AF789569

Springer Proceedings in Physics 27

Springer Proceedings in Physics

Managing Editor: H. K. V. Lotsch

Volume 1 *Fluctuations and Sensitivity in Nonequilibrium Systems* Editors: W. Horsthemke and D. K. Kondepudi

Volume 2 *EXAFS and Near Edge Structure III* Editors: K. O. Hodgson, B. Hedman, and J. E. Penner-Hahn

Volume 3 *Nonlinear Phenomena in Physics* Editor: F. Claro

Volume 4 *Time-Resolved Vibrational Spectroscopy* Editors: A. Laubereau and M. Stockburger

Volume 5 *Physics of Finely Divided Matter* Editors: N. Boccara and M. Daoud

Volume 6 *Aerogels* Editor: J. Fricke

Volume 7 *Nonlinear Optics: Materials and Devices* Editors: C. Flytzanis and J. L. Oudar

Volume 8 *Optical Bistability III* Editors: H. M. Gibbs, P. Mandel, N. Peyghambarian, and S. D. Smith

Volume 9 *Ion Formation from Organic Solids (IFOS III)* Editor: A. Benninghoven

Volume 10 *Atomic Transport and Defects in Metals by Neutron Scattering* Editors: C. Janot, W. Petry, D. Richter, and T. Springer

Volume 11 *Biophysical Effects of Steady Magnetic Fields* Editors: G. Maret, J. Kiepenheuer, and N. Boccara

Volume 12 *Quantum Optics IV* Editors: J. D. Harvey and D. F. Walls

Volume 13 *The Physics and Fabrication of Microstructures and Microdevices* Editors: M. J. Kelly and C. Weisbuch

Volume 14 *Magnetic Properties of Low-Dimensional Systems* Editors: L. M. Falicov and J. L. Morán-López

Volume 15 *Gas Flow and Chemical Lasers* Editor: S. Rosenwaks

Volume 16 *Photons and Continuum States of Atoms and Molecules* Editors: N. K. Rahman, C. Guidotti, and M. Allegrini

Volume 17 *Quantum Aspects of Molecular Motions in Solids* Editors: A. Heidemann, A. Magerl, M. Prager, D. Richter, and T. Springer

Volume 18 *Electro-optic and Photorefractive Materials* Editor: P. Günter

Volume 19 *Lasers and Synergetics* Editors: R. Graham and A. Wunderlin

Volume 20 *Primary Processes in Photobiology* Editor: T. Kobayashi

Volume 21 *Physics of Amphiphilic Layers* Editors: J. Meunier, D. Langevin, and N. Boccara

Volume 22 *Semiconductor Interfaces: Formation and Properties* Editors: G. Le Lay, J. Derrien, and N. Boccara

Volume 23 *Magnetic Excitations and Fluctuations II* Editors: U. Balucani, S. W. Lovesey, M. G. Rasetti, and V. Tognetti

Volume 24 *Recent Topics in Theoretical Physics* Editor: H. Takayama

Volume 25 *Excitons in Confined Systems* Editors: R. Del Sole, A. D'Andrea, and A. Lapiccirella

Volume 26 *The Elementary Structure of Matter* Editors: J.-M. Richard, E. Aslanides, and N. Boccara

Volume 27 *Competing Interactions and Microstructures: Statics and Dynamics* Editors: R. LeSar, A. Bishop, and R. Heffner

Volume 28 *Anderson Localization* Editors: T. Ando and H. Fukuyama

Competing Interactions and Microstructures: Statics and Dynamics

Proceedings of the CMS Workshop,
Los Alamos, New Mexico, May 5–8, 1987

Editors: R. LeSar, A. Bishop, and R. Heffner

With 160 Figures

Springer-Verlag Berlin Heidelberg New York
London Paris Tokyo

Dr. Richard LeSar
Dr. Alan Bishop
Dr. Robert Heffner
Los Alamos National Laboratory, Los Alamos, NM 87545, USA

ISBN-13: 978-3-642-73500-4 e-ISBN-13: 978-3-642-73498-4
DOI: 10.1007/978-3-642-73498-4

Library of Congress. Library of Congress Cataloging-in-Publication Data. CMS Workshop (1987: Los Alamos, N.M.) Competing interactions and microstructures: statics and dynamics: proceedings of the CMS Workshop, Los Alamos, New Mexico, May 5–8, 1987 / editors, R. LeSar, A. Bishop, and R. Heffner. p.cm.– (Springer proceedings in physics; v. 27) Includes index. 1. Condensed matter – Congresses. 2. Phase transformations (Statistical physics) – Congresses. 3. Dynamics – Congresses 4. Statics – Congresses. I. LeSar, R. (Richard), 1953–. II. Bishop, A. (Alan), 1947–. III. Heffner, R. (Robert), 1942–. IV. Title. V. Series. QC173.4C65C58 1987 530.4'1 – dc19 88-6472

This work is subject to copyright. All rights are reserved, whether the whole or part of the material is concerned, specifically the rights of translation, reprinting, reuse of illustrations, recitation, broadcasting, reproduction on microfilms or in other ways, and storage in data banks. Duplication of this publication or parts thereof is only permitted under the provisions of the German Copyright Law of September 9, 1965, in its version of June 24, 1985, and a copyright fee must always be paid. Violations fall under the prosecution act of the German Copyright Law.

© Springer-Verlag Berlin Heidelberg 1988
Softcover reprint of the hardcover 1st edition 1988

The use of registered names, trademarks, etc. in this publication does not imply, even in the absence of a specific statement, that such names are exempt from the relevant protective laws and regulations and therefore free for general use.

Printing: Weihert-Druck GmbH, D-6100 Darmstadt
Binding: J. Schäffer GmbH & Co. KG., D-6718 Grünstadt
2154/3150-543210

Preface

Many macroscopic properties of materials, such as strength and response, are determined primarily by inhomogeneous structures and textures. These intermediate-scale "mesoscale" structures most often arise from competing or, sometimes, cooperating interactions, which stem from interactions within a material that operate on different length scales or in opposing (or cooperating) manners. Our understanding of such phenomena has increased substantially with the identification and theoretical description of solid-state materials with incommensurate and long-period modulated phases – ferroelectrics, charge-density-wave compounds, epitaxial layers, polytypes, molecular order/disorder, etc. Furthermore, the experimental diagnosis of inhomogeneous ground states and metastable phases has advanced such that these are now well-accepted phenomena, at least in specific solid-state and condensed-matter communities. As conference organizers, we felt that it was timely to bring together a diverse group of physicists *and* materials scientists to review developments in this area and to examine possible future directions; in particular, how the microscopic understanding emerging in "bench-top" solid-state systems can be applied to materials science. To that end, the workshop "Competing Interactions and Microstructures: Statics and Dynamics" was held at the Los Alamos National Laboratory, May 5–8, 1987.

The workshop can be loosely divided into three areas: the physics of competing interactions; how the structures that arise from these interactions affect material properties (especially phases and phase transitions); and the dynamics of such phenomena (the study of which is far less advanced, and thus appears to us to be an important direction for the future).

The balance between interactions in condensed phases can lead to very complex structures. The theoretical understanding of these structures is complicated by the fact that they are often of an intermediate scale, with characteristic lengths on the order of hundreds of angstroms to micrometers, whereas the interactions that lead to these structures are short ranged, extending over only a few neighbors. Much progress has been made in this area by studying phenomenological models, such as the ANNNI model, which build competition between structures into the Hamiltonian describing the systems in relatively simple ways. While these basic models are quite successful in describing such phenomena as domain structure, glassy response, etc., the physics of the real systems is rich, and only beginning to be explored. In Part I, four articles, which are by nature reviews and which address these fundamental questions, are given. The introduction to the physics of competing interactions by V. Heine sets the stage for many of the discussions to follow. The article by G. van Tendeloo does the same for the experimentally important technique of electron diffraction. H. Aaronson then gives a materials scientist's view of phase transitions, while J. Krumhansl presents a physicist's view.

In Part II, the discussions center on the intermediate-scale structures caused by the microscopic interactions, which determine to a large extent the macroscopic properties of materials. Of particular interest are "precursor structures" (or "pretransformation microstructures") which are known to be important for technologically relevant materials such as martensites (perhaps even including the newly discovered oxide superconductors), ferroelectrics, omega phase materials, and polytypes. These newly emerging and unifying concepts are examples of the interface between microscopic and macroscopic approaches to materials. Also included are discussions of the roles of competing interactions in quasi-two-dimensional systems – atoms on or between layers of graphite – as well as superlattices and liquid crystals.

The study of the large-scale dynamics of these structures is limited to poorly communicating fields and models, such as ideal- and disordered-Frenkel-Kontorova models, random-field spin models, spin glasses, weakly pinned charge-density waves, etc. In fact, these various models share many characteristics (e.g. hysteretic and "glassy" dynamics), and it seems reasonable now to expect that underlying systematics and phenomenology will emerge if we study them side by side. In Part III, we have a series of papers that discuss these dynamics, from charge-density waves to spin glasses to grain growth to more abstract neural networks.

Sponsorship of this workshop was provided by the Center for Materials Science at the Los Alamos National Laboratory. We are thankful for the expert secretarial help from the Center for Materials Science: Stella Taylor and Bettye McCulla. We also thank the Los Alamos National Laboratory for use of their excellent conference facilities and organizational staff, in particular Luz Woodwell.

The prospect of developing a microsopic understanding of the macroscopic properties of materials is an exciting one, and one that needs a synthesis of different perspectives. We hope that this series of articles – representing the work of physicists, chemists, and materials scientists – can help to promote this synthesis.

Los Alamos
November 1987

R. LeSar
A.R. Bishop
R.H. Heffner

Contents

Part III **Dynamics**

Part I

Introductory Surveys

The Microscopic Understanding of Modulated Structures and Polytypes

V. Heine

Cavendish Laboratory, Madingley Road,
Cambridge, CB3OHE, UK

Abstract

Incommensurate structures and polytypes show a variety of phenomena. These result from the interplay of various forces in solids, sometimes cooperation and sometimes competition between them. A general understanding now exists of the origin of modulated structures and polytypes. Computer simulation on some simple archetypal materials validates the general mechanisms and improves our understanding of the details.

1. Introduction

The relevance of incommensurate (IC) structures and polytypes to the present workshop is that they represent a certain kind of texture. The texture is regularly periodic but that is not the main point. Scientifically the interesting question about these materials is "Why do they do it?" When we understand that we understand more about the combinations of forces in solids which can lead to various texture phenomena. There is nothing very interesting about a piece of rock salt. It is cubic, and that is about about all it does. That does not get us much further in understanding the interplay of solid state forces. By contrast IC structures show a wealth of phenomena. Some have a purely sinusoidal modulation whereas others form sharp antiphase boundaries as in polytypes. Some are incommensurate while others have a wave vector that locks on to rational fractions of a reciprocal lattice vector to form long period superlattices. Some exist as equilibrium phases only over a narrow temperature range while other are stable apparently to 0 K. We shall therefore be concerned in this opening paper with the origin and stability of IC structures and polytypes, with the squaring up of the modulation wave form, and with the wave vector locking on to commensurate rational values. All these phenomena have their counterpart in more general non-periodic textures.

The present state of the subject is that there is a general theoretical understanding of several microscopic mechanisms that can give rise to IC structures [1]. A brief summary and classification of these will be given in Sect. 2. In some materials one can identify more or less plausibly the microscopic mechanism, whereas in others it is far from clear. The next step is to study a few archetypal cases in detail by computer simulation and quantitative theory. That serves to validate the general mechanisms and to add extra insights. I shall draw on the work in Cambridge [2-18] on $NaNO_2$, biphenyl and the SiC polytypes in Sects. 3, 4 and 5 respectively. To reference fully every point which I touch on would turn this brief survey into a review article such as I am preparing for publication elsewhere [1], and I apologise for the many omissions.

2. Mechanisms

There is no tidy and complete list of mechanisms for IC structures. Rather there are several general ideas which can be elaborated and subdivided in various ways. Thus any list is to some extent a matter of personal taste. In studying any real material several of these ideas may need to be combined, as will be illustrated by my remarks on SiC. One mechanism may ride on the back of another. For example in the noble metal alloys it is generally accepted that the Fermi surface mechanism gives oscillatory interplanar interactions. These can be fed into an ANNNI type of model to discuss the effect of temperature on the sharpening up of the modulation to form long period superlattices. In some cases a specific microscopic mechanism can be described in alternative ways with one mechanism being mapable mathematically onto another. Each may be illuminating in different ways. The following list [1,5] must therefore be taken as a set of ideas to be developed and possibly combined in understanding specific materials.

2.1 The Fermi Cut-Off in Metals

This mechanism is well documented particularly when there are flattened or nesting regions of Fermi surface. But the Fermi cut-off gives rise to relatively long ranged and irregularly alternating interactions even for more general band structures [19]. In either case the incommensurateness arises from the beats between the wavelength λ_F of the interaction and the lattice constant of the crystal structure. The effect can be destroyed at high temperature or by impurities either by the Fermi function fuzzing the sharpness of the cut-off in energy, or elastic scattering doing the same in k-space. Little quantitative work has been done on that, or on the role of the electronic structure among other factors determining the shape of the modulation. In some cases the entropy of disorder at the antiphase boundaries is essential to the stability of the structure.

2.2 Cooperating Order Parameters of Different Symmetryl

This mechanism goes back at least to LEVANYUK and SANNIKOV [20] as a phenomenological model and has been developed more physically by McCONNELL and myself in a variety of ways [21, 2 - 4]. Its features will be explained with reference to $NaNO_2$ in Sect. 3. In brief, it depends on the cooperation between two processes of different but related symmetry in the crystal. It may turn out to be the commonest mechanism in complex materials where two order parameters with the related symmetries are likely to exist among the many degrees of freedom. Examples include mullite Al_2 (Al_{2+2x} Si_{2}-

$_{2x}$) O_{10-x} [8], the (001) surface of molybdenum [17], thiourea, quartz, $ThBr_4$, the cooperative Jahn Teller system $K_2PbCu(NO_2)_6$ [6], and the intermediate plagioclase feldspars [21]. The fact that the mechanism involves the cooperation between the two order parameters means that some of these IC structures are extremely stable. For example mullite melts before it disorders, and $ThBr_4$ remains IC down to the lowest temperatures measured (about 1 K).

2.3 Competing Short Ranged Interactions J_1 and J_2

Consider a line of entities with some internal degree of freedom. An entity may be a magnetic atom whose moment is free to rotate in the xy plane, or an Ising site for A/B ordering; it can be a biphenyl molecule which can twist by an angle ϕ in the positive or negative sense about the central bond, or a rotating BX_4 tetrahedron in the family of A_2BX_4 compounds. In a one-dimensional model our 'entity' is conceived as a whole xy plane of such atoms or molecules. For example in SiC each atomic double layer may be stacked in two possible orientations on the layer below in what is the (111) direction in the zinc blende structure. With different stacking sequences an infinite number of random and regular structures can be constructed, of which several dozen regular polytypes have been reported.

We have therefore a row of entities in the z-direction with first and second neighbour couplings J_1 and J_2. Now suppose the second neighbour interaction tends to order entities in opposite sense. We may call this 'antiparallel' and describe it by an 'antiferromagnetic' negative J_2. Whatever the sign, positive or negative of the nearest neighbour interaction J_1, its effect via two nearest neighbour hops is to align second neighbours parallel. There is therefore a frustration or conflict between the effects of J_1 and J_2. The intersite energy is minimised by an incommensurate modulation if $|J_1|<4|J_2|$. Thus IC structures are found in biphenyl, some magnetic insulators, and several A_2BX_4 compounds such as K_2SeO_4.

Of course the model as such has nothing to say about the origin of the couplings J_1 and J_2. In metallic alloys it is the Fermi cut-off as mentioned in Sect. 1.1, whereas in biphenyl it is simply steric hindrance (see Sect. 3). In SiC it is presumably a remnant of the Fermi mechanism and in K_2SeO_4 an unknown mixture of electrostatic interactions and steric hindrance. Clearly there is no difficulty in extending the model to include third or further neighbour couplings.

The intersite coupling tends to give an IC modulation as we have seen: but what actually happens depends strongly also on the intrasite potential and the interactions within an xy plane. In biphenyl the twist of a single molecule including the cage effect of the neighbours (Sect. 3) is described by a rather flat-bottomed single-well potential. In consequence the IC modulated structure persists down to the lowest temperatures measured (about 1K) and presumably absolute zero, at least under modest pressure. At the opposite extreme one has in SiC a whole atomic double layer oriented in one of the two possible directions so that in the polytypes the modulation has sharpened up into a series of discrete antiphase boundaries between bands of two or three atomic double layers. These boundaries are atomically flat over large areas as any irregularity would imply broken bonds with a very high energy cost. The situation is different again in some alloys and the Axial Next Nearest Neighbour Ising (ANNNI) model where the entropy of disorder at the antiphase boundaries is crucial to observing the IC modulation. At low temperature the ANNNI model settles into one of three possible ground states, namely the 'ferromagnetic', the 'antiferromagnetic' or the fourfold superlattice of 'two up, two down' usually denoted by $\langle\infty\rangle$, $\langle 1\rangle$ and $\langle 2\rangle$ respectively. This transition to a simple commensurate state is due to the discreteness of an Ising spin, in contrast to the potential well for the twist angle ϕ as a continuous variable for the biphenyl molecule.

2.4 Combining Unrelated Length Scales L_1 and L_2

Several examples will be given later in this conference where L_1 is the lattice constant of a graphite layer and L_2 the lattice constant of an intercalate such as cobalt chloride or of a surface adsorbed monolayer of krypton. Many other combinations of structural units are known, arranged either in layers or columns. If both of the structural units are rather rigid then they just sit side by side forming a structure which is effectively IC or has a very long repeat period.

But very interesting phenomena occur if one of the units is relatively easily deformable. Consider a rigid periodic potential V(x) with lattice periodicity L_1, the so-called 'washboard'. On it lay an infinite string of particles connected by springs of natural length L_0 and spring constant λ. The density of particles can be varied by changing their chemical potential in the system, modelled by a tension T in the string. The particles will tend to sit in the hollows of the washboard potential V(x) depending on the depth D of the washboard corrugations in relation to strength λ of the springs. If the V(x) is purely sinusoidal we have the original FK/FM model [22] but more general potentials have been considered [23]. Let the mean spacing between the particles be L_2. Then as T is varied, the ratio L_1/L_2 goes in jerks through <u>all</u> rational fractions m/n, in big jerks at simple fractions and ever finer jerks at higher order fractions. The totality of all rational fractions is called a devil's staircase with an infinite number of steps but occupying a finite range of T. If V(x) is a smooth (mathematically well behaved) function, then there are only these periodic groundstates and the staircase is termed 'complete'. But if V(x) has cusps or worse singularities then in addition to the periodic ground states with $L_1/L_2 = m/n$, the staircase is 'incomplete' with genuinely incommensurate states in between consecutive commensurate ones as shown by AUBREY [23]. There are always in addition a large number of metastable periodic and aperiodic states.

2.5 Long Range Convex Repulsive Interactions J(n)

Consider an infinite line of sites spaced L_1 apart, and a density of particles with mean spacing L_2 (greater than L_1). The particles have to sit on sites and there is an interaction energy J(n) between particles n sites apart. If the J(n) are repulsive, the particles will try to space themselves out as evenly as possible and we have a situation similar to that in Sect. 2.4. If also J(n) extends to infinity and decreases monotonically with

$$J(n) > \tfrac{1}{2}\,[J(n-1) + J(n+1)] \text{ for all } n, \qquad (1)$$

then BAK and BRUINSMA [24] have shown that a complete devil's staircase results. The condition (1) is called convexity. The structures have fascinating mathematical properties and can be constructed quite simply by laying down a line of particles at equal spacing L_2 and moving each one to the nearest allowed lattice site. Some charge transfer salts and alkali metals intercalated in graphite are considered examples of this mechanism. The Coulomb interaction, partially screened in the case of the graphite intercalates, provides the long range repulsion J(n).

2.6 Bilz's Periodons

In the shell model of an atom often used in the calculation of lattice dynamics, the atom is represented by a rigid spherical core plus a spherical shell of valence electrons. Core and shell are connected by a potential V(u-v) where u and v are the displacements of core and shell respectively. For some atoms

$$V(w) = bw^2 + cw^4 \qquad (2)$$

can be quite anharmonic, particularly for oxygen ions where b may be considered negative. Core and shell can then interact with neighbouring cores and shells by harmonic forces. The behaviour of such a strongly anharmoic model has not been investigated very far. However BILZ et al. [25] have shown that some unique exact periodic solutions exist with frequencies $\omega(q)$ which they termed periodons. These show a strong anomaly at q around g/3 and by suitable choice of parameters can give a soft mode and IC structure at around this value of q. The q is not exactly g/3 because there is no special symmetry at this rational value.

The status of these ideas is uncertain. Since the model is non-linear, a periodon for given q has a fixed finite amplitude and solutions of the equations cannot be superposed. It is therefore not entirely clear what relation, if any, the periodons have to a system in thermal agitation. We may suppose that the latter bears some locally fluctuating resemblance to the former. The important point about this model is that it gives a natural explanation of why q should be near g/3 as found in quite a few materials such as K_2SeO_4. The force from differentiating (2) contains $4cw^3$ and when one substitutes

$$w \sim \exp(iqx - i\omega t)$$

one naturally generates third harmonics and by interference the anomaly around g/3. Personally I regard this as probably significant because when I draw threefold superlattices they don't seem to make any structural 'sense', in fact rather the reverse: and yet there seem to be too many examples of them for it to be entirely accidental.

2.7 General Conclusions

Several general remarks can be drawn from the above brief survey. Firstly there are always two length scales, two order parameters, two interaction, two (or more) somethings. Secondly these two effects may cooperate with each other to give some very stable IC structures, or they may compete against each other to give rather unstable ones, or they may just interact to give a devil's staircase. Thirdly IC structures occur in all types of solids: metals and insulators, organic and inorganic, magnetic materials and Jahn Teller systems. Fourthly there is no finite set of precise models onto which all materials can be mapped. Rather there is a set of broad ideas and phenomena which can be elaborated and combined in various ways (which other authors may indeed classify in different ways). For example BRUINSMA and ZANGWILL [26] have considered a set of alloy phases rather like a devil's staircase. The basic interaction between antiphase boundaries was considered to arise from the Fermi cut-off mechanism (Sect. 2.1), giving roughly a J_1, J_2 model to account for the approximate periodicity of the antiphase boundaries (Sect. 2.3). But these authors had to invoke elastic effects mapped mathematically onto the FK/FM model [22] (Sect. 2.4) to account for the stability of the superlattice structures at low temperature. Fifthly our list is not even complete. For example the origin of the IC structure in propyl ammonium manganese chloride is probably more transparent than that in any other material [27], but it does not conveniently fit into any of the above categories. In addition there are non-equilibrium mechanisms such as spinoidal decomposition giving similar textures very relevant to this conference. Finally incommensurate structures have been put very firmly on the map. It used to be thought that all true equilibrium structures had to be commensurate, i.e. had to have some long but rational period superlattice: certainly it was thought this had to be so for the ground states at T = 0°K. However AUBREY'S discussion [23] of washboard model has shown that this need not be so. Certainly the lock-on energy was undetectably small in some computer studies with the J_1,J_2 model and soft on-site wells

3. Co-operating Order Parameters and Gradient Interaction in Sodium Nitrite.

The first point to make is that a modulation always contains two separate but related symmetries. This can be seen from Fig. 1 where the order parameter ψ represents the degree of alignment of the polar NO_2^- ions in $NaNO_2$ as shown schematically by arrows and some unit cells in the lower part of the figure. Note the symmetry around A and C were the ions are more or less uniformly aligned. Now consider the symmetry around B. It is different, but uniquely defined by the 'up' alignment to its left and 'down' on its right. Whereas at A we have a vertical mirror plane, at B this has become a twofold rotation axis perpendicular to the page. Any Bloch wave, any modulation, automatically has <u>two</u> such related symmetries in it [4].

Now to the physics. The gradient in the degree of alignment around B creates a 'force field' which will interact with any deformation or ordering in

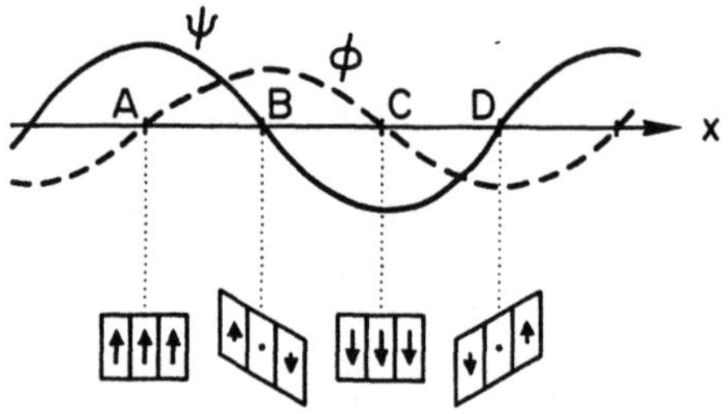

Fig. 1. The spatial modulation of the order parameter ψ describing the alignment of NO_2^- ions, shown below as arrows, and the local shear ϕ. (After Ref. [3].

the crystal with that second symmetry. Such a process is a second order parameter ϕ also shown in Fig. 1. In the case of $NaNO_2$ the ϕ is a shear of the unit cell.

The two order parameters ψ, ϕ cooperate to produce a lower free energy than either alone. This can be illustrated in an oversimplified way by considering the bilinear free energy expression

$$G = A\psi^2 = 2H\,\psi\phi + B\phi^2. \qquad (3)$$

We can minimise this expression for fixed ψ by setting ϕ equal to $-H\psi/B$ which on substitution into (3) gives

$$G = A_{eff}\,\psi^2 = (A - H^2/B)\psi^2. \qquad (4)$$

This clearly gives a lower free energy than (3) with ψ alone. The argument is oversimplified because it leaves out the q-dependence of all the symbols in (3) and much else, but the gist of it is correct [3].

The IC modulation arises because we have a gradient interaction between ψ and ϕ. In Fig. 1 the ϕ has maximum amplitude where ψ has maximum gradient. If there is no modulation, then there is no gradient in ψ anywhere and nothing for ϕ to interact with, so that the argument of (3), (4) falls down. To be precise, the fact that we have a gradient interaction introduces a factor of iq into H(q), thus making H = O in the uniform limit of q = O.

A gradient interaction is not some new magical force but the result of steric hindrance and other normal atomic interactions. Its origin can be demonstrated in atomic terms using what we have called the exaggerated gradient ploy (Fig. 2). Since the shear ϕ is caused by a gradient in the alignment of the NO_2^- groups, we highlight the interaction by creating as sharp a gradient as possible. This is achieved at an antiphase boundary where the polarisation changes from totally 'up' to totally 'down' (with respect to the b axis) from one atomic plane to the next. $NaNO_2$ has a body centred structure of alternate layers stacked in the a direction shown as clear and shaded in Fig. 2.

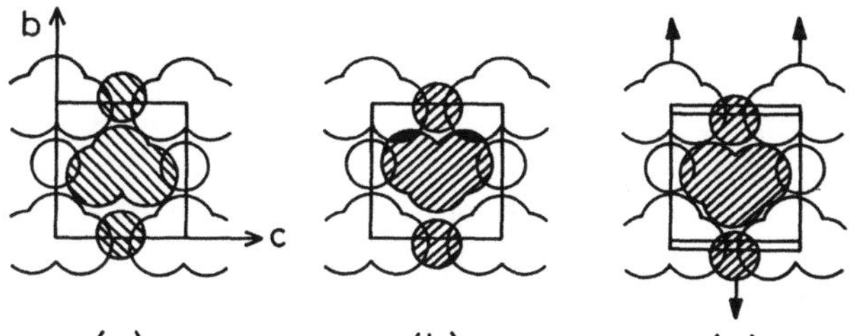

Fig. 2 (a) Ordered structure of uniform (ferroelectric) ordering of $NaNO_2$ showing alternate atomic planes. (b) Atomic misfit if one plane is reversed to create a strong gradient in the ordering parameter. (c) Restoration of good atomic fit between the planes by a relative displacement of one with respect to the other, equivalent to a shear of the unit cell. (After Ref. [7].)

The layers clearly fit snugly on top of one another if all the NO_2^- groups are aligned parallel as in Fig. 2a depicting the low temperature ferroelectric phase. In Fig. 2b we have reversed one layer as at an antiphase boundary: it no longer fits so well onto the one below. However if we displace one layer with respect to the other, then it fits well again as in Fig. 2c. Such relative displacement constitutes a shear of the unit cell between the two layers. We conclude that the gradient ploy demonstrates the gradient interaction with the second order parameter ϕ [2,3,7] and can indeed be used to discover the nature of ϕ in new cases [6,8,17]. By making a more quantitative drawing in terms of ionic radii, reading off the required displacement in the sense of Fig.2c and scaling to a realistic gradient, we were able to arrive at a quantitative estimate of the shear which agreed to within a factor of 1.5 with experiment [7]. I believe this confirms the general validity of the picture. The gradient ploy has been used in several of the examples listed in Sect. 2.2 [6,8,17].

Molecular dynamic simulation adds a further insight [7]. As usual in discussing second order phase transitions, we start mentally with the disordered state at high temperature with its local fluctuations that are microscopic precursors of the phase transition. As the temperature is lowered, these fluctuations grow in amplitude and condense at the transition temperature T_{IC} into a macroscopic ordering. Figure 3 shows the local fluctuations in the degree of NO_2^- ordering and in the shear ϕ in the sense of Fig. 2c. There are three points to note Firstly Fig. 3c shows that there is a substantial degree of correlation (ξ of order one half) over most of the Brillouin zone at temperatures well above T_{IC}. The gradient interaction demonstrated in Fig. 2 is strong and effective locally at all temperatures so that fluctuations of ψ are accompanied by correlated fluctuations in ϕ. Even at high T they are not independent modes as one might have supposed. Secondly Figs. 3a and 3b show the rapid growth in the magnitudes of both fluctuations as T_{IC} is approaching. But the growth in intensity is only at the value of q_{IC} which becomes the wave vector of the IC structure. Note there is little growth in ψ_q at $q = 0$. This is surprising since the

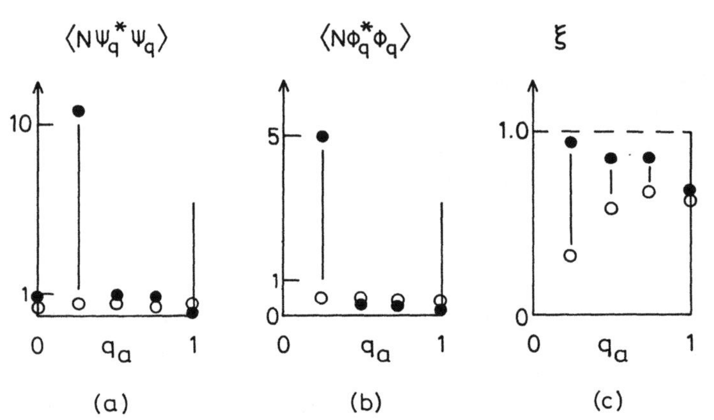

Fig. 3. Intensities of the fluctuations in $NaNO_2$ at two temperatures T in the disordered phase, shown as a function of q along the a* axis in the Brillouin zone. Open circles: T about 200° above the phase transition at T_{IC}. Filled circles: T a few degrees above T_{IC}. (a) The fluctuations of ψ_q describing the alignment of the NO_2^- ions. (b) The fluctuations of the shear ϕ_q in the sense of Fig. 2c. (c) The degree of amplitude and phase correlation between the fluctuations of ψ_q and ϕ_q, where $\xi = 1$ means complete correlation. The results are derived from Molecular Dynamics Simulation. (After Ref. [7].

material has a lock-in commensurate transition to the uniform $q = 0$ ferroelectric phase at a temperature T_c only about 2° below T_{IC}. (The fact that this interval is so small is due to some special features of $NaNO_2$.) Thirdly note in Fig. 3c that the correlation between the ψ_q and the ϕ_q fluctuations becomes 100% at q_{IC} where the amplitude grows. We can say that the NO_2^- ordering fluctuations ψ_q and the shear fluctuation ϕ_q are condensing together at T_{IC} in a correlated or united way to form the modulated structure incorporating both. For T in the range $T_c < T < T_{IC}$ the IC structure is better ordered., has lower entropy, than a uniform ferroelectric ordering because the former orders both the NO_2^- orientations and the local shears whereas the latter would leave the atomic displacements completely disordered. We can express it another way in terms of the temperature T_0 at which uniform ferroelectric ordering would occur if there were no coupling to the displacements. The T_0 is less than T_{IC} so that between T_0 and T_{IC} the IC structure is thermodynamically stable while the ferroelectric ordering is still unstable.

So what have we learnt from the study of NaNO2 ? Firstly the mechanism of cooperating ordering effects can be clearly seen at work in the molecular dynamic simulations. We have given reasons in Sect. 2.2 for expecting the effect to be of increasing importance as one studies more complex materials. Secondly to get an IC structure one needs a gradient interaction but there is nothing mysterious about this. The gradient ploy shows how it arises from simple atomic forces, and the idea can be applied to other materials. Thirdly the gradient coupling also produces strongly correlated fluctuations in the disordered phase so that one may expect it to be relevant to the formation of textures in non-equilibrium or non-uniform conditions.

4. Computer Modelling of Biphenyl

Biphenyl is an excellent example of the J_1, J_2 coupling mechanism of Sect. 2.3. As a result of our intensive study over the last two years [9-14] it must now be the best understood IC structure from the point of view of the basic microscopic interactions. Moreover the phase diagram (Fig. 4) and other behaviour shows a varied richness that tests our understanding. The single molecule (Fig. 4a) would lie flat but for steric hindrance between the four central hydrogens which are rather close together. As a result, the free molecule has a twist of $2\phi \approx 40°$. In the solid state (phases II and III) the twist angle 2ϕ is modulated with an amplitude of about 11°. For a review of the experimental results see Ref. [28].

The inherent tendency of an individual molecule to twist is described by an intramolecular potential

$$v(\phi) = -A\phi^2 + B\phi^4 \qquad (5)$$

of double well form. In the solid the molecules are packed in a herring-bone pattern in ab layers with molecular axis roughly in the c direction (Fig. 5a). As one molecule twists and the phenyl ring rotates, it will tend to rotate the neighbouring phenyl rings simply because the molecules are in contact. Moreover the neighbour will be turned in the opposite direction, making both J_1 and J_2 negative ('antiferromagnetic' in the sense of Sect. 2.3). Remember from Sect. 2.3 that J_2 has to be negative to give frustration and an IC structure.

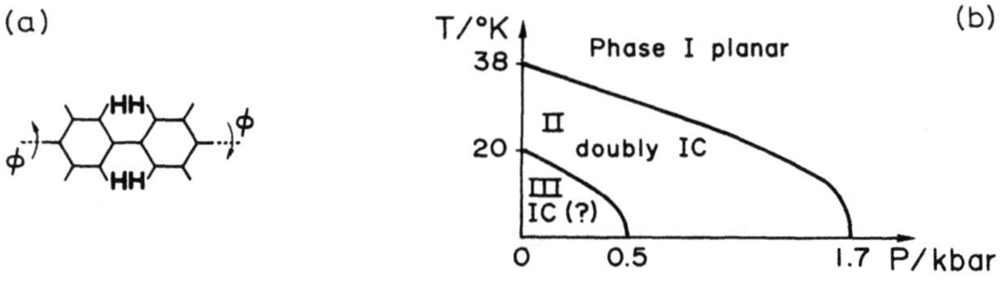

Fig. 4. (a) Individual biphenyl molecule, consisting of two joined phenyl (benzene) rings. The central hydrogens are indicated for emphasis. (b) Phase diagram of crystalline biphenyl under modest pressure. (After Ref. [28].) Phase II is doubly incommensurate in the sense that the modulation wave vector (10) has a strong component along the unique b* axis and a weak component in the a*c* plane. Phase III is observed as incommensurate with modulation only along b* but the theoretical analysis indicates such a structure must be metastable. The true ground state in the phase III region is the simple commensurate 'two up, two down' superlattice.

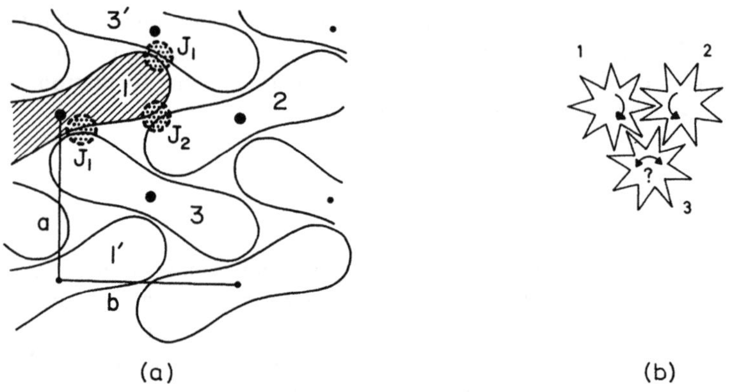

Fig. 5. (a) An ab layer of biphenyl molecules in the crystal. The section cuts roughly through the centre of one of the phenyl rings, with twist axis approximately perpendicular to the paper. The interaction zones (schematic) with first (J_1) and second (J_2) neighbours are shown. (b) The enmeshed cogwheel analogy. The rotation of one cogwheel (phenyl ring) induces an opposite rotation in a neighbour in contact. This results in locking or frustration in the case of three wheels in contact. (After Ref. [9].)

How the frustration arises is very obvious here by considering a set of three molecules in contact, or three cogwheels [9]. Suppose phenyl ring 1 turns clockwise (Fig. 5). It will tend to rotate rings 2 and 3 both anticlockwise. However an anticlockwise rotation 2 will drive ring 3 clockwise. Thus ring 3 is caught between opposing influences. This translates directly into interplanar interactions for modulation along the b* direction: molecules 1, 1′,... form one plane, molecules 3, 3′,... the nearest neighbour plane, and molecules 2, 2′,... the second neighbour plane.

By computer modelling, the interaction between all sets of neighbours has been 'measured' [9, 12]. For a given pair of molecules i and j, the interaction potential can be written in terms of four constants [12]:

$$U_{ij}= ½ \ J_{ij} \ (\phi_i + \phi_j + \gamma_{ij})^2 + ½ \ K_{ij} \ (\phi_i - \phi_j + \delta_{ij})^2. \qquad (6)$$

when U_{ij} is summed over a centro-symmetric structure the angles α_{ij} and δ_{ij} cancel out so that we will drop them for simplicity. Then U_{ij} can be rewritten

$$U_{ij} = (J_{ij} + K_{ij}) \ (\phi_i^2 + \phi_j^2) + K_{ij} \ - \ J_{ij}) \ \phi_i \ \phi_j \ . \qquad (7)$$

Each term has a physical significance. We have called the first term the cage effect [9]. It adds to the quadratic term in (5) with positive sign, i.e. giving an effective intramolecular well

$$v_{eff}(\phi) = A_{eff} \ \phi^2 + B\phi^4 \qquad (8)$$

$$A_{eff} \quad = \Sigma(J_{ij} + K_{ij}) - A \qquad (9)$$

which is a much shallower double well or, as here suggested, actually a rather flat bottomed single well with A_{eff} small and positive. Anyway it is clear that the first term of (7) tends to keep the molecules more nearly planar because they are hemmed in by the cage formed by their near neighbours. Under pressure this term increases as the cage gets tighter and the IC phases disappear (Fig. 4b) [13]. The second term in (7) is the net interaction between the molecules characterised by the coupling constant $(K_{ij} - J_{ij})$. The interaction constants in (6) were 'measured' with a computer modelling programme by evaluating the energies for various combinations of $\phi_i, \phi_j = 0$ or $\pm$ 11° and fitting the form (6), and this was repeated for a range of intermolecular distances to obtain the pressure effects.

The whole phase diagram can be calculated once the $(K_{ij} - J_{ij})$ coupling constant have been obtained [11, 13]. We used mean field theory with a correction for the usual deviation of the correct transition temperature T_{IC} from mean field theory. Another correction takes into account that biphenyl is to some degree a quasi one-dimensional material with the coupling between molecules 1 and 2 in Fig. 5a in the b direction being an order of magnitude larger than any other. The only adjustment was a little fine tuning on A and B in (5) which are not well known from the free molecule, particularly at the low values of ϕ relevant in the solid. The results for both phase transitions in Fig. 4 are then in good agreement with experiment. The exact value of the IC vector

$$q_{IC} = (\tfrac{1}{2} \pm q_b)b^* \pm (q_a a^* - q_c c^*) \qquad (10)$$

depends on the coupling to translations and librations of the molecules which were only treated fully partially. In particular the values of q_a and q_c depend on tiny energy differences which were not obtained wholly correctly. Nevertheless the overall agreement with experiment is sufficiently impressive that it leaves no room to doubt the general correctness of the model.

Several observations emerge from the work. Firstly as already remarked it leaves no doubt about the model of competing J_1, J_2 interactions due to enmeshed rotation of the molecules, as the origin of the incommensurateness. Secondly at pressures between 0.5 and 1.7 Kbar the genuine IC phase II exists down to the lowest temperatures measured (1°K) and any lock-in to a high commensurate wave vector would appear to be very weak. Certainly there need be no lock-in to one of the simple ground-states phases which one has in the ANNNI model. Thirdly the energy of a quilted pattern involving all four wave vectors (10) is higher than that of striped domains using pairs of wave vectors. The reason is that the quilted pattern has more regions where the order parameter goes through zero. Fourthly the intermolecular forces turned out to be extremely close to harmonic over the range of angles $|\phi| \leq 11°$ relevant in the solid. Only the intramolecular potential (5) is appreciably anharmonic. It is probably true fairly generally that in systems like quartz and thiourea where one expects some double well feature, all the forces may be taken as harmonic except for the one special feature After all, one knows from specific hears and coefficients of linear expansion that all forces in all straight-forward solids are quite close to harmonic. One therefore expects models based on a single anharmonic feature to be a good representation of reality. Fifthly one can obtain a creditable model of terphenyl by doubling (5) on account of two joining bonds per molecule and multiplying all the intermolecular couplings by three because of three rings per molecule. This explains the much higher transition temperature. Also the v_{eff} (8) now becomes a double well, explaining the observed order-disorder nature of the phase transition in terphenyl, in contrast to the single well and soft mode transition in biphenyl [11,12].

The phase III at low T and P calls for special comment. Experimentally it has been described as an IC phase with only the q_b component in the IC wave vector q_{IC}. However our calculations show that the equilibrium structure must be the simple superlattice denoted by <2> in the ANNNI model It is shown in two symmetry-related forms on the left and right of Fig. 6. When the IC phase II tries to make a lock-in transition to the true ground state <2>, the q_a, q_c modulation is easily lost because the energy involved in them is tiny. However the modulation q_b in (10) in the b* direction is not lost easily and one observes what must be a metastable regular array of solitons (alias anti-phase boundaries) such as the one shown in Fig. 6. The activation energy to move the soliton one step to the right (from molecule 1 to molecule 2 in Fig. 6) turns out to be surprisingly high. We calculate 160°K at P=O [14], nearly an order of magnitude higher than the transition temperature of 20°K between phases II and III. The reason for the difference is that the transition temperature is largely a measure of the <u>inter</u>molecular coupling whereas the activation energy of moving the soliton is principally determined by the effective <u>intra</u>molecular potential (8) because in Fig. 6 it involves twisting molecule 1 before untwisting molecule 2. The strongest experimental evidence for our view of phase III is the observation of a strong NMR relaxation mechanism with activation energy of 172°K, in good agreement with our calculation [14].

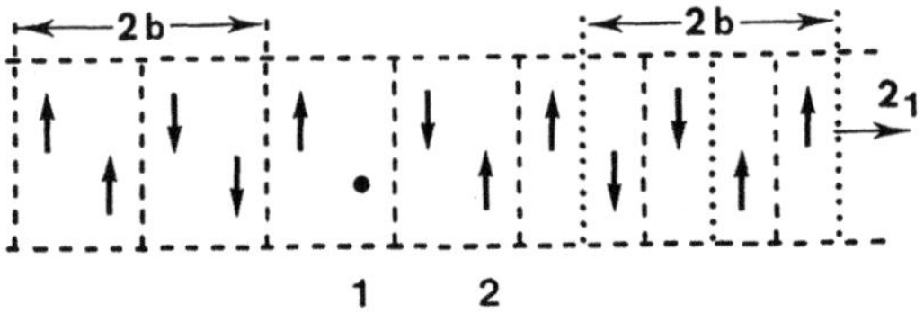

Fig. 6. A soliton in phase III of biphenyl. The commensurate equilibrium structure below 20°K with lattice constant 2b is shown in two symmetry-related forms on the left and right sides. Up and down arrows denote molecules with right and left handed twists. Molecule 1 at the centre of the soliton has zero twist. Note that the centre of the soliton lies on a molecule, not between two molecules, because the intramolecular potential (8), (9) is a single well and not a double well. (After Ref. [14].)

<u>5. Multiphase Degeneracy in silicon carbide polytypes</u>

This is a difficult problem and an unfinished story [29, 30, 15, 16]. The ordinary cubic form of SiC can be viewed as hexagonal atomic double layers of Si and C atoms, stacked on top of one another in the (111) crystallographic direction. Now each double layer (just called 'layer' for short hereafter) can be placed on the one below in two distinct possible ways by rotating it by 180° in its plane. Thus an infinite number of regular and random stacking sequences are possible and several dozen regular periodic polytypes have been observed. What microscopic mechanism causes these polytypes ? The simpler ones now appear to be equilibrium phases under some conditions and the others must also be so or at least very close to being equilibrium phases or they would hardly be observed so frequently with such regular structure.

The current picture involves two distinct ingredients, [15, 16] of which the first can be called an interplanar interaction model with a multiphase degeneracy. It is convenient to specify the two stacking orientations of each layer as the two states $\sigma = \pm 1$ or 'up' and 'down'. The total energy or free energy of the system can be written

$$G = \Sigma\,(-J_1\,\sigma_n\,\sigma_{n+1} - J_2\,\sigma_n\,\sigma_{n+2} - J_3\,\sigma_n\,\sigma_{n+3}) \qquad (11)$$

where the J_i are i'th neighbour interplanar interactions, truncated arbitrarily after J_3. If we set $J_3 = 0$ then (11) has a multiphase degeneracy for $J_1 = 2|J_2|$: all structures have equal energy consisting of bands of two or more consecutive 'up' or 'down' layers, i.e. consisting only of what we call n-bands where $n \geq 2$. The J_2 has to be negative ('anti ferromagnetic') for this, and the degeneracy results from the frustration already discussed in the J_1, J_2 mechanism (Sect. 2.3). It is the same as in the well known ANNNI model at $T = O$ [29, 30]. For $J_3 < 0$ the degeneracy point becomes two multiphase degeneracy lines (Fig. 7), one of which

$$J_1 + 2J_2 - 3J_3 = 0 \qquad (12)$$

consists of all structures of 2-bands and 3-bands which includes almost all the observed polytypes except the cubic one $<\infty>$.

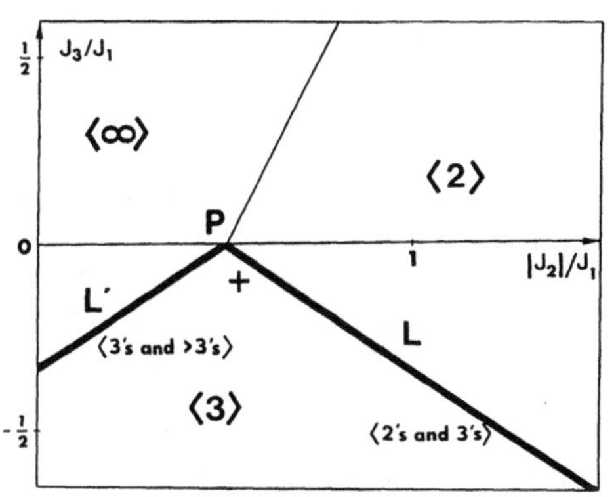

Fig. 7. Phase diagram for the interplanar interaction model, including one ordinary phase boundary (light line) and two multiphase degeneracy lines L and L' (heavy) radiating from the degeneracy point P where $J_1 = 2|J_2|$ and $J_3 = 0$. This line (12) is L, and the cross represents the values of J_1, J_2, J_3 found in the calculations for T = 0°K. (After Ref. [16].)

We have made ab initio calculations of the total energy of five simple polytypes and analysed them in terms of (11) to obtain the J_1 [15, 16]. These calculations involved self-consistent solutions of the Schrodinger equation in the local density approximation to density functional theory using norm-conserving pseudopotentials from calculated free atoms. The results represent energies at T = O of course. In order to obtain the small differences between polytypes the energies had to be calculated to an accuracy of about 1 in 10^6, with special precautions about sampling the electron states in k-space to ensure identical basis functions for all the polytypes [16].

The results show that (11) represents the energy very well, even for a very different and artificial polytype, with J_4 very small. The sign of J_2 is negative as required, and the value of $J_1/|J_2|$ comes out at 1.97 which is very close to the multiphase degeneracy ratio of 2. The J_3 is quite small and the representative point in a J_1, J_2, J_3 phase diagram is very close to the degeneracy line (12) as shown in Fig. 7. The phase <3>, i.e. 'three up, three down' stacking, is one of the two commonest polytypes and comes out with the lowest energy by 4 x 10^{-4} eV per SiC pair of atoms.

So far so good. The system is clearly near the multiphase degeneracy line (12) where nearly all the polytypes would be equilibrium phases. Our calculations at T = 0°K give the system near the line but not exactly on it. However if we include the phonon contributions to the free energies of the polytypes, then G and the J_i in (11) and (12) become temperature-dependent free energies. It would require only a 0.03% difference in average phonon frequencies between the phases <3> and <2> to shift the representative point on Fig. 7 from our calculated value at T = O°K onto the multiphase degeneracy line (12) at 2000°C. Such a very small difference in phonon frequency is entirely plausible in view of the large differences (up to a factor of two) in the band gaps of different polytypes. The movement of the representative point onto the multiphase degeneracy line at high temperature would allow all those degenerate phases to form as equilibrium structures. However it would also allow the equilibrium formation of all random stacking sequences of 2-bands and 3-bands, and this is not what we want because there are infinitely more random sequences than regular ones so that the latter would never be observed in practice.

We must therefore assume as a second ingredient of our theory that there is some additional mechanism which splits the multiphase degeneracy line (12) slightly. On the phase diagram it would be replaced by a fan of very many very narrow discrete phases. As the representative point moves on the phase diagram it would cut through these phases each one being the equilibrium structure at slightly different temperatures. The polytype structures would then be retained metastably after cooling. The most important point is that all known mechanisms for such splittings eliminate the random stackings. The large number of observed polytypes suggests a devil's staircase of all rational long period superlattice lengths.

What mechanism could cause such splitting of the multiphase degeneracy [16]? The mechanism in the ANNNI model is implausible for SiC because it involves local little kinks on the stacking planes which are not observed in electron micrographs in SiC and would have very high energy. Another possibility might be thermal lattice vibrations [31] which have a large amplitude at the high temperatures of polytype formation. The phonon effects can be surprisingly long range in diamond-type structures. Another possibility is an adaptation of the BRUINSMA and ZANGWILL [26] mechanism which depends on a distance dependence of the J_i and consequent longitudinal relaxation of the interlayer distances. The theory maps mathematically more or less onto the FK/FM model [22] (Sect. 2.4). Such inequalities of bond lengths are certainly found experimentally in the <3> polytype.

1. For review see V. Heine: Solid State Physics, to be published.
2. V. Heine, J.D.C. McConnell: Phys.Rev.Lett. 46, 1092 (1981)
3. V. Heine, J.D.C. McConnell: J.Phys.C. 17, 1199 (1984)
4. J.D.C. McConnell, V. Heine: Acta Cryst. A40, 473 (1984)
5. V.Heine: In Statics and Dynamics of Non-Linear Systems, ed by H. Bilz, G. Benedek and R. Zeyher (Springer, Berlin, Heidelberg 1983).
6. J.D.C. McConnell, V. Heine: J.Phys.C 15, 2387 (1982)
7. V. Heine, R.M Lynden-Bell, J.D.C. McConnell, I.R. McDonald: Z.Phys.B. Cond.Matter 56, 229 (1984)
8. J.D.C. McConnell, V. Heine: Phys. Rev. B. 31, 6140 (1985)
9. V. Heine, S. L. Price: J.Phys. C 18, 5259 (1985)
10. C. Benkert, V. Heine, E.H. Simmons: Europhys. Lett. 3, 833 (1987)
11. C. Benkert, V. Heine: J.Phys. C. to appear.
12. C. Benkert, V. Heine: J.Phys. C. to appear
13. C. Benkert, J.Phys.C to appear
14. C. Benkert, V. Heine: Phys. Rev. Lett. 58, 2232 (1987)
15. C. M. Cheng, R.J. Needs, V. Heine, N.I. Churcher: Europhys. Lett. 3, 475 (1987)
16. C.M. Cheng, R.J. Needs, V. Heine: submitted to J.Phys.C.
17. V. Heine, J.J.A. Shaw: Surf.Sci. to be published
18. V. Heine, J.J.A. Shaw: Surf.Sci to be published
19. V. Heine: Solid State Phys. 35, 1 (1980) especially p.104. V.Heine, J.H. Samson: J.Phys.F. 10, 2609 (1980)

20. A.P. Levanyuk, D.G. Sannikov: Fiz. Tverd. Tela, 18, 423 and 1927 (1976) Translation Sov. Phys. Solid St. 18, 245 and 1122 (1976)

21. J.D.C. McConnell: Zeif.f.Kristall. 147, 45 (1978)

22. Y. I. Frenkel, T. Kontorova: Zh. Eksp. Teor. Fiz. 8, 1340 (1938). F.C. Frank, J.H..van der Merwe: Proc. Roy. Soc. A198, 205 and 216 (1949)

23. S. Aubry: J.Phys.C 16, 2497 (1983)

24. P. Bak, R. Bruinsma: Phys. Rev. Lett. 49, 249 (1982)

25. H. Bilz, H. Buttner, A. Bussmann-Holder, W. Kress, U. Schroder: Phys. Rev. Lett. 48, 264 (1982)

26. R. Bruinsma, A. Zangwill: Phys. Rev. Lett. 55, 214 (1985)

27. R. Kind, P. Muralt: In Incommensurate Phases in Dielectrics: 2 Materials, ed. by R. Blinc, A.P.Levanyuk, p.301 (North Holland, Amsterdam 1986)

28. H. Cailleau In Incommensurate Phases in Dielectrics: 2 Materials, ed. by R. Blinc, A.P.Levanyuk, p.71 (North Holland, Amsterdam 1986)

29. J.D.C. McConnell, V. Heine: Eurphysics Conf. Abstracts 6A, 172 (1982)

30. J. Smith, J. Yeomans, V. Heine: In Modulated Structure Materials, ed. by T. Tsakalakos, p.95 (Martinus Nijhoff, Dordrecht 1984)

31. J. Yeomans: private communication

Electron Microscopy of Static and Dynamic Phenomena

G. Van Tendeloo, J. Van Landuyt, and S. Amelinckx

University of Antwerp, R.U.C.A.,
Groenenborgerlaan 171, B-2020 Antwerp, Belgium

1. Introduction

Due to the unique combination of real space and reciprocal space information, electron microscopy is an extremely powerful technique for revealing and analysing static or dynamic structural features. With modern electron microscopes structural details of the order of 0.2 nm can be directly observed so that structural defects can be studied down to an atomic scale. For a more general introduction on electron microscopy and the possibilities of the technique we refer e.g. to [1,2,3]. In this contribution we will treat four case studies taken from different types of materials; they are meant to illustrate the possibilities of electron microscopy combined with electron diffraction in the study of static and dynamic phenomena.

2. Static or dynamic ?

Using X-rays or neutron diffraction techniques it is relatively easy although not always unambiguous to decide whether diffuse streaking in reciprocal space is due to static or dynamic phenomena. Using electron diffraction only this distinction is almost impossible because electron diffraction is no longer kinematical but is complicated by dynamical scattering events. High energy electrons interacting with the material undergo multiple scattering even for very small thicknesses of the material. Local thickness changes and slight differences in orientation furthermore complicate a straightforward interpretation. However electron diffraction combined with high resolution real space information allows in a number of cases to separate both contributions. We will consider two related examples here : <110>* diffuse scattering in $Ni_{1+x}Al_{1-x}$ and in Cu-Zn-Al [4]. Electron diffraction patterns of these two materials along [001] are shown in fig. 1b and 2b respectively, while their corresponding high resolution pictures are reproduced in fig. 1a and 2a. To find out whether the information incorporated in the electron diffraction pattern is also reflected in the image one uses optical diffraction whereby the negative acts as an optical grating for the laser beam in order to obtain an optical fourier transform of the imaged intensities (fig. 1c and 2c). Comparing the original electron diffraction patterns with their optical analogue learns that for Ni-Al the <110>* streaking is still present in the optical diffraction while for Cu-Zn-Al it is almost completely absent. This indicates that for Ni-Al the "tweed" modulation which is also visible in the high resolution image is at least static during the time to record the picture which is typically around 1 sec. For Cu-Zn-Al no such effect is observed; the diffuse intensity has vanished and as Dvorack has pointed out for Cu-Zn-Ni, using single crystal X-rays, it has to be mainly attributed to thermal diffuse scattering [5]. In another example of lead orthovanadate [$Pb_3(VO_4)_2$] diffuse streaking, appearing as a pretransition effect for the $\gamma \to \beta$ transition, has been found to be completely of dynamic origin; in the high resolution image no effect at all has been observed although the wavelengths involved (~ 0.6 nm) were large enough to be resolved. Structural details of the order of 0.25 nm are easily resolved in the image but no extra modulation is present. The electron diffraction pattern shows

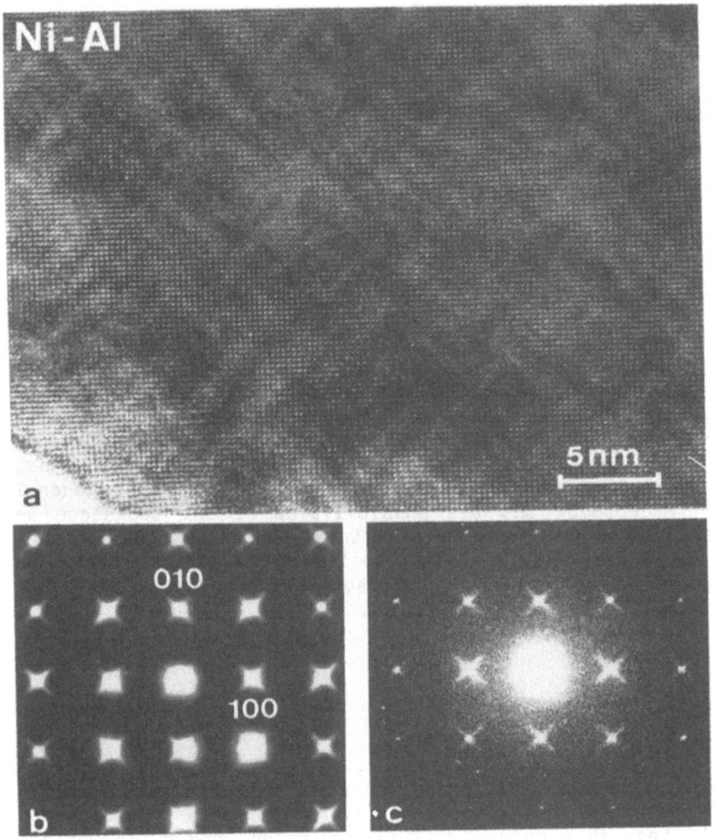

Fig. 1. a) High resolution image of $Ni_{62}Al_{38}$ along the [001] zone. The basic lattice is clearly resolved as white dots separated by 0.4 nm but coarse irregular modulations (tweed [8]) are observed along (110) and (1$\bar{1}$0).
b) [001] electron diffraction pattern showing diffuse streaking along different <110>* directions.
c) Optical diffraction pattern obtained from the high resolution image a).

pronounced satellites in the c-zone of the trigonal γ-phase while the optical diffraction is devoid of any diffuse intensity [6][7].

3. α-β Phase Transformation in SiO_2 Quartz

The α-β phase transformation at 843K in quartz, although known for more than 100 years has been more frequently studied than any other transformation; however it still revealed unexpected aspects in recent studies [9-14]. Using dark field electron microscopy the presence of an incommensurate columnar triangular domain structure was observed in a narrow temperature interval between the high temperature β-phase and the low temperature α-phase [9]. More recently neutron and electron microscopy studies have identified the incommensurate phase as stable structure within a temperature range of about 1.3K; the $\bar{q}$-vector is in the range of 0.06 nm^{-1} and its orientation is approximately along the y-axis [10,11,12]. In the present report we would like to show evidence of :

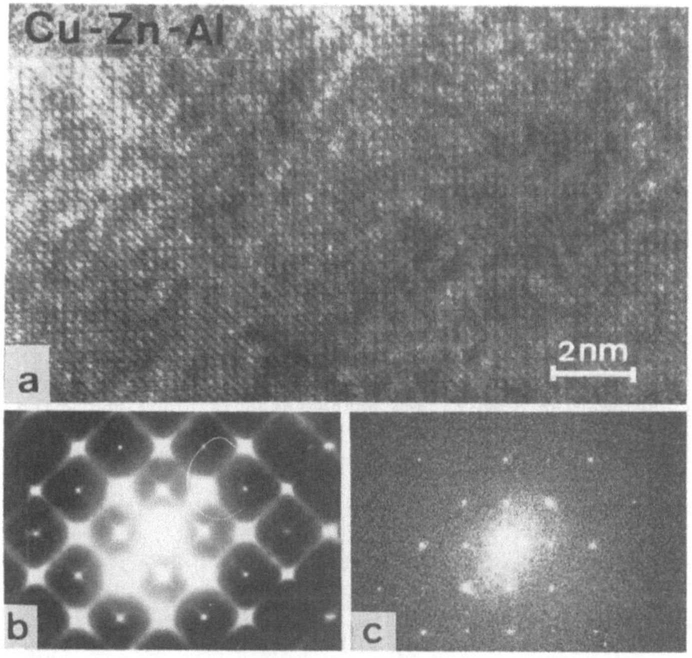

Fig. 2. a) High resolution image of Cu-Zn-Al. The white dots of the basic lattice are at a distance of 0.2 nm.
b) Corresponding [001] electron diffraction pattern; pronounced <110>* streaking is present.
c) Optical diffraction pattern obtained from the high resolution image a); note the complete absence of diffuse intensity.

a) the existence of the incommensurate phase
b) the dependence of magnitude $|q|$ and direction of $\bar{q}$ with temperature
c) the existence of macrodomains
d) the effect of stress upon the 3q state.

The $\beta \to \alpha$ phase transformation is a structural phase transition as a result of which the Si-ions undergo displacements as represented in fig. 3a. The oxygen tetrahedra which surround the Si-ions and which undergo synchronous rotations have been omitted in the figure. The order parameter η can be taken to have a magnitude equal to the magnitude of the displacement of a given Si-ion; it is positive if the displacement is in the direction indicated and negative if the displacement is in the opposite direction. Obviously $\eta=0$ in the high temperature β-phase. The Dauphiné twins in the α-phase which in the electron micrographs appear as bright or dark triangular domains correspond to areas with $\eta>0$ and areas with $\eta<0$ respectively. These domains are related by a 180° rotation and evidently have the same free energy. From symmetry considerations it is evident that the free energy per unit area of domains should have a minimum at a position $\phi \neq 0$ where ϕ is the angle the domain has rotated about the z-as, away from the (x,z) orientation [12]. Actually α dauphiné domains being rotated over $+\phi$ cluster together and form macro-domains separated from areas where all α-domains are rotated over $-\phi$. Such actual configuration is shown in fig. 4 where the white lines separate $+\phi$ and $-\phi$ macrodomains. In fig. 3b two such macrodomains are represented schematically. These observations can be performed in situ in the electron microscope by heating the material till close to the transition and then triggering the transformation by controlling the intensity of the electron beam.

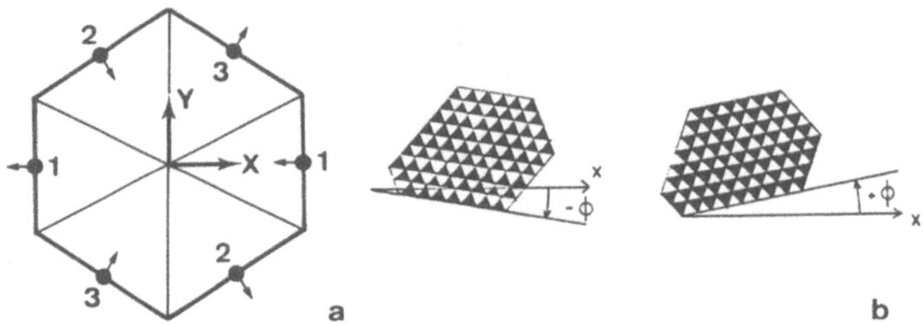

Fig. 3. a) Wigner Seitz unit cell of SiO_2-quartz. Only the Si-ions are represented. Arrows indicate ion displacements in the low temperature α-phase.
b) Two macrodomains with parallel c-axis rotated respectively over +ϕ and -ϕ away from the X-axis. α_1 and α_2 triangular domains are represented in black and white.

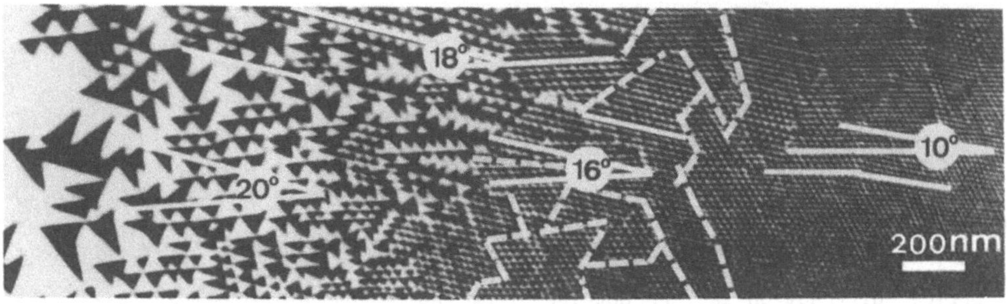

Fig. 4. $30\bar{3}1$ dark field of quartz heated inside the electron microscope till 843K The right part is at a higher temperature; the left part at a lower temperature. Macrodomains marked by dotted lines are clearly visible. The angle 2ϕ between neighbouring macrodomains is also indicated; it clearly increases with decreasing temperature.

The strong black-white contrast in the electron microscope images between the Dauphiné twin variants α_1 and α_2 is obtained using the dark field mode; for imaging a diffracted beam is selected for which the structure factor for α_1 is appreciably different from that of α_2. The present observations are all $30\bar{3}1$ dark field images for which the kinematical intensity from α_1 is 30x larger than from α_2. The recorded configurations are extremely sensitive to small temperature changes and the domains constantly vibrate at a frequency detectable by the eye. Since the electron beam necessarily introduces a temperature gradient the behaviour as function of temperature can be easily recorded (see fig. 5). At the cooler side (top part) individual α_1-α_2 Dauphiné twins can be recognized as black or white domains. The size of the domains decreases towards the bottom of the picture where the temperature is highest. The wavelength of modulations decreases from q^{-1}= 62 nm till q^{-1} = 14 nm. This necessarily introduces "dislocations" in the domain configuration as visible e.g. in the inset corresponding with q^{-1}= 22 nm. At the same time the angle 2ϕ between macrodomains varies from about 20° down to about 10° (see fig. 4). This is in agreement with theoretical calculations of the free energy of a domain as function of its orientation [13]. Dauphiné twins smaller than 14 nm are never observed because the resolution in the image is limited by the dynamical character of the Dauphiné domain walls which are constantly vibrating and changing position. The contrast fades out (see bottom of fig. 5) and the material transforms into the β-phase.

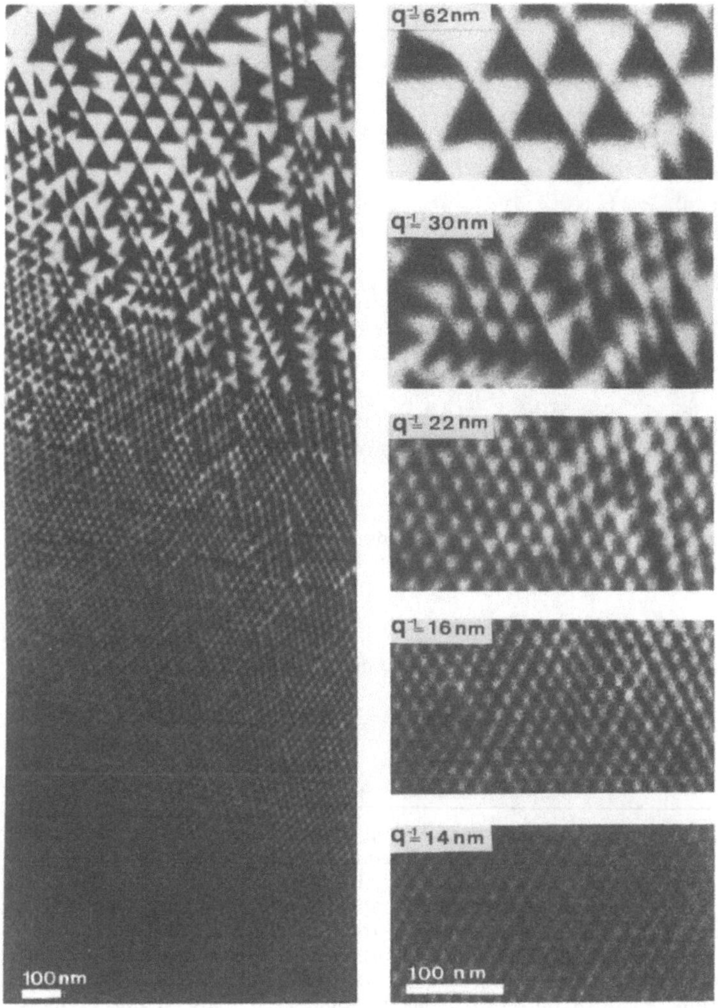

Fig. 5. Dark field image of quartz where the variation of domain size with temperature is clearly visible. No domain smaller than 14 nm has been observed.

When quartz thin foil single crystals are heated in the electron microscope cracks are formed when cycling through the transition due to thermal expansion and the volume change associated with the α-β transformation. This effect is due to local strains inside the material which could be the reason why a 3q state is not always observed. In some areas as in fig. 6 the transformation is clearly more of the 1q state; the transformation front no longer consists of equilateral prisms as it is the case in fig. 4 or 5. Since the direction of such strain field cannot be controlled one has to be lucky to have a strain field perpendicular to one of the $\bar{q}$ vectors. Fig. 6 is a clear example of such a configuration under stress. Only one q vector is prominent and the needle-shaped domains at the transformation front are far from being equilateral triangles. A 16mm movie has been made from real-time observations in the electron microscope; it shows clear evidence for the movement and the shape changes of the α_1-α_2 domains with temperature [14]. The material can be cycled several times through the transition

Fig. 6. Dark field image of quartz transformed under stress. Only a line pattern is observed indicating the formation of a one-dimensional (1q) modulated structure.

and no memory phenomena are observed except when one of the domains is pinned to a small precipitate or defect cluster.

4. Orthorhombic-tetragonal transformation in $Ba_2YCu_3O_{7-\delta}$

The recent discovery of superconductors with transition temperatures above 77K has stimulated great interest in the structure and microstructure of these materials. It is now well established that in the case of $Ba_2YCu_3O_{7-\delta}$ the superconducting phase is orthorhombic and that the superconducting properties strongly depend on parameters such as non-stoichiometry, heat treatment microstructure etc. [15][16]. In view of the subject of the present meeting we will not treat all the aspects of a microscopy study; we refer to [17], but restrict ourselves to those phenomena associated with dynamic aspects of the transitions in these materials.

The superconducting phase of $Ba_2YCu_3O_{7-\delta}$ is based on the perovskite structure and has an orthorhombic, pseudo tetragonal unit cell consisting of three perovskite cubes stacked along the c-axis. The corners of the cubes are occupied by Cu-atoms, the edge centers by oxygen or vacancies and the body centers by Y and Ba, in a sequence Y Ba Ba Y Ba Ba ... along the c-axis. The non stoichiometry is accommodated by vacancies on the oxygen sublattice. A structural model is reproduced in fig. 7. The symmetry reduction from tetragonal to orthorhombic is solely introduced by the ordered arrangement of the vacancies. Such vacancies in the basal plane form one-dimensional chains along the $\bar{b}$-axis. The difference in lattice parameters between a_0 and b_0 is extremely small : 0.382 compared to 0.389 nm. The structure becomes orthorhombic when all vacancies are filled up by oxygen or when the vacancies disorder on the O-sublattice.

In the superconducting state the structure is orthorhombic and the material shows frequent twinning i.e. areas where the a and b axis are permuted; the c-axis remaining common. The coherent twin plane is the (110) plane. Such defects can be easily identified by electron microscopy. Initially all twin boundaries are strictly parallel to (110) (or $1\bar{1}0$) and their average separation is around

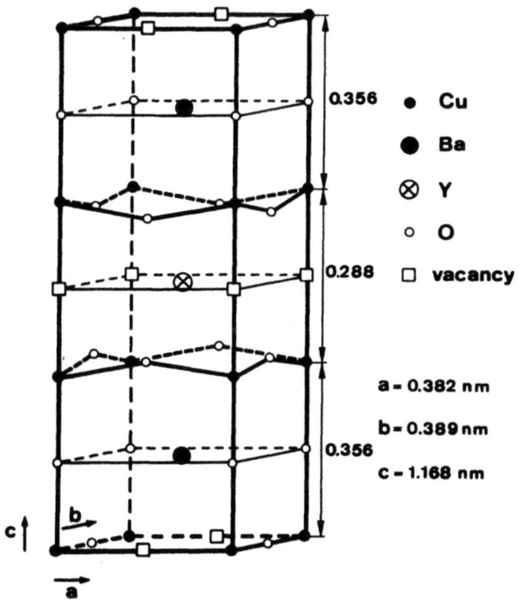

Fig. 7. Structure model of $Ba_2YCu_3O_{7-\delta}$.

100 nm. When imaged in the diffraction contrast mode as in fig. 8a these defects can be easily recognized by the difference in background contrast on both sides of the interface. Under high resolution imaging conditions which reveal the projected heavy metal configuration, the slight change in orientation of the (100) or (010) planes over the twin boundary can be observed (fig. 8b). This orientation change is caused by the permutation of the $\bar{a}$ and $\bar{b}$ axes which are slightly different in length. Optical diffraction from the H.R. images such as fig. 8cde allow to select areas down to 20 nm and clearly attribute the spot splitting observed in the electron diffraction pattern to (110) or ($1\bar{1}0$) twinning at the observed defects.

In the electron microscope a local heating can be introduced by focussing and defocussing the electron beam on the specimen. Since the material at room temperature is an insulator it will heat up under the influence of the electron beam. Upon extreme heating above 750°C one passes the O→T transformation; as a consequence the twins disappear; which upon cooling however re-establish, at different positions [17]. After in situ heating the material above the orthorhombic-tetragonal transition and subsequently observing the same area at room temperature or slightly above, a series as reproduced in fig. 9 can be obtained as function of time. Immediately after the heat pulse the crystal is in the tetragonal phase (fig. 9b); no twin interfaces are observed. The corresponding electron diffraction pattern (fig. 10b) also fails to show any spot splitting. After having left the crystal during an interval of time of the order of 5 minutes at microscope temperature the twin interfaces reappear (fig. 9c) as well as the spot splitting in the diffraction patterns (fig. 10c) albeit with a smaller magnitude of splitting as in the original configuration (fig. 10a). The twin interfaces are vague at first and not strictly located along (110) but gradually they sharpen up again (fig. 9d) and the spot splitting in the diffraction pattern increases to the magnitude before heating.

In fig. 10b it is further noticed that weak streaks of diffuse scattering are visible along the [100]* and [010]* rows of spots. After some time the diffuse streaks have shortened somewhat into elongated diffuse superstructure spots centered at positions 1/2 0 0 and 0 1/2 0. The high resolution images exhibit short segments of extra bright rows of dots along the [100] or [010] directions with an average spacing which is twice the normal (100) or (010) spacing. The directions of the doubly spaced dot rows differ by about 90° in orientation in

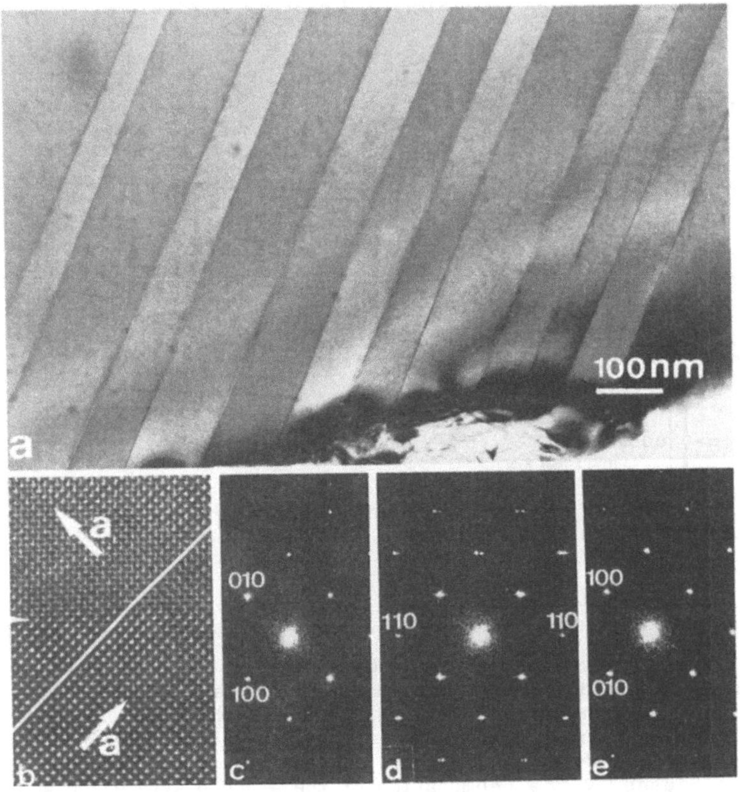

Fig. 8. a) Twin boundaries in $Ba_2YCu_3O_{7-\delta}$ parallel with the (110) plane imaged in the diffraction contrast mode.
b) A single twin boundary imaged under high resolution conditions; over such twin boundary the $\bar{a}$ and $\bar{b}$ axis are interchanged.
c)d)e) optical diffraction patterns from the high-resolution image of b selecting an area of 20 nm diameter.
c) from the left part
d) over the twin boundary; the splitting parallel with the $[1\bar{1}0]^*$ row is clearly visible
e) from the right part.

successive twin bands (fig. 11a). This is even more pronounced when recording optical diffraction patterns from successive twin bands (fig. 11b,c); the diffuse streaking has clearly also rotated over 90°. A similar doubling of the (010) periodicity has already been observed in specimens slowly furnace cooled between 750°C and 600°C [18] so that the effect described here is certainly not an artifact introduced by in situ electron microscope heating. All the observed phenomena can consistently be explained as resulting from an order-disorder sequence taking place on the oxygen sublattice. The disordering is further accompanied by some loss of oxygen resulting in an excess in the oxygen vacancy concentration. Following [19, 20] we assume that the oxygen atoms in the Cu-O-planes (i.e. the CuO layer common to the two barium-containing cubes) are the most loosely bound ones in the structure of $Ba_2YCu_3O_{7-\delta}$ since their thermal agitation ellipsoids are largest. These are the oxygen atoms which disorder first on heating and eventually are lost.

Since the structure is orthorhombic only because of the ordering of the vacancies, it becomes tetragonal when the vacancies are randomly distributed; it would

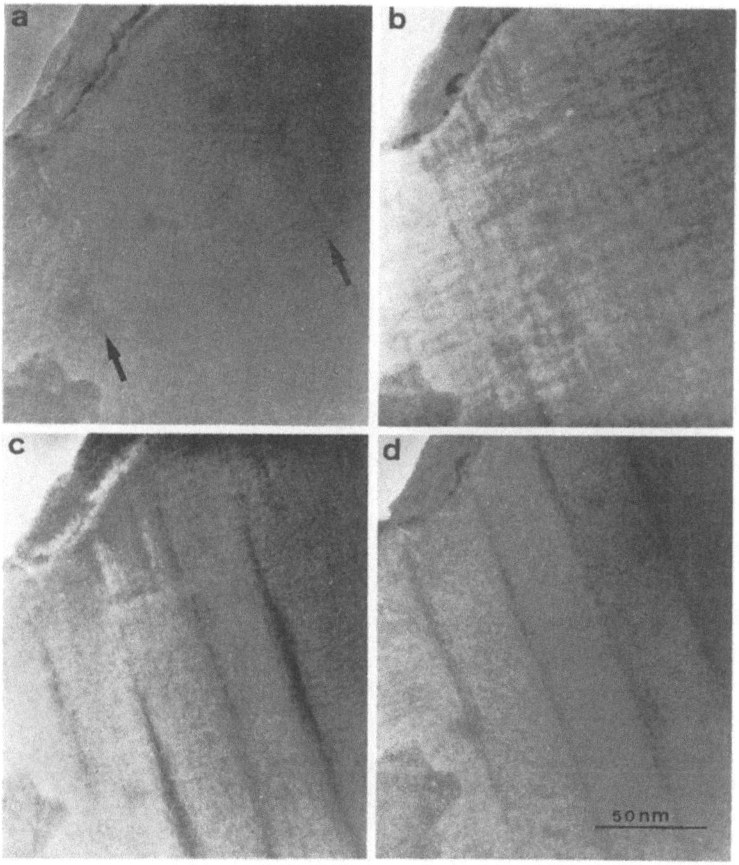

Fig. 9. High-resolution images of the same area all recorded at the same microscope temperature.
a) Starting configuration; only two twin boundaries are present.
b) Immediately after heat pulsing the twin interfaces have disappeared; only a mottling is present.
c) The twin interfaces start forming but they remain vague.
d) After about five minutes the twin interfaces have sharpened; a configuration different but similar to a) is realised.

also become tetragonal if the Cu-O-□ layer would become completely devoid of oxygen, i.e. become pure Cu-□-layers; the composition would then become $BaYCu_3O_6$, which is no longer a superconductor.

During heat pulsing the oxygen is apparently mainly disordered and only to a small extent lost, since an orthorhombic structure is recovered after a few minutes at microscope temperature. A model for the structure of this Cu-O- plane is represented [17]. The present model is at variance with one of the models proposed previously [17]; it is in a sense "complementary" to that model. The difference is based on the fact that we now convince ourselves that the structure must accommodate a deficiency of oxygen rather than an excess.
The decrease of the spot splitting due to twinning is a direct consequence of the decrease in the orthorhombic deformation. This in turn is consistent with the fact that in the superstructure the density of Cu-O-Cu-O rows or strings along the b_0 direction is only half of that present in the normal structure.

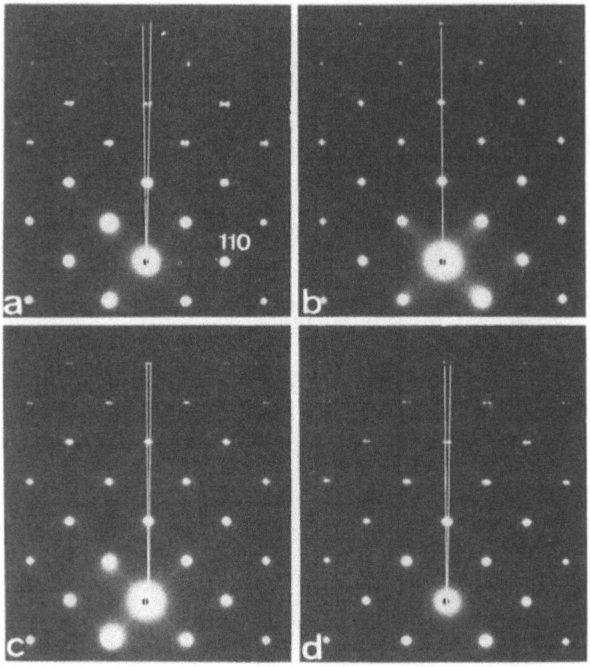

Fig.10. Diffraction patterns along [001] corresponding to the same area as fig. 9 and recorded immediately after the recording of the images of fig. 9a-d. a) Initial state; b) immediately after pulsing in the tetragonal phase; c) after about 3 min.; d) after five minutes.

Fig.11. a) High-resolution image along [001] after repeated heat pulses. The dot rows parallel with b_0 acquire a double period.
b,c) Optical diffraction pattern from successive twin bands exhibit weak spots due to a one-dimensional superstructure. The two variants b) and c) are clearly the result of twinning.

The fact that the superstructure is only one-dimensional and hence that we still have O-Cu-O-Cu linear strings can be concluded from the presence of the extra bright dot rows in the high-resolution image and from the corresponding optical diffraction patterns (fig. 11).
Actually such a doubling of the unit cell along a_0 or b_0 has been suggested by de Fontaine and Moss [21]. The structure proposed however is a conservative one with respect to the vacancy concentration of the "ideal" structure.

References

1. P.B. Hirsch, A. Howie, R.B. Nicholson, D.W. Pashley and M.J. Whelan, "Electron Microscopy of Thin Crystals", Butterworths, London 1965.
2. Diffraction and Imaging Techniques in Materials Science, Eds. S. Amelinckx, R. Gevers, J. Van Landuyt, N. Holland Publ. Co. (1978).
3. J.C.H. Spence, "Experimental High-Resolution Electron Microscopy", Clarendon Press, Oxford 1981.
4. G. Van Tendeloo, S. Amelinckx, Scripta Met. 20, 335 (1986).
5. M.A. Dvorack and H. Chen, Scripta Met. 17, 131 (1983).
6. C. Manolikas, G. Van Tendeloo and S. Amelinckx, Solid State Communications 60, 749 (1986).
7. C. Manolikas, E. Paloura, G. Van Tendeloo and S. Amelinckx, Mat. Res. Bull. 21, 695 (1986).
8. L.E. Tanner, Phil. Mag. 14, 111 (1966).
9. G. Van Tendeloo, J. Van Landuyt and S. Amelinckx, Phys. stat. sol. (a) 33, 723 (1976).
10. G. Dolino, J.P. Bachheimer, B. Berge, C.M.E. Zeyen, J. Physique 45, 361 (1984).
11. G. Dolino, J.P. Bachheimer, B. Berge, C.M.E. Zeyen, G. Van Tendeloo, J. Van Landuyt, S. Amelinckx, J. Physique (Paris) 45, 901 (1984).
12. J. Van Landuyt, G. Van Tendeloo, S. Amelinckx, M.B. Walker, Phys. Res. B31, 2986 (1985).
13. M. B. Walker, Phys. Rev. B28, 6407 (1983).
14. J. Van Landuyt, G. Van Tendeloo and S. Amelinckx, Phase transitions between α- and β-phases in quartz, Scientific Films, Institut Wiss. Film Götingen, 1984, ref. E2699.
15. D.G. Hinks, L. Soderholm, D.W. Capone II, J.D. Jorgensen, I.K. Schuller, C.U. Segre, K. Zhang and J.D. Grace, Phys. Rev. Letters (in the press).
16. R.J. Cava, B. Batlogg, R.B. Van Dover, D.W. Murphy, S. Sunshine, T. Siegrist, J.P. Remeika, E.A. Rietman, S. Zahurak and G.P. Espinosa, Phys. Rev. Letters, 58, 1676 (1987).
17. H.W. Zandbergen, G. Van Tendeloo, T. Okabe and S. Amelinckx, Phys. stat. sol. (a) (in the press), 1987.
18. G. Van Tendeloo, H.W. Zandbergen and S. Amelinckx, Solid State Comm. 63, 603 (1987).
19. J.J. Capponi, C. Chaillout, A.W. Hewat, P. Lejay, M. Marezio, N. Nguyen, B. Raveau, J.L. Soubeyroux, J.L. Tholence and R. Tournier, Europhys. Letters 3, 1301 (1987).
20. M.A. Beno, L. Soderholm, D.W. Capone, D.G. Hinks, J.D. Jorgenson, Y.K. Schuller, C.K.Segre, K.Zhang, J.D. Grace, Appl. Phys. Letters (in the press).
21. D. de Fontaine and S.C. Moss, Phys. Rev. Letters (to be published).

Some Problems for Physicists in First Order Diffusional Phase Transformations in Crystalline Solids

H.I. Aaronson and R.V. Ramanujan

Department of Metallurgical Engineering and Materials Science,
Carnegie-Mellon University,
Pittsburgh, PA 15213, USA

Although the homogeneous nucleation kinetics of precipitates formed during first order diffusional phase transformations now appear to be well understood when the precipitate has the same crystal structure and orientation as the matrix, analysis of homogeneous nucleation involving a significant change in crystal structure has just begun. The core problem is evaluation of the chemical interfacial energy as a function of boundary orientation with a fixed lattice orientation relationship for coherent interphase boundaries. Quantitative understanding of nucleation at dislocations requires analysis of strain energy interactions between faceted nuclei and the dislocations, using anisotropic, inhomogeneous elasticity theory. Diffusional growth of precipitates whose crystal structure differs from that of their matrix can be qualitatively understood in terms of interphase boundaries which have a partially coherent structure. These boundaries are immobile in the direction normal to themselves and can migrate only by means of the ledge mechanism. Theory of the origin of ledges, and also of kinks on the risers of ledges, is the main gap to be filled in order to place this approach to diffusional growth on a quantitative basis. Theory of morphological instability and of growth of the unstable products at the point of instability appears applicable in the large to precipitation from solid solution but must be reworked in terms of the ledge mechanism in order to make such applications more realistic.

There have been increasing indications in recent years that physicists are becoming seriously interested in problems involving first order phase transformations in crystalline solids which heretofore have mainly concerned physical metallurgists. The purpose of this talk is to encourage the further development of such interest. A stronger and deeper theoretical base will surely lead not only to a better understanding of existing observations but also to the insight needed to make important new discoveries.

We will discuss first problems in diffusional nucleation. Then we will move on to current interests in the area of diffusional growth. This choice of topics is quite selective. Thus, nothing will be said of either martensitic transformations or of continuous phase transformations such as spinodal decomposition and spinodal ordering. Nonetheless, the subjects which are considered are applicable to very wide ranges of crystalline materials and classes of transformations taking place within them.

1. NUCLEATION

1.1 Homogeneous Nucleation

Metallurgists have the impression, quite pronouncedly, that physicists are unhappy with nucleation theory. Perhaps this is because GIBBS' [1] mathematics leads to very important results through the agency of algebra and calculus of the simplest types! But then there are also more substantial experimentally based reasons for such displeasure. Several studies of homogeneous nucleation of one liquid within another [2-5], or of a vapor within a liquid [6], have revealed important apparent disagreements between nucleation theory and experiment. The most striking of these is that the nucleus:matrix interfacial energy back-calculated from experimental measurements and homogeneous nucleation theory is higher than that independently measured. Whereas a lower back-calculated value strongly implies that nucleation was heterogeneous, a higher interfacial energy indicates that there is a serious deficiency in the theory. BINDER and STAUFFER [7] seem to have been the first to recognize that these disagreements arose from defects in the design of the experiments. Evaluating nucleation kinetics by means of the "cloud point"--the undercooling at which homogeneous nucleation apparently first becomes copious--actually involves, they realized, a melange of nucleation, growth and coarsening. Pursuing this lead further, LANGER and SCHWARTZ [8] concluded that: (a) effectively all of the available data is consistent with a complex analysis involving all three of these processes, and (b) sorting out nucleation kinetics on this approach is feasible only at such small undercoolings that nucleation rates may then be negligible.

Because metallurgists usually work with crystalline solids in which diffusivities are much smaller than in liquids and in which transformations processes may be terminated prior to completion by quenching to room temperature, they have in principle the opportunity to do a better job of studying separately these processes. KIRKWOOD and co-workers [9-11], however, obtained data on the homogeneous nucleation kinetics of ordered fcc Ni_3Al within disordered fcc α Ni-Al by means of transmission electron microscopy which turned out to fall almost entirely within the coarsening regime. SERVI and TURNBULL [12] organized better fortune with the homogeneous nucleation of fcc Co-rich precipitates in fcc Cu-rich Cu-Co matrices through the use of resistivity measurements and a complicated four-level analysis. However, the indirectness of this approach left some concerns as to the accuracy of the nucleation rate data derived through its application. Recently, LE GOUES and AARONSON [13] redetermined homogeneous nucleation kinetics in Cu-Co alloys by employing the direct technique of counting particle number densities with the aid of TEM (transmission electron microscopy). They showed that there is a rather narrow "window", between ca. 0.5 and 1 A/O Co, within which nucleation and growth rates are neither too slow nor too rapid. Hence, as shown in Figs. 1a and 1b, particle number density could be evaluated at sufficiently early stages of reaction so that overlap of the diffusion fields of adjacent precipitates remained negligible. In the one composition used in both their work and that of SERVI and TURNBULL, excellent agreement was obtained between the two sets of nucleation rate data as a function of reaction temperature.

To account for the measured nucleation kinetics, three different forms of nucleation theory were used. One was classical homogeneous nucleation theory, wherein a discrete lattice plane, nearest neighbors, regular solution approach analysis [14] was used to compute the polar γ-plot, from which the Wulff construction was then derived [15]. The results are shown in Fig. 2 as a function of T/T_c, where T_c is the critical temperature of the regular solution miscibility gap. These shapes are geometrically similar to the critical nucleus shape; knowing this, all shape-dependent variables in the classical equation for the nucleation rate are then readily derived. The second was CAHN-HILLIARD [16] continuum non-classical nucleation theory and the third was COOK-DE FONTAINE [17, 18] discrete lattice point theory [19, 20]. Discrete lattice point theory permits the activation free energy for critical nucleus formation, ΔF^*, to be evaluated satisfactorily at large as well as at the small undercoolings appropriate to classical theory, and additionally permits the anisotropies of both interfacial energy and strain energy to be taken readily into account.

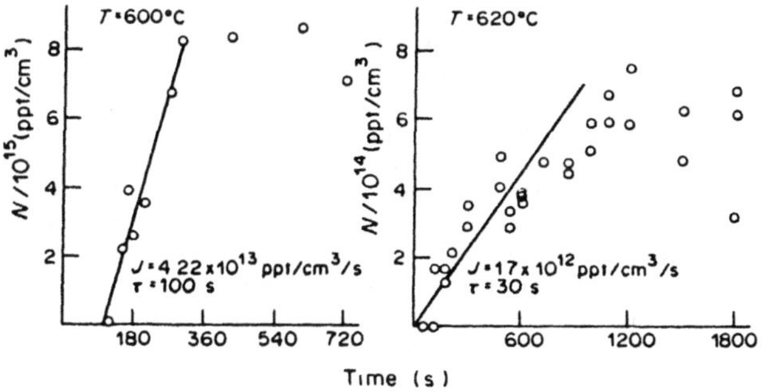

Fig.1 Representative plots of precipitate number density vs. isothermal reaction time in Cu-1 A/O Co at two different temperatures (13)

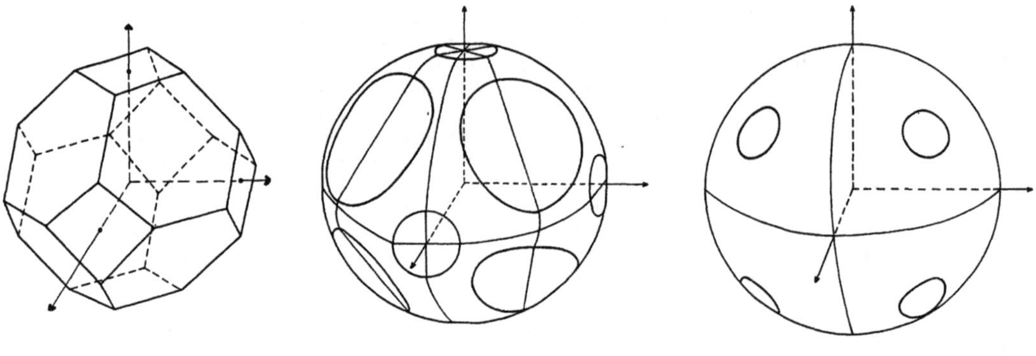

Fig.2 Wulff constructions performed on polar γ-plots of fcc:fcc chemical interfacial energy at (a) 0°K, (b) 0.25 T_c, and (c) 0.50 T_c for Cu-Co alloys (15)

Figs. 3a-c show that Cook-DeFontaine theory accounts remarkably well for the nucleation rate data in all three alloys. Since it is not easy to incorporate the volume strain energy attending transformation in Cahn-Hilliard theory, the nucleation kinetics calculated from all three theories are compared in the absence of strain energy in Fig. 3d for one of the alloys studied. The three theories are seen to yield nearly the same results within the temperature range studied. The larger differences with respect to the experimental data now exhibited are due to the (deliberate) failure to take account of strain energy. Thus classical theory is viable over a wider temperature range than anticipated [16]--and all three theories look good. Hence homogeneous nucleation theory really does have firm experimental support.

We are now attempting an undertaking which, at least in the past, might not have interested physicists. We seek to understand homogeneous nucleation during a diffusional phase transformation involving a change in both crystal structure and composition. The fcc->hcp transition was deliberately chosen for this enterprise as the simplest one involving a significant change in structure usually encountered in metallic alloys [21]. It should be noted, though, that transformations involving a change in crystal structure and composition are extremely common in alloy systems. A basic assumption underlying this study is that nucleation is coherent [22]. This may be justified in two different ways. The first is thermodynamic. On the considerations of VAN DER MERWE [23], precipitates in the size range typical of critical nuclei, i.e., <~ 10 nm, will have lower energy boundaries when these are fully rather than partially coherent, particularly when misfit is small. The second is kinetic. From the standpoint of the acquisition kinetics of misfit dislocations, particularly in a well annealed matrix, and again in the presence of relatively small misfits, say of the order of a few pct., it would be very difficult to acquire the necessary misfit dislocations <u>and</u> organize them into minimum energy arrays on the interphase boundaries in the usually very short time when an embryo is "growing" through the size regime from just below to just above the critical nucleus size [24, 25].

When the $\{111\}_{fcc}//\{0001\}_{hcp}$, $\langle 110\rangle_{fcc}//\langle 11\bar{2}0\rangle_{hcp}$ interface is considered, full coherency can be achieved simply by assuming the appropriate lattice parameter ratio. Using this interface for our initial studies, we have employed the discrete lattice plane approach to examine some of the problems involved when there is a change in crystal structure across the interphase boundary [26]. Under these circumstances, different bond energies are required in each phase. Unlike the results obtained from the CAHN-HILLIARD [16] continuum treatment and the discrete lattice plane analysis of fcc:fcc interfaces [14], interfacial energy does not vary in a simple manner with temperature. Fig. 4 shows that the concentration profile is again diffuse, as in the situations previously investigated [14, 16], but that it is also asymmetric. The magnitude of the asymmetry was found to be a function of the ratio of the regular solution constants assumed for the fcc and hcp phases.

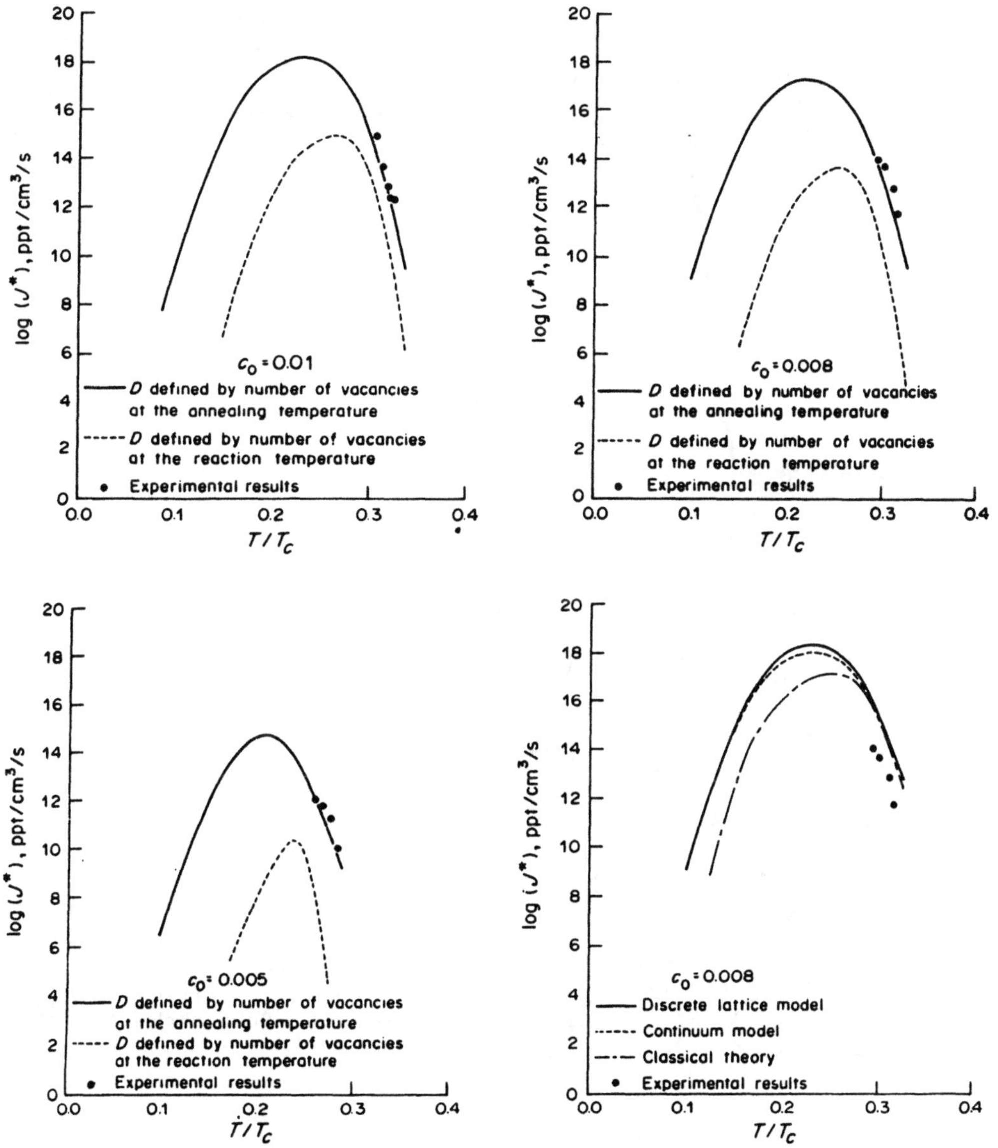

Fig.3 Comparison of calculated and measured nucleation rates vs. T/T_c in Cu-Co alloys containing: (a) 1% Co, (b) 0.8% Co and (c) 0.5% Co. (d) Comparison of experimental nucleation kinetics in Cu-1% Co with kinetics calculated by means of discrete lattice point (18), continuum (16) and classical nucleation theories (15), all in the absence of volume strain energy (13)

At other orientations of the fcc:hcp boundary, assuming that the same lattice orientation relationship is present, matching between the fcc and hcp lattices is not nearly as good. Fig. 5 illustrates the situation at an interface which we know, from observations made during growth [27], can be partially coherent--and hence ought to be fully coherent during nucleation. However, an accurate general method is lacking for calculating the chemical interfacial energy as a function of boundary orientation when the boundary is

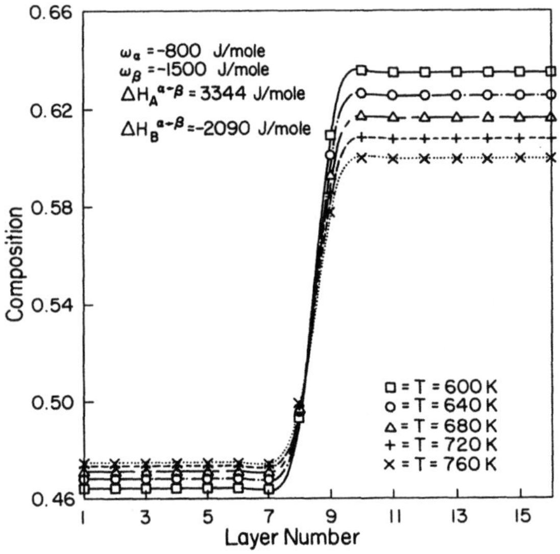

Fig.4 Concentration profile normal to the fcc:hcp structural interface (layers 8 and 9) at five temperatures. $\omega_\alpha(\omega_\beta)$ is the regular solution constant in the fcc (hcp) phase $\Delta H_A^{\alpha\to\beta}$ is the enthalpy of transformation of A from $\alpha\to\beta$ (26)

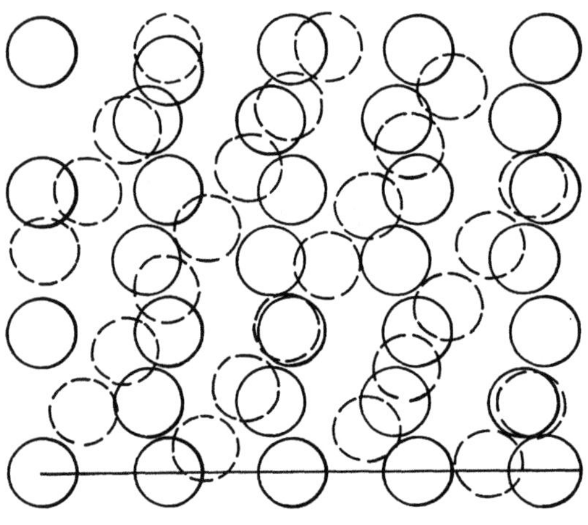

Fig.5 Atom patterns at $(1\bar{1}3)_{fcc}//(2\bar{1}2)_{hcp}$, $\langle 110\rangle_{hcp}//\langle 112\rangle_{fcc}$ (solid line). Solid circles are hcp, dashed are fcc

coherent. The strains needed to develop coherency at individual orientations can be evaluated, particularly when misfit is not large. Even in this situation, the problem arises of balancing the coherency strains applied to various interfaces so as to ensure that the critical nucleus as a whole is in elastic equilibrium. At the present time, we are using an approximate approach to this problem [28]. The nearest-neighbor version of the broken-bond discrete lattice plane model is being employed. Entropy is again assumed to be of the ideal configurational type only. On this analysis, no adjustments are made in the positions of the atoms so as to improve coherency at individual interfaces. This is, instead, accomplished macroscopically through application of the ESHELBY [29] method, as further developed by LEE et al [30], to an oblate ellipsoid of revolution. Since preliminary calculations, on a low temperature approximation of the chemical interfacial energy analysis, indicate that the critical nucleus should be a thin, nearly circular plate, probably faceted round the rim, this approximation should not be an inaccurate one, since it is qualitatively similar to the morphology of hcp γ' observed during the early stages of growth in an fcc α Al-Ag matrix [27]. However, a check of atom pattern matching across the interfaces of the Wulff construction, i.e., the critical nucleus shape, will be made before and after the application of coherency strains to ascertain the levels of improvement in matching which have been accomplished.

Of course, one could do a computer simulation of atom matching across many different fcc:hcp interfaces as a basis for constructing a more precise polar γ-plot, but aside from the immense amount of computer time required, the need to utilize different interatomic potentials for individual alloy systems, and the uncertainty as to which one is most appropriate in any system, suggests that this approach might not be much of an improvement.

Beyond the problem of calculating the chemical interfacial energies of fcc:hcp interfaces, there lie those of computing the chemical interfacial energy of the presumably still less satisfactorily matching fcc:bcc and bcc:hcp interfaces. Many other pairs of crystals may subsequently merit consideration in this manner. Achieving solutions to the problems of finding a more general method for evaluating these energies for pairs of infinite planes (in the presence of coherency strain energy), and also for the situation in which the planes are of small extent and interactive as on a critical nucleus at large undercooling, is critical for developing homogeneous nucleation theory in crystallographically realistic form.

1.2 Heterogeneous Nucleation

Balancing the needs and problems of theory against those of experiment, perhaps the simplest case of heterogeneous nucleation is that of forming a precipitate nucleus at an isolated straight edge or straight screw dislocation. This problem has, in fact, been undertaken many times by excellent theorists for both coherent and incoherent nucleation [31-37]. For the reasons previously given, only coherent nucleation will be considered here. The mathematical difficulties involved are sufficiently severe so that the critical nucleus shape usually assumed is a sphere [36] or some other unfaceted morphology [37]. However, such critical nucleus shapes usually imply that the nucleus:matrix boundaries have a high-energy disordered or incoherent structure. Even when nucleation is assumed to occur at a disordered grain boundary (actually, a grain face) in the matrix phase, and the critical nucleus is taken to consist of two abutting spherical caps (whose spherical surfaces correspond to disordered nucleus:matrix interfaces), calculated nucleation rates at the low undercoolings at which nucleation can be experimentally observed are of the order of $10^{-1,000,000}$/cm.^{2}-sec. [38, 39]! Comparable rates may be anticipated at individual dislocations under these circumstances. Hence it is clear that more realistic critical nucleus shapes must be employed, as discussed in the previous sub-section.

In addition to taking account of the boundary orientation-dependence of interfacial energy in constructing the shape of the critical nucleus at a given undercooling, it is also necessary to incorporate in the critical nucleus shape the influence of interaction with the dislocation. GOMEZ-RAMIREZ and POUND [37] have sketched some plausible shapes for critical nuclei formed on dislocations (Fig. 6). However, a rigorous solution to this problem has yet to be reported.

The primary mechanism through which a dislocation reduces ΔF^* for nucleation is evidently diminution of the volume strain energy attending formation of the nucleus. At least two major items of experimental evidence support this statement. (i) The matrix plane with respect to which the broad face of a precipitate plate is parallel is known as its "habit plane". When the lattice orientation relationship between the matrix and precipitate phases permits multiple, crystallographically equivalent habit planes, that particular habit plane is chosen whose normal lies most nearly parallel to the Burgers vector of the dislocation [40, 41]. This normal, which is also known as the "misfit vector" of the precipitate, is taken to denote the direction in which misfit between precipitate and matrix lattices is a maximum; since misfit between the two lattices is usually at or near a minimum parallel to the habit plane, this direction of maximum misfit seems to be at least reasonably precise. Hence this choice of habit plane achieves the maximum feasible reduction in the volume strain energy associated with critical nucleus formation. (ii) HORNBOGEN and ROTH [42] have shown that while ordered fcc Ni_3Al nucleates on dislocations in fcc α Ni-Al at low undercoolings, when the misfit between

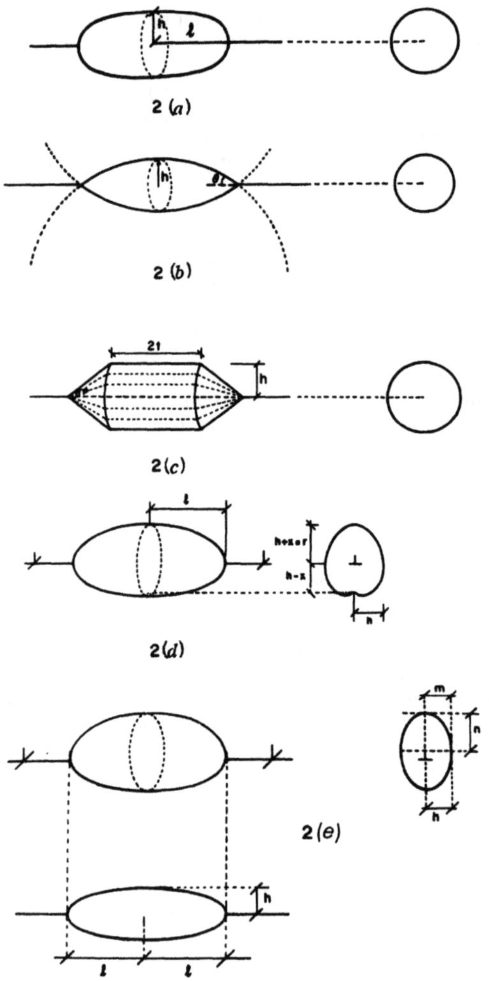

Fig.6 Critical nucleus shapes at dislocations. GOMEZ-RAMIREZ and POUND (37)

the two phases is made very small through addition of a suitable proportion of an appropriate third element, dislocations are no longer preferred nucleation sites.

Thus the main additional theoretical problem which must be solved is that of the interaction of the strain fields of the matrix dislocation and of the critical nucleus. Further, these calculations must be undertaken with anisotropic elasticity, and a form of calculation must be employed which permits different elastic constants to be utilized for the matrix and precipitate phases. As an illustration of the effects of employing anisotropic elasticity and different elastic constants for the matrix and precipitate phases in considerably simpler circumstances, Fig. 7 shows the variation of the strain energy of a homogeneously nucleated intragranular precipitate of hcp γ $AlAg_2$ (in fcc Al) with the morphology of an ellipsoid of revolution as a function of the aspect ratio, β, of the ellipsoid [30]. The variously numbered curves in the plot correspond to the lattice orientation relationships listed in Table I. W_o is the volume strain energy when both phases have the same elastic constants. Considerable differences in behavior are seen to attend changes in orientation relationship, and the result characteristic of cubic-type metals, i.e., that strain energy passes through either a maximum or a minimum at β = 1 (i.e., a sphere) is no longer obtained. Since even calculation of the strain energy associated with a homogeneously nucleated precipitate which is partially faceted and

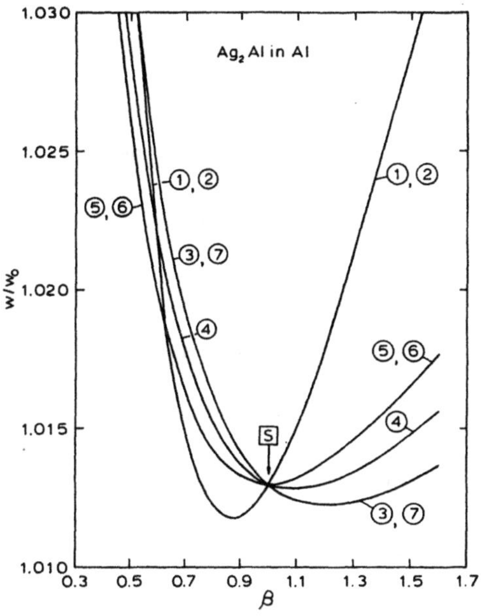

Fig.7 Relative dilatational strain energy vs. aspect ratio for ellipsoids of revolution calculated for hcp $AlAg_2$ in Al matrix assuming the orientation relationships in Table I. LEE et al (30)

otherwise spherically curved (e.g., Figs. 2b and c) is a formidable problem which probably cannot be solved analytically [43], deduction of the strain energy interaction between a critical nucleus with such a morphology and that of the strain field of, say, an isolated straight edge dislocation is readily seen to represent a substantial escalation in complexity.

Once nucleation at isolated dislocations has been analyzed, it will then be necessary to consider the critical nucleus shape and ΔF^* calculations at arrays of dislocations, spaced progressively closer together and corresponding to grain boundaries produced by successively larger misorientation of the crystals forming them. At large-angle grain boundaries, where dislocation arrays may be replaced by random close-packed polyhedra [44, 45] (Fig. 8), it is possible that the nature of the calculations required will change. Particularly if the critical nucleus dimensions in contact with the grain boundary are comparable to those of a polyhedron, interfacial energy as well as strain energy contributions will have to be considered. Macroscopic concepts may at this point no longer be useful, and computer calculations would thus be required instead.

Similar considerations are needed concerning nucleation at point defects, ranging from one or two excess vacancies on up to small clusters of vacancies and their successors such as dislocation loops and tetrahedra. Nucleation at point defects and aggregations thereof, it should be noted, is an excellent area toward which the modern experimentalist can profitably direct his efforts; radiation damage is a familiar method for increasing markedly the number density of such defects. These defects ought to be overwhelmed during the earliest stages of growth, if not during the nucleation process itself. The only way presently available in which to investigate their influence upon nucleation seems to be to investigate the point defects present in a thin foil with atomic resolution TEM prior to aging. Employing a transformation which had previously been shown not to be affected significantly by taking place within a thin foil, precipitation would then be permitted to occur in situ. The same areas as were previously studied prior to transformation would again be examined in order that individual defects might be related to the specific details of the precipitates which had formed at and replaced them. Whether analysis of nucleation at point defects would be based upon strain energy interactions alone, or upon a combination of strain energy and interfacial energy reductions, remains to be investigated. Significant differences may well be anticipated amongst the different types of point defect developed.

Table 1.

ORIGINS OF GROWTH LEDGES (25)

Source	Experimental Evidence
Two-dimensional nucleation	Θ' Al-Cu, Al-Au (70)
Dislocation with out-of-plane component emerging from ppt.	Same
Edges or corners of plates Mg_2Si (73)	γ Al-Ag (72), Θ' Al-Cu (70),
Stepped misfit dislocations	γ Al-Ag (71,72)
Intruder dislocations which create an unhealable ledge	None
Volume change distorts paths of interphase boundary	α Fe-C (67)
Junctions between allotriomorphs and sideplates	Same
Anti-phase boundaries in ppt.	Θ Al-Cu (70)
Impurity ppts. in contact with interphase boundary	Θ Al-Cu, Al-Au (73)

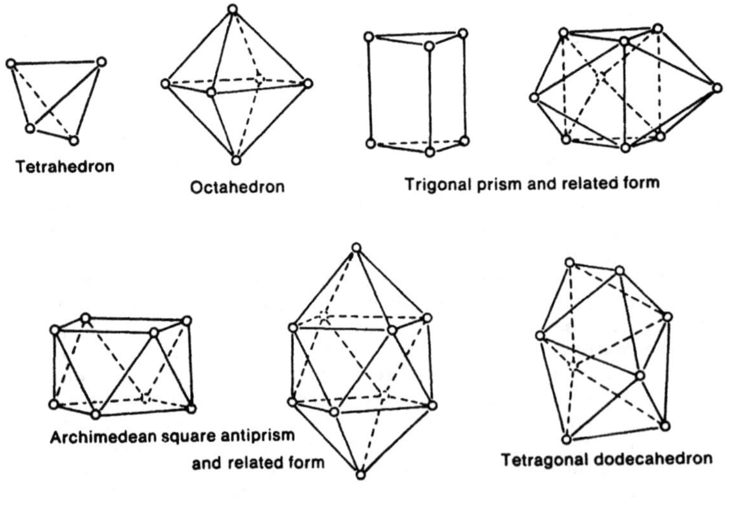

Fig.8 Random close-packed polyhedra: possible structural components of large-angle grain boundaries. POND, SMITH and VITEK (44)

1.3 On the Applicability of Macroscopic Values of Various Energies to Near Atomic-Scale Critical Nuclei

A final topic to be noted is the old question of the extent to which data on macroscopic values of interfacial energy, volume free energy change and strain energy apply to the atomistic scale upon which nucleation takes place. This has been for many years an important form of criticism of nucleation theory, particularly of the classical variety. Considering first the most important question, that of the applicability of macro-scale interfacial energies to critical nuclei, theory appears to be badly divided upon the subject. Some investigators have concluded that interfacial energy increases at small particle sizes [46-48] whereas others claim the reverse [49-54]. It would be helpful indeed for those who are oriented toward comparison of theory and experiment if significant further advances in the theory of this problem could be made reasonably soon. The good results achieved with classical theory (Figs. 3a-d) suggest, but certainly do not prove, that macro-scale interfacial energies may remain accurate until critical nuclei are composed of a quite small number of atoms.

In respect of both the volume free energy change and the volume strain energy, it is of interest to note the results of EGELHOFF and TIBBETTS [55]. These investigators studied Cu, Ni and Pd films formed atop graphite and amorphous carbon. They employed both Auger and photoemission spectroscopy. Their findings were that the electronic structure of these metals approached that of the bulk material at 1/4 - 3/4 of a monolayer. Only a few monolayers were needed to attain full bulk properties. These results hint that as long as the critical nucleus is larger than a few atoms, all three properties under consideration may be closely similar to those of bulk material.

2. DIFFUSIONAL GROWTH

2.1 A General Theory of Precipitate Morphology [24, 56]

Although this theory is presently qualitative, it has nonetheless been quite successful in accounting for the main features of the morphologies developed by precipitate crystals during growth driven by a chemical free energy change. If theoretical needs (to be introduced here) can be satisfied, it should become possible to place the theory on a quantitative basis. In its original form, the basis of this theory was the assumption that for a given orientation relationship between the lattices of the matrix and precipitate phases, matching across the interphase boundary was good enough at only one or a very few boundary orientations to permit these interfaces to be partially coherent, i.e., composed of misfit dislocations periodically spaced along an otherwise coherent boundary. The remaining boundary orientations were taken to have a disordered or incoherent structure. Those boundaries having the latter type of structure were thus free (in the absence of a solute drag-like effect [57, 58] or of a diffusion short circuit [59, 60]) to migrate at rates controlled by the non-interface structural factors of the chemical driving force, represented usually as the difference between the equilibrium interface composition and the far field composition, the inter-diffusivity in the matrix phase and the local radius of curvature of the interphase boundary. The partially (or fully) coherent areas of the boundary, on the other hand, are considered to be entirely immobile in the direction normal to themselves when there is a significant difference in crystal structure across the interphase boundary. Such boundaries can be displaced only by means of GIBBS' [1] ledge mechanism, schematically illustrated in Fig. 9a. This concept was introduced by GIBBS to explain atomic attachment to close-packed surfaces and was subsequently adapted to the analogous problem of displacing partially and fully coherent interphase boundaries between phases differing in crystal structure [24, 56]. Unlike small-angle grain boundaries describable in an analogous manner, this immobility does not derive from difficulties in displacing the interfacial dislocations. Instead, immobility arises from the kinetic infeasibility of changing the stacking sequence across the boundary one atom at a time when atom pattern matching across the boundary is very good. Such changes would have to be accomplished by placing substitutional atoms temporarily in interstitial sites. Because of the dense packing characteristic of the coherent areas of interphase boundaries, the probability that such sites will be occupied should be vanishingly small. On the other hand, ledges were proposed to furnish sites for easy atomic attachment by

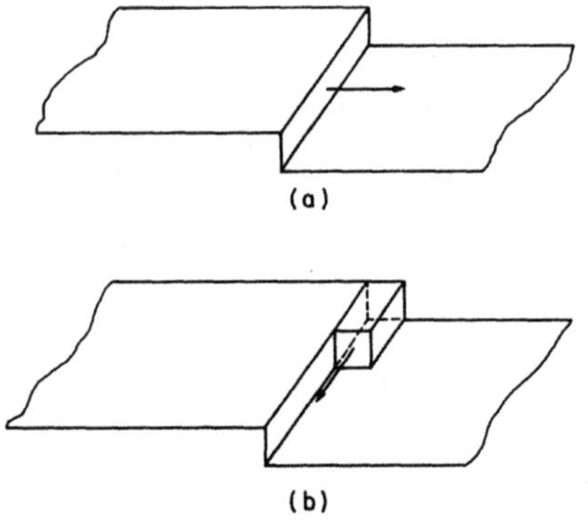

Fig.9 Schematic drawings of: (a) a growth ledge, and (b) a ledge-on-ledge or a kink-on-ledge growth configuration

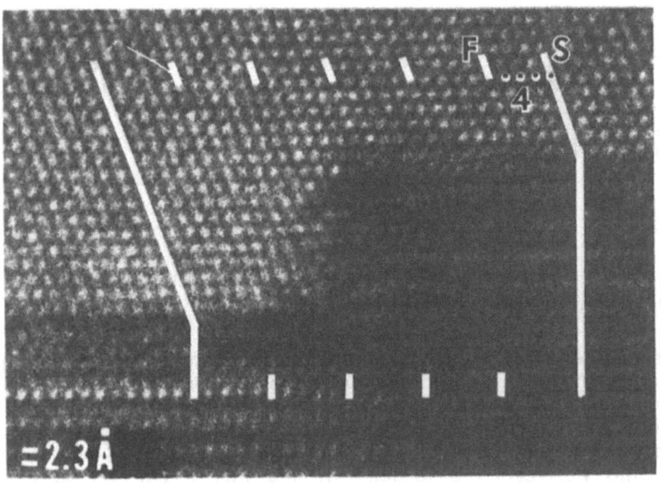

Fig.10 A multi-layer growth ledge on the broad face of an $AlAg_2$ plate formed at 350[{]C in an Al-4.2 A/O Ag alloy. HOWE et al (66)

virtue of their risers, assumed to have a disordered interfacial structure. Thus the overall migration rate of partially coherent boundaries in the direction normal to themselves would be determined by the migration rate of ledges and the spacing between them [61]. Ledge height, incidentally, does not enter this kinetic picture, since the lateral migration rate of a ledge is inversely proportional to its height [62].

Now putting together the picture of the interphase boundaries enclosing a particular precipitate crystal, nucleated say within a crystal of the matrix phase, we have two different types of interphase boundary structure, each migrating by a different mechanism with different kinetics under a given driving force. Unless the growth ledges are quite closely spaced and there has been minimal depletion of supersaturation by the prior passage of ledges [63, 64], the disordered areas of the interphase boundaries will grow more rapidly than the partially coherent ones. If the average inter-ledge spacing at one boundary orientation is particularly large, the evolution of the plate morphology with broad faces parallel to a particular plane in the matrix phase follows automatically. So, also, does the increasing tendency for the formation of anisotropic precipitate morphologies with decreasing temperature, i.e., with increasing undercooling below the equilibrium temperature.

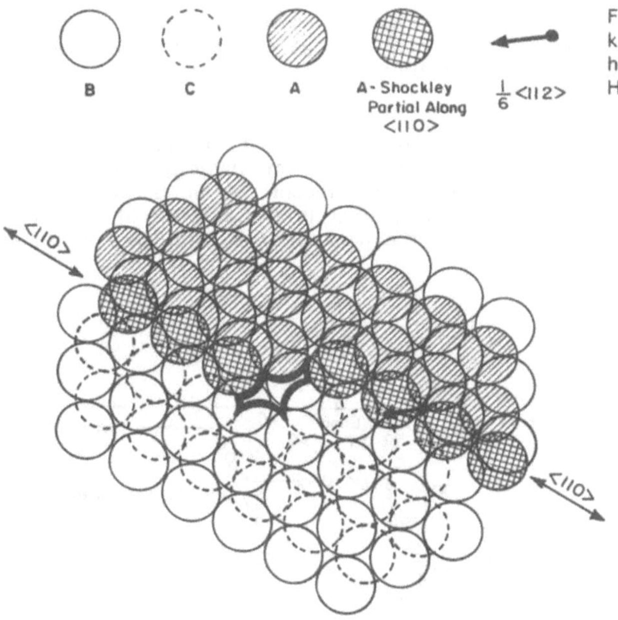

Fig.11 Hard sphere construct of a kink on a riser during growth of an hcp precipitate into an fcc matrix. HOWE et al (69)

This basically simple picture has undergone at least one major change since it was first developed. Increasingly, it appears that partial coherency can develop at a sufficiently large number of boundary orientations, for a given orientation relationship between two crystals differing significantly in structure, to enclose completely a precipitate crystal [24, 65]. This consideration extends even to the risers of ledges. Contrary to Gibbsian expectation, growth ledges at interphase boundaries are often at least several atom layers high, as illustrated in Fig. 10 [66], and can even be microns in height [67]. Risers are being found repeatedly, by direct TEM observation [66, 68], to be partially coherent. Atomic attachment must thus be achieved at ledges or kinks on the risers [24]. As shown in Fig. 11 from the work of HOWE [69], sufficient atomic disorder can be created at an atomic kink to make the necessary jumps across the boundary energetically feasible. Thus, as shown schematically in Fig. 9b, the ledge-on-ledge or kink-on-ledge mechanism probably replaces the conventional ledge mechanism in many alloy systems--and perhaps in nearly all of them.

Evolution of anisotropic precipitate morphologies thus depends upon a particular boundary orientation having not only a particularly large inter-ledge spacing but also an especially wide inter-kink spacing. And now a temperature dependence of morphological anisotropy requires a temperature-dependence of inter-ledge and/or of inter-kink spacings.

2.2 Problems in Ledgemanship

Table I summarizes the origins of ledges so far proposed and the alloy systems, if any, in which evidence supporting these origins has been found [25]. No doubt still more origins will be postulated or experimentally discovered. Some of these mechanisms, such as that of intruder dislocations from the matrix phase, are not particularly amenable to prediction in advance of experimental observations. Others, however, such as the junctions between grain boundary allotriomorphs and sideplates, should be quantitatively describable. No explicit theory is as yet available, though, for any of these mechanisms. Unsurprisingly, theory for kinks is also unavailable, unless kinks are taken to be thermally activated. On the other hand, there is now a substantial body of theory for the migration kinetics of ledges [62-64,74-77]. Very little has been done, however, to predict the modifications required when the atomic growth process is based upon kinks rather than upon risers [24]. Hence, accounting for the growth kinetics of ledged precipitates requires that measurements be provided of both the inter-ledge and the inter-kink spacings operative. Theories of ledge and of kink formation by various mechanisms are thus urgently needed.

In order to encourage such theoretical developments, it may be useful to summarize some of the successes which the theories of precipitate morphology and of ledge migration kinetics have achieved. Fig. 12 shows by means of hot-stage TEM and thermionic electron emission microscopy, respectively, that in the absence of ledges but in the presence of a driving force sufficient for the growth of a disordered interphase

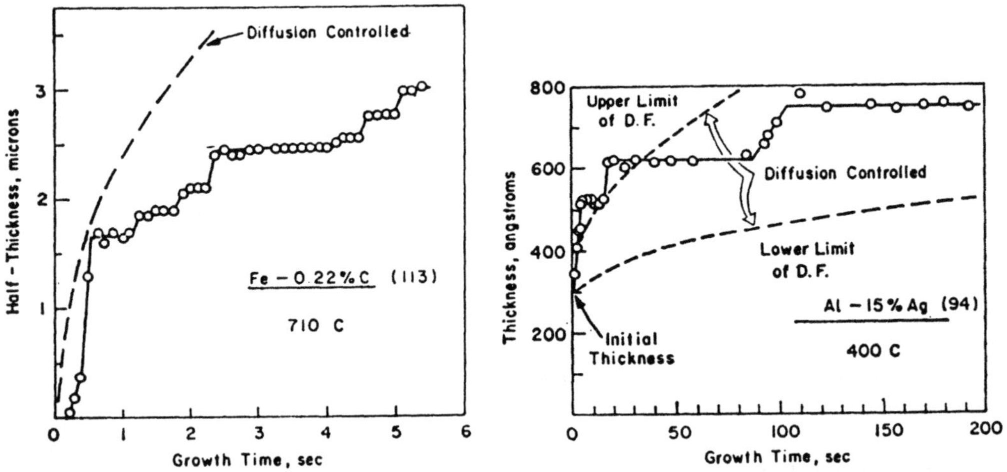

Fig.12 (a) Thermionic electron emission microscopy data on the ledgewise thickening of a bcc ferrite plate in the fcc austenite matrix of an Fe-C alloy (67); (b) Hot-stage TEM data on the ledgewise thickening of an hcp $AlAg_2$ plate in an fcc Al-Ag matrix (71)

Fig.13 Two-dimensional atomic resolution micrograph of a hcp $AlAg_2$ plate in an fcc Al-Ag matrix, showing that the terrace of a ledge on the plate is atomically flat. HOWE et al (66)

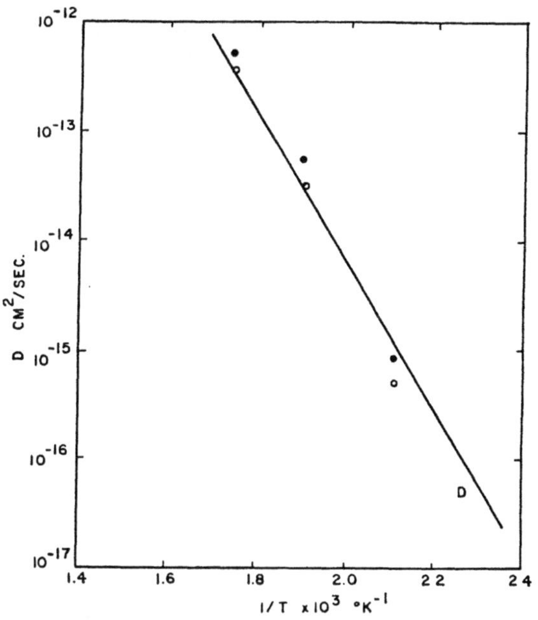

Fig.14 Effective interdiffusivities in α Al-Cu back-calculated from data on thickening kinetic of and inter-ledge spacings on Θ' plates in Al-4% Cu and the ZENER (100) and JONES-TRIVEDI (62) analyses

Table 2.

COMPARISON OF CALCULATED AND MEASURED LEDGE VELOCITIES IN Al-15 W/O Ag

Temp.(°C)	Calc x 10^5 cm/s Jones & Trivedi (62)	Meas. x 10^5 cm/s Laird and Aaronson (71)
400	2.23 - 0.64	- 2.95 - 1.18
425	0.62 - 0.08	1.76 - 0.08

boundary at a goodly pace, neither γ $AlAg_2$ plates nor α Fe-C plates will thicken at all. Fig. 13 [66] demonstrates by means of lattice fringe imaging that a γ $AlAg_2$ plate is atomically flat, i.e., that normal migration of the terraces of ledges is not occurring by atomic attachment directly to them. Fig. 14 shows that when the average inter-ledge spacing is measured as a function of reaction time on the broad faces of Θ' Al-Cu plates, the diffusivity back-calculated from two different analyses of ledge growth kinetics agrees well with the conventional measurement interdiffusivity extrapolated downwards from the α Al-Cu solid solution region [70]. Table II [71] shows the agreement achieved between the experimentally measured growth kinetics of individual ledges on γ $AlAg_2$ plates, expressed in terms of interdiffusivities back-calculated from JONES-TRIVEDI [62] theory, and extrapolated conventional interdiffusivities in α Al-Ag [78].

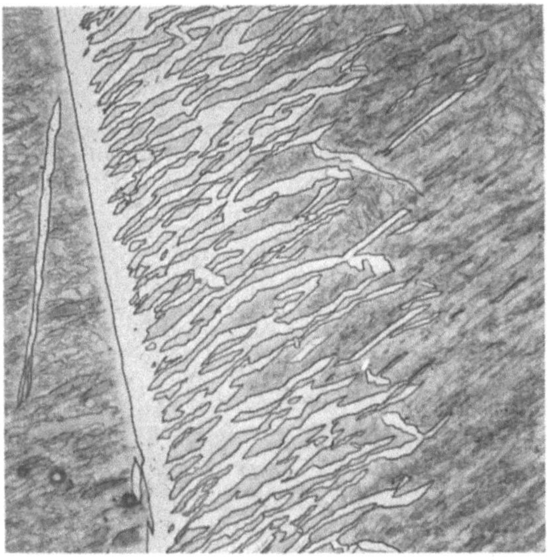

Fig.15 Degenerate proeutectoid ferrite sideplates in steel containing 0.29% C, 0.76% Mn and 0.25% Si

With these successes in hand, we briefly note, in Fig. 15 [56] a considerably more complex morphological situation for which further development of these theories may some day permit an explanation. Sideplates of ferrite formed in a dilute Fe-C base alloy are seen to exhibit repetitive degeneracy or deviations from the usual morphology of parallel, narrow isosceles triangles as seen on a plane of polish. Only a limited number of types of degeneracy has been observed. Each type is reproducible at different isothermal reaction times and over a range of isothermal reaction temperatures. Clearly, complicated diffusional interactions amongst adjacent ledges must be occurring here--but the important feature in all of this complexity is its overall reproducibility, and hence its presumed amenability to rational understanding.

2.3 Optimization Problems in Diffusional Phase Transformations

We have so far concluded that growth of single phase precipitates by the ledge mechanism is ubiquitous when the crystal structures of the matrix and precipitate phases are significantly different. The same conclusion is now appearing to be equally applicable to two-precipitate phase products of eutectoid decomposition [79-81]. (This is the solid-solid analogue of eutectic decomposition, with the liquid matrix being replaced by a single phase solid matrix.) In both single phase and two-phase precipitation, important problems of morphological stability and of optimizing the growth kinetics of the product of morphological instability when this condition prevails have attracted the attention of theoreticians for many years [82-86]. Let us first consider the additional complexities which result when the ledge mechanism intervenes in the case of a single phase transformation product.

When the driving force for growth, achieved by undercooling below the equilibrium temperature, becomes sufficiently high and crystallographic conditions are suitable, grain boundary allotriomorphs--precipitates which nucleate at matrix grain boundaries and then grow preferentially and more or less smoothly along them [87]--appear to become morphologically unstable. They develop sideplates or sideneedles, as illustrated in Fig. 16. Some efforts have been made to account for quantitative aspects of this phenomenon, such as the inter-sideplate spacing [88] and the undercooling below the equilibrium temperature needed to develop sideplates [89], through application of MULLINS-SEKERKA [83, 84] theory; these have been rewarded with mixed success. However, these accountings assumed that the interphase boundaries of the allotriomorphs have a primarily disordered structure. It has been recently shown, though, that their structure is composed

Fig.16 Secondary ferrite sideplates evolved from grain boundary allotriomorphs of the same phase in an 0.29% C, 0.76% Mn, 0.25% Si steel

of complex arrays of misfit dislocations and ledges [90]; on considerations of nucleation theory, this result is likely to be general. Hence it now becomes useful for theory to redevelop the analyses of MULLINS and SEKERKA, their peers [82, 85] and their successors [86], for ledged interphase boundaries. Perhaps one may legitimately hazard the prediction that such theoretical development will show that when the ledges are widely spaced, instability is inhibited, but that it will again become feasible, under more restrictive conditions, as the ratio of inter-ledge spacing to ledge height diminishes.

Once instability appears and sideplates or sideneedles form, many experiments have shown that these anisotropic morphologies lengthen at a constant rate at a given reaction temperature [91-93]. The eutectoid counterpart of a well-formed eutectic structure, called "pearlite", is composed of alternating plates of the two precipitate phases which also lengthen at a constant isothermal growth rate. Sideplates and sideneedles are characterized by a constant radius of edge or tip curvature [94] and pearlite by a constant inter-lamellar spacing [95], again at a particular reaction temperature. Earlier views to the contrary, the edges of plates [27, 95-97] and the tips of needles [98] are being shown to be densely ledged. Recently, HACKNEY and SHIFLET [79-81] have found that the edges of the alternating lamellae comprising pearlite grow by means of ledges shared between the two product phases! Thus these phases are constrained to grow in "lock step". Both the radius of the edges or tips of the single phase plates or needles and the inter-lamellar spacing of pearlite are characteristic of a given temperature [94, 95]. In the case of pearlite, this spacing can even be recovered if growth is caused to take place temporarily at a different temperature and then allowed to resume at the original reaction temperature [99]! It has yet to be determined experimentally whether or not these statements are also applicable to the ledge structures through which these transformation products grow. However, it is difficult to understand how their characteristic growth rates would be linked to their characteristic dimensions if the average inter-ledge spacing were not also particular to a given reaction temperature and alloy composition. Thus the question arises: what factor(s) determine these characteristic dimensions?

To date, ZENER'S [100] ad hoc suggestion, made in 1946, that the edge or tip radius or the inter-lamellar spacing will be that which maximizes the growth rate has been most frequently applied to evaluate these dimensions. Other proposals have been made for the ruling optimization principle, e.g., the minimum rate of entropy production [101]. These also have been questioned [102]. A particular favorite at the present time is Langer's [103] suggestion that "certain shapes propagate under conditions of marginal stability, i.e., under growth conditions such that small change of growth parameters in one direction will produce instability; presumably, changes in the opposite direction are compensated by non-linear effects that are not completely understood" [104]. Since the shapes which succumb to instability are removed from further consideration, those which survive do so at the very margin of stability.

The marginal stability criterion quantitatively explains the lengthening kinetics of rod-shaped dendrites during the growth of succinonitrile from its liquid [105] and the growth of ice into water [106] whereas the maximum growth rate hypothesis gives a quite poor accounting for these precisely determined experimental data. However, TRIVEDI [107] has recently shown that when the sides of a plate or rod are constrained against sidebranching--as obtains during precipitate plate, needle and pearlite growth as a result of the presence of partially coherent boundaries at the sides of the growing crystals--then marginal instability reduces to maximum growth rate.

Insofar as we are aware, marginal instability has yet to be applied to pearlite. However, further developing our suggestion in respect of the evolution of sideplates from grain boundary allotriomorphs, it is now suggested that pearlite be treated on the basis of the HACKNEY-SHIFLET [79-81] shared ledge mechanism. Similarly, the lengthening of precipitate places and needles can profitably be reconsidered on a ledgewise basis. Maximum growth rate can now be more safely applied. However, this hypothesis must henceforth be used in a situation where closely spaced ledges, mounted on a sharply curved interface, are so narrow that diffusion must be examined in three dimensions, probably in a moving coordinate system. Incorporation of inter-ledge (and inter-kink) spacings in the optimization process ought also to be attempted--though the physical legitimacy of these variables remains to be established.

SUMMARY

The status of theories of diffusional nucleation and diffusional growth in crystalline solids is reviewed from the standpoint of important unsolved problems located at the theory:experiment interface. After establishing that experiment has now shown that three different forms of nucleation theory can indeed predict the kinetics of homogeneous nucleation in the absence of a crystal structure change (during fcc->fcc precipitation in Cu-Co alloys) [12, 13], problems involved in the central question of evaluating the critical nucleus shape during homogeneous nucleation in the presence of a significant difference in crystal structure between the precipitate and matrix phases are discussed. Calculation of the chemical interfacial energy as a function of boundary orientation for a fixed lattice orientation relationship while maximizing coherency across the interphase boundary is the core of this problem. Computation of the critical nucleus shape during heterogeneous nucleation at a dislocation requires a further escalation in complexity: interaction of the strain field of realistically shaped, e.g., faceted critical nuclei with the strain field of the dislocation must be calculated using inhomogeneous, anisotropic elasticity. These calculations should then be extended to interaction with arrays of dislocations at small-angle grain boundaries, with the atomic polyhedra which appear to be the basic structural units of large-angle grain boundaries [44, 45], and with various point defects. Such calculations should be repeated for various crystal structure pairs. When these critical nucleus shapes are utilized to compute nucleation rates, the old question put to applications of classical nucleation theory arises: are macroscopic values of interfacial energy, volume free energy change and volume strain energy applicable to critical nuclei composed of small numbers of atoms? This question now needs to be revived and further investigated with some degree of urgency.

A qualitative general theory of precipitate morphology [24, 56], applicable when the crystal structures of matrix and precipitate are significantly different, is briefly reviewed and is shown to be supported by available experimental evidence. The central aspect of this theory is the immobility of partially coherent boundaries between crystals with different structures. These boundaries can migrate only by the ledge mechanism; on an atomic scale, the risers of ledges may often have to be displaced by means of kinks, i.e., a ledge-on-ledge mechanism. Placement of this theory on a quantitative basis requires not only prediction of interphase boundary structure as a function of boundary orientation for a given orientation relationship but also prediction of the kinetics of ledge and kink generation under these circumstances. A number of different mechanisms for ledge formation is noted; quantitative analysis of these mechanisms is now badly needed.

Although the theory of morphological instability [83, 84] is usually applied to disordered interfaces, there is now a major requirement to extend this theory to the densely ledged partially coherent interphase boundaries which represent the broad faces of grain boundary allotriomorphs [90]. Evolution of secondary Widmanstatten sideplates [88] from allotriomorphs appears to be a direct consequence of such instability. Further, when a precipitate plate develops and lengthens by the ledge mechanism, or alternating lamellae of two precipitate phases (known as "pearlite") undergo coupled growth by means of shared ledges [79-81], the radius of the plate edge and the interlamellar spacing of pearlite are characteristic of reaction temperature and alloy composition. These characteristic dimensions appear to be determined by an optimization principle. Maximum growth rate [100] has been used effectively for this purpose but its theoretical justification remains incomplete, despite an important recent advance [107]. Marginal stability [103] is now favored as the controlling principle. Analyses conducted on this basis should next be undertaken with the explicit recognition that both the lengthening of single phase plates and the lengthening of stacks of plates of two alternating phases , i.e. pearlite, take place by the ledge mechanism. It may even be desirable to incorporate the inter-ledge (and perhaps also the inter-kink) spacing, as well as the characteristic "macroscopic" dimensions of these microstructures, in the optimization analyses.

Acknowledgments

Support for the preparation of this review by Grant DMR86-15997 from the Division of Materials Research of the National Science Foundation and by Grant AFOSR84-0303 from the Air Force Office of Scientific Research is gratefully acknowledged.

References

1. J. W. Gibbs: Collected Works, Vol. 1 (Yale University Press, New Haven, CT 1948).
2. B. E. Sundquist and R. A. Oriani: Jnl. Chem. Phys. 36, 2604 (1962).
3. R. B. Heady and J. W. Cahn: Jnl. Chem. Phys. 58, 896 (1973).
4. R. G. Howland, N. C. Wong and C. M. Knobler: Jnl. Chem. Phys. 73, 522 (1980).
5. A. J. Schwartz, S. Krishnamurthy and W. I. Goldburg: Phys. Rev. A21, 1331 (1980).
6. J. S. Huang, W. I. Goldburg and M. R. Moldover, Phys. Rev. Lett. 34, 639 (1975).
7. K. Binder and D. Stauffer: Advances in Physics 25, 343 (1976).
8. J. S. Langer and A. J. Schwartz: Phys. Rev. A21, 948 (1980).
9. D. H. Kirkwood: Acta Met. 18, 563 (1970).
10. A. W. West and D. H. Kirkwood: Scripta Met. 10, 681 (1976).
11. T. Hirata and D. H. Kirkwood: Acta Met. 25, 1425 (1977).
12. I. Servi and D. Turnbull: Acta Met. 14, 161 (1966).
13. F. K. LeGoues and H. I. Aaronson: Acta Met. 32, 1855 (1984).
14. Y. W. Lee and H. I. Aaronson: Acta Met. 28, 539 (1980).
15. F. J. LeGoues, H. I. Aaronson, Y. W. Lee and G. J. Fix: In Proceedings of an International Conference on Solid-Solid Phase Transformations (TMS-AIME, Warrendale, PA 1983) p. 427.
16. J. W. Cahn and J. E. Hilliard: Jnl. Chem. Phys. 31, 688 (1959).
17. H. E. Cook, D. De Fontaine and J. E. Hilliard: Acta Met. 17, 765 (1969).
18. H. E. Cook and D. De Fontaine: Acta Met. 17, 915 (1969); 19, 607 (1971).
19. F. K. LeGoues, Y. W. Lee and H. I. Aaronson: Acta Met. 32, 1837 (1984).
20. F. K. LeGoues, H. I. Aaronson and Y. W. Lee: Acta Met. 32, 1845 (1984).
21. R. V. Ramanujan: Ph.D. thesis research in progress, Carnegie-Mellon Univ. (1987).
22. H. I. Aaronson and K. C. Russell: Proceedings of an International Conference on Solid-Solid Phase Transformations (TMS-AIME, Warrendale, PA 1983) p. 371.
23. J. H. Van der Merwe: Jnl. App. Phys. 15, 73 (1967).
24. H. I. Aaronson, C. Laird and K. R. Kinsman: In Phase Transformations (ASM, Metals Park, OH 1970) p. 313.
25. H. I. Aaronson: Jnl. of Microscopy 102, 275 (1974).

26. R. V. Ramanujan, J. K. Lee, F. K. LeGoues and H. I. Aaronson: to be submitted to Acta Met.
27. C. Laird and H. I. Aaronson: Acta Met. 15, 73 (1967).
28. R. Cadoret: Phys. Stat. Solidi (B) 46, 291 (1971).
29. J. D. Eshelby: Prog. Sol. Mech. 2, 88 (1961).
30. J. K. Lee, D. M. Barnett and H. I. Aaronson: Met. Trans. 8A, 963 (1977).
31. J. W. Cahn: Acta Met. 5, 169 (1957).
32. D. M. Barnett: Scripta Met. 5, 261 (1971).
33. C. C. Dollins: Acta Met. 18, 1209 (1970).
34. F. S. Ham: Jnl. App. Phys. 30, 915 (1959).
35. F. Montheillet and J. M. Haudin: Phys. Stat. Sol. 154, 271 (1979).
36. F. C. Larche: In Dislocations in Solids (North Holland, Amsterdam 1979) p. 137.
37. R. Gomez-Ramirez and G. M. Pound: Met. Trans. 4A, 1563 (1973).
38. W. F. Lange III, M. Enomoto and H. I. Aaronson: Met. Trans., in press.
39. M. R. Plichta, J. H. Perepezko, H. I. Aaronson and W. F. Lange III: Acta Met. 28, 1031 (1980).
40. G. Thomas and J. Nutting: In The Mechanism of Phase Transformations in Metals (Institute of Metals, London 1956) p. 57.
41. H. B. Aaron and H. I. Aaronson: Met. Trans. 2, 23 (1971).
42. E. Hornbogen and M. Roth: Zeit. Metallkunde 58, 842 (1967).
43. D. M. Barnett: Stanford Univ., private communication.
44. R. C. Pond, D. A. Smith and V. Vitek: Scripta Met. 12, 699 (1978).
45. M. F. Ashby and F. Spaepen: Scripta Met. 12, 193 (1978).
46. A. G. Walton: Jnl. Chem. Phys. 39, 3162 (1963).
47. A. G. Walton and D. R. Whitman: Jnl. Chem. Phys. 40, 2722 (1964).
48. R. A. Oriani and B. E. Sundquist: Jnl. Chem. Phys. 38, 2082 (1963).
49. R. C. Tolman: Jnl. Chem. Phys. 16, 758 (1948).
50. R. C. Tolman: Jnl. Chem. Phys. 17, 333 (1949).
51. J. G. Kirkwood and F. P. Buff: Jnl. Chem. Phys. 17, 338 (1949).
52. J. G. Kirkwood and F. P. Buff: Jnl. Chem. Phys. 18, 991 (1950).
53. G. C. Benson and E. A. Flood: In The Solid-Gas Interface (Marcel Dekker, New York 1967) p. 243.
54. I. W. Plesner: Jnl. Chem. Phys. 40, 1510 (1964).
55. W. F. Egelhoff,Jr. and G. G. Tibbetts: Phys. Rev. B19, 5028 (1979).
56. H. I. Aaronson: In Decomposition of Austenite by Diffusional Processes (Interscience, NY 1962) p. 387.
57. K. R. Kinsman and H. I. Aaronson: In Transformation and Hardenability in Steels (Climax Molybdenum Co., Ann Arbor, MI 1967) p. 39.
58. H. I. Aaronson: In The Mechanism of Phase Transformations in Crystalline Solids (Institute of Metals, London 1969) p. 270.
59. H. B. Aaron and H. I. Aaronson: Acta Met. 16, 789 (1968).
60. A. D. Brailsford and H. B. Aaron: Jnl. App. Phys. 40, 1702 (1969).
61. J. W. Cahn, W. B. Hillig and G. W. Sears: Acta Met. 12, 1421 (1964).
62. G. J. Jones and R. Trivedi: Jnl. App. Phys. 42, 4299 (1971).
63. M. Enomoto: Acta Met. 35, 935 (1987).
64. M. Enomoto: Acta Met. 35, 947 (1987).
65. H. I. Aaronson: Trans. Indian Inst. Metals 32, 1 (1979).
66. J. M. Howe, H. I. Aaronson and R. Gronsky: Acta Met. 33, 649 (1985).
67. K. R. Kinsman, E. Eichen and H. I. Aaronson: Met. Trans. 6A, 303 (1975).
68. P. R. Howell and R. W. K. Honeycombe: In Proceedings of an International Conference on Solid-Solid Phase Transformations (TMS-AIME, Warrendale, PA (1983)) p. 399.
69. J. M. Howe, U. Dahmen and R. Gronsky: Phil. Mag., in press.
70. R. Sankaran and C. Laird: Acta Met. 22, 957 (1974).
71. C. Laird and H. I. Aaronson: Acta Met. 17, 505 (1969).
72. C. Laird and H. I. Aaronson: Jnl. Inst. Metals 96, 222 (1968).
73. G. C. Weatherly: Acta Met. 19, 181 (1971).
74. G. J. Jones and R. Trivedi: Jnl. Crystal Growth: 29, 155 (1975).
75. C. Atkinson: Proc. Roy. Soc. A378, 351 (1981).
76. C. Atkinson: A384, 167 (1982).
77. M. Enomoto, H. I. Aaronson, J. Avila and C. Atkinson: Proceedings of an International Conference on Solid-Solid Phase Transformations (TMS-AIME, Warrendale, PA 1983) p. 567.

78. T. Heumann and S. Dittrich: Zeit. Elektrochem. 61, 1138 (1957).
79. S. A. Hackney and G. J. Shiflet: Scripta Met. 19, 757 (1985).
80. S. A. Hackney and G. J. Shiflet: Acta Met. 35, 1007 (1987).
81. S. A. Hackney and G. J. Shiflet: Acta Met. 35, 1019 (1987).
82. C. Wagner: J. Electrochem. Soc. 99, 369 (1952).
83. W. W. Mullins and R. F. Sekerka: Jnl. App. Phys. 34, 323 (1963).
84. W. W. Mullins and R. F. Sekerka: Jnl. App. Phys. 35, 444 (1964).
85. V. V. Voronkov: Sov. Phys. Sol. State 6, 2378 (1965).
86. R. F. Sekerka: In Encyclopedia of Materials Science and Engineering (Pergamon Press, New York 1986) p. 3486.
87. C. A. Dube, H. I. Aaronson and R. F. Mehl: Rev. de Met. 55, 201 (1958).
88. R. D. Townsend and J. S. Kirkaldy: Trans. ASM 61, 605 (1968).
89. P. Krahe, K. R. Kinsman and H. I. Aaronson: Acta Met. 20, 1109 (1972).
90. T. Furuhara, A. M. Dalley and H. I. Aaronson: unpublished research, Carnegie-Mellon University (1987).
91. M. Hillert: unpublished research, reproduced in L. Kaufman, S. V. Radcliffe and M. Cohen: Decomposition of Austenite by Diffusional Processes (Interscience, NY 1962) p. 313.
92. R. F. Hehemann: In Phase Transformations (ASM, Metals Park, OH 1970) p. 397.
93. E. P. Simonen and R. Trivedi: Report to U.S.A.E.C., Iowa State Univ., Ames, IA (Aug., 1972).
94. E. P. Simonen, H. I. Aaronson and R. Trivedi: Met. Trans. 4, 1239 (1973).
95. R. F. Mehl and W. C. Hagel: Prog. in Met. Phys. 6, 74 (1956).
96. C. Laird and H. I. Aaronson: Trans. TMS-AIME 242, 1393 (1968).
97. H. J. Lee and H. I. Aaronson: Jnl. Mat. Sci., in press.
98. K. Chattopadhyay and H. I. Aaronson: Acta Met. 34, 695 (1968).
99. J. W. Cahn and W. C. Hagel: In Decomposition of Austenite by Diffusional Processes (Interscience, NY 1962) p. 131.
100. C. Zener: Trans. AIME, 167, 550 (1946).
101. J. S. Kirkaldy: In Decomposition of Austenite by Diffusional Processes (Interscience, NY 1962) p. 29.
102. J. W. Cahn and W. W. Mullins: In Decomposition of Austenite by Diffusional Processes (Interscience, NY 1962) p. 123.
103. J. S. Langer: Rev. Mod. Phys. 52, 1 (1980).
104. R. F. Sekerka: In Phase Transformations and Material Instabilities in Solids (Academic Press, NY 1984) p. 147.
105. M. E. Glicksman, R. J. Schaefer and J. D. Ayres: Met. Trans. 7A, 1747 (1976).
106. J. S. Langer, R. F. Sekerka and T. Fujioka: Jnl. Crystal Growth 44, 414 (1978).
107. R. Trivedi: In Proceedings of an International Conference on Solid-Solid Phase Transformations (TMS-AIME, Warrendale, PA 1983) p. 477.

Competing Displacive Interactions, Phonon Anomalies, and Structural Transitions Which Do Not "Soften"

J.A. Krumhansl

Department of Physics,* Cornell University,
Ithaca, NY 14853, USA

ABSTRACT Displacive structural phase transitions (i.e. diffusionless) comprise a significant fraction of all phase transformations. Interest among materials scientists was greatly stimulated by the idea put forth by Cochran and by Anderson that some phonons "softened" to $\omega_s^2 \rightarrow 0$ at a transition temperature, and thereby the transformed material adopted the displacement patterns of that soft mode. This simple, but powerful, idea has stimulated many experiments over the past 30 years. However, after a few early successes (A-15, perovskites, K_2SeO_4) in which a special mode did soften, more careful and extensive study has shown that hardly any materials transform in this manner. Although they have phonon anomalies, and product structures are related to those, it is clear that a different explanation is needed. This paper suggests an alternative mechanism, taking into account nonlinear effects leading to perhaps weak but definitely first order transition.

I. COMPETING DISPLACIVE INTERACTIONS

We begin with a few illustrative examples of how competing atomic and electronic interactions can lead to anomalous phonon dispersion spectra and structural instabilities.

Body Centered Cubic Lattices: Many crystals, both metallic and insulating have the so called β-cubic (body centered) structure. In the long wave (elastic regime) they show startling anisotropy in their shear wave dispersion spectra; they are much more compliant to $[1\bar{1}0]$ shear transverse to the [110] direction than to any other strain. That elastic behavior was known long ago (Zener[1]); but with the advent of inelastic neutron spectroscopy it was further found that phonons propagating in the {110} direction had further anomalies in the form of dips at $\underline{q}$ regions out from the origin in the Brillouin zone.[2,3,4] While in the great majority of cases the frequencies there changed only slightly with temperature, nonetheless resulting structures with modulations characteristic of the q-anomalies were frequently found, suggesting incipient instabilities.

In essence, what Zener pointed out was that for closed shell (or nearly) ion cores, considered as hard spheres, in lowest order the planes normal to [110] could slide without resistance along $[\bar{1}10]$, and that it was the ion-electron gas interactions which stabilized the system overall. More quantitative calculations[5,6,7] have confirmed this basic concept, and in fact usually for closed shell systems the short-range contribution to the potential energy for shear is negative, and unless coulomb interactions are large enough[6] or, at higher temperature anharmonic corrections are taken into account,[7] the β-cubic form is unstable (i.e. in harmonic language ω^2 is negative).

The harmonic phonon spectra of many transition metals have been studied: Nb, Mo, Zr, as well as alloys in that series.[8] It becomes clear from these studies that Zener's original idea appears in many manifestations; both lattice instabilities and phonon anomalies are related to competition between short range (unstable) ionic or bonding forces and longer range coulomb or electron gas non-bonding forces.

Attention should be called in passing to the likelihood that the frequent attribution of phonon anomalies to the Kohn effect is generally unjustified as has been noted in ref. 8.

Spin Systems: Since it is not straightforward in most of the calculations referred to above to identify the competing interactions and their consequences, it is pedagogically amusing to examine a simpler model system, that of the helical spin systems[9] discussed many years ago.

Consider a one dimensional line of equally spaced spins which have ferromagnetic nearest neighbor exchange interactions and anti-ferromagnetic next nearest neighbor exchange. The energy is given by

$$U = -2J[\sum_p \underline{S}_p \cdot (\underline{S}_{p+1} - \varepsilon \underline{S}_{p+2})] \tag{1}$$

i.e. $J_1 = J$, $J_2 = -\varepsilon J$, i.e. a $1-d$ ANNNI model; ε measures the strength of the competing AF exchange.

Assume that there are two possible states: (a) ferromagnetic, with all $\underline{S}_p$ in the Z direction, (b) a helical state in which adjacent spins twist mutually through an angle θ in the plane perpendicular to the line of spins. For the latter case the energy is

$$\bar{U} = -2JNS^2(\cos\theta - \varepsilon\cos 2\theta) \tag{2}$$

whence $(\frac{\partial U}{\partial \theta}) = 0$ leads to the condition that

$$\theta = \pm\cos^{-1}(1/4\varepsilon) \tag{3}$$

It is apparent that if $\varepsilon > \frac{1}{4}$ the ferromagnetic state is unstable against helix formation, otherwise no real solution for θ exists, except $\theta = 0$ which is just the simple ferromagnetic state.

Now it is instructive to relate this to the spin wave dispersion in the ferromagnetic (i.e. "untransformed") state. One easily finds (for small amplitude spin waves):

$$\hbar \mid \omega \mid = 4JS[\sin^2(\frac{ka}{2}) - \varepsilon\sin^2(ka)] \tag{4}$$

Expanding for small ka, to terms quadratic in ka,

$$\hbar \mid \omega \mid = 4JS[(\frac{1}{4} - \varepsilon)k^2a^2] \tag{5}$$

whence it is seen that if $\varepsilon > \frac{1}{4}$, the condition for instability of the ferromagnetic state against the helical state, the dispersion relation near $k = 0$ goes "soft". For larger k values the quartic term contributes a dip (i.e. anomaly) in the spin wave spectrum.

In similar fashion we may expand Eq. (2) up to terms in θ^4, and get a "Landau" energy:

$$U = U_0 + (1 - 4\varepsilon)\frac{\theta^2}{2} - \frac{1}{4}(1 - 16\varepsilon)\theta^4 + \cdots \tag{6}$$

For $(1/16) < \varepsilon < (1/4)$ the minimum is at $\theta = 0$. For $\varepsilon > (1/4)$ the point $\theta = 0$ is a maximum of $\bar{U}$, i.e. unstable; in this case there are two stable minimum, approximately as given by Eq. (3).

To summarize: competing interactions can give rise to (a) structural instability, (b) a "soft" ($\omega \to 0$) spin wave mode, and (c) a (static) Landau free energy compatible with (a).

<u>Competing Interactions in Insulators</u>: Continuing to illustrate that the general features under discussion are found in many kinds of materials[10] it is particularly instructive to cite studies of the orthorhombic ferroelectric insulators, on which extensive diffraction and neutron studies have been made, stimulated initially by the discovery that K_2SeO_4 showed almost complete softening at the temperature of transition to an incommensurate structure. A comprehensive and timely review has been written by Axe, Iizumi, and Shirane (1986).[11]

In passing we (again) call attention to two points: (a) these are good insulators, so neither charge density waves nor the Kohn effect can occur here, although many authors invoke both to explain their results; (b) to quote (p. 3 of ref. 11): "It is a puzzling fact that although many other examples of similar (i.e. to K_2SeO_4) phase transformations have been found in isostructural materials, none of these materials have shown clear evidence of well defined soft mode behavior." So again we have it; the widely attributed soft mode explanation is irrelevant, in reality. We return to this point later; for now we summarize the analysis of the origin of phonon anomalies as reviewed in ref. 11.

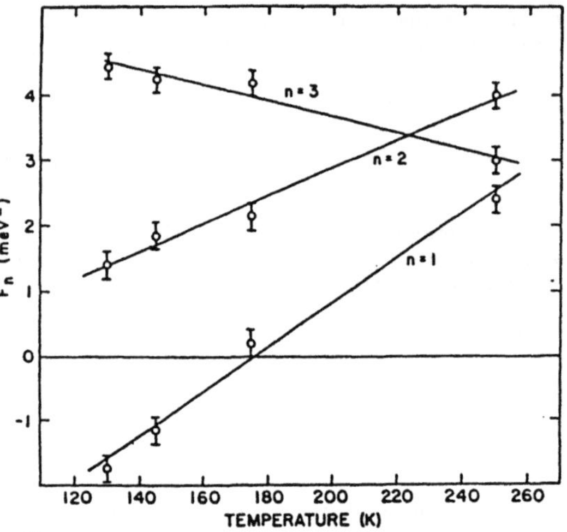

Fig. 1

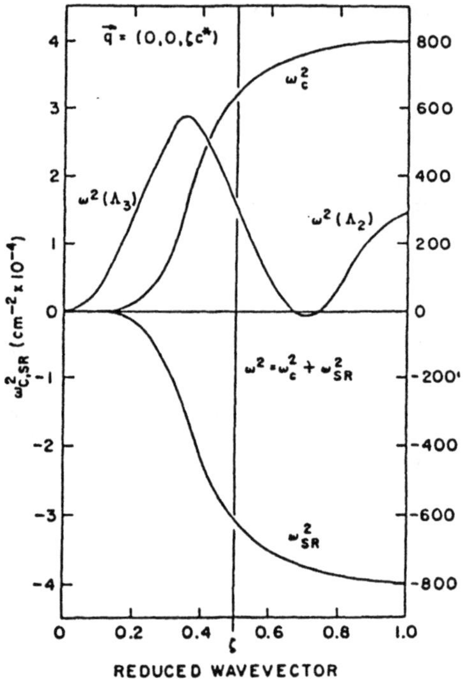

Fig. 2

1. Contributions of first, second, and third neighboring interplanar force constants to phonon dispersion for K_2SeO_4 [from Ref. 12].
2. Competition between coulomb (ω_c^2) and short range (ω_{sr}^2) contributions to phonon dispersion as discussed in Ref. 11.

Iizumi et al.[12] found that the neutron phonon dispersion curves could be fit by a set of interplanar force constants F_n whose temperature behavior is shown in Fig. 1 leading to the dispersion relation

$$\omega^2 = \sum_n F_n(1 - cos n\pi\varsigma) \tag{7}$$

where n denotes the n-th neighboring plane. It is noteworthy that the interaction 3 planes away is as large as it is; the underlying microscopic reason is unclear from just this analysis. [Parenthetically, Iizumi also applied this analysis to b.c.c. Thallium[13] and to b.c.c. to h.c.p. martensitic transformations.]

Haque and Hardy[12] shed further light on the nature of the interactions which dominated the anomalous mode, by carrying out an extensive analysis of the lattice dynamics and separating the short range near-neighbor contribution ω_{sr}^2 from the coulomb contribution ω_c^2, to give the actual frequency

$$\omega^2 = \omega_{sr}^2 + \omega_c^2 \tag{8}$$

The remarkable competition between these two is shown in Fig. 2 from Haque and Hardy[12] (Fig. 10 of ref. 11). Again from these two sets of calculations it is clear that there is significant competition between stabilizing and destabilizing forces, and both the positions of phonon anomalies, and whether significant softening occurs, are strongly affected by the concellation of relatively large and opposite interactions. Very similar behavior is found in metals (alloys)[4,5,6,7,8] so we doubt that Kohn anomalies are generally present; rather, the physics is actually one form or another of Zener's original ideas of competing interactions.

II. PHONON ANOMALIES AND DISPLACEMENT CORRELATIONS

It is the thesis of the present discussion that phonon anomalies are related, but not in the simple Cochrane-Anderson sense, to most structural transformations.

Some particularly interesting phenomena, other than the transformation itself, are the pretransitional effects, characterized by strong diffuse scattering in reciprocal space near the would-be product positions, or real space modulated structures in diffraction or electron real space imaging. We can gain insight into these by considering the displacement-displacement correlation functions associated with an anomalous phonon branch, starting say above a transition temperature where the phonons are still moderately well defined. In this regime we can obtain the correlation functions from the phonon Green's functions; they in turn determine the diffraction pattern, the distribution of thermal displacement fluctuations, and the displacive response of the lattice to imposed forces (e.g. defects).[13]

While the complete calculation requires a knowledge of all the phonon branches (frequencies and polarization vectors), there are strongly distinguishing features of the class of materials under study that: (1) the transition is driven by only one or a few "phonons" at particular wave vectors $\underline{q}_s$ (i.e. the "star" of $\underline{q}_s$), as discussed in the models developed by Cowley and Bruce;[14] (2) a very important distinguishing feature of many of these systems is that the dominant dispersion surfaces and displacements are extremely anisotropic in $\underline{q}$ space (magnitude ratios of $10-10^2$ are possible). Morii and Iizumi[15] have pointed out qualitatively, and we demonstrate more formally below, that the result of this great anisotropy is a quite different picture of the atomic displacements from the soft mode model.

The phonon Green's function[13] is given by (we use the classical form which is adequate at high temperature for the present):

$$\underline{\underline{G}}(\underline{R},\underline{R}';\omega)=\sum_{\underline{q}}\frac{\underline{e}(\underline{q})\underline{e}^*(\underline{q})}{N}\;\frac{exp[i\underline{q}\cdot(\underline{R}-\underline{R}')]}{\omega^2(\underline{q})-\omega^2} \tag{9}$$

where $\underline{R},\underline{R}'$ are site indices, $\underline{e}(\underline{q})$ phonon polarization vectors, and $\omega^2(\underline{q})$ and ω^2 are mode and measurement frequencies respectively. From $\underline{\underline{G}}$ we can obtain: (1) the time average of the displacement correlation function $<\underline{U}(\underline{R})\underline{U}(\underline{R}')>_t$ by setting $\omega=0$, or (2) the displacement $\underline{u}(\underline{R})$ due to an applied force $\underline{F}(\underline{R}')$. Since x-ray or neutron diffraction patterns are determined by the former we address that here, assuming for purposes of illustration (see Fig. 3) the simplest situation discussed by Cowley and Bruce,[14] modified to introduce strong anisotropy (Fig. 3b). Specifically we assume that near $\pm\underline{q}_s$,

$$\omega^2(\underline{q})=\omega_s^2(T)+a_\ell^2(T)(q_\ell-q_s)^2+a_t^2(T)(q_t-q_s)^2 \tag{10}$$

Here $a_\ell(T)$, q_ℓ refer to the soft direction along q_s, similarly for "t" transverse to the soft direction, and the temperature dependence T implies (anharmonically) renormalized parameters. In reference 14 $a_\ell \simeq a_t$; but here $a_\ell \ll a_t$. The polarization vector is assumed constant and purely transverse. Approximating the dispersion near q_s simply by two ellipsoidal families at $\pm q_s$, it is possible to compute $\underline{G}$ formally in several limiting cases.

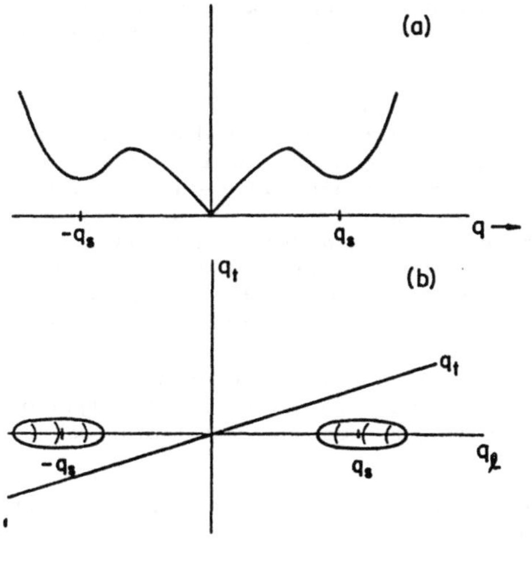

3. (a) Model phonon dispersion along anomalous direction $\pm\underline{q}_s$; (b) Surfaces of constant ω in q-space.

The main contributions to $\stackrel{G}{=}$ come from two regions: (i) the modes near $\underline{q}=0$, described by conventional (long wave) elasticity theory, (ii) additional (significant) dynamical correlations from the phonon anomaly around $\underline{k}=\underline{q}_s$. Thus, approximately, $\stackrel{G}{=} = \stackrel{G}{=}_{elastic}+\stackrel{G}{=}_{anomalous}$ and there are two dominant contributions to the correlation function. Thus, diffuse x-ray scattering (streaking, in the case of strong anisotropy) will be seen both near to the Bragg peaks i.e. the conventional elastic thermal diffuse or Huang scattering related to $\stackrel{G}{=}_{elastic}$, and, near $\underline{k}=\underline{q}_s$ from $\stackrel{G}{=}_{anomalous}$, having a characteristic lattice modulation $\sim \cos(q_s x)$ along the $\hat{\underline{q}}_s$ direction. This latter anomalous diffuse scattering cannot be found from conventional elastic theory, so we have evaluated the correlation function on the basis of the model dispersion relation, Eq. (10).

Defining x to be along $\underline{q}_s$, y to be transverse, the following time average correlation functions are found from $\stackrel{G}{=}_{anomalous}$:

$$< u(x,y)u(o,y) > \sim \quad (a_t^2 a_\ell)^{-1} \cos(q_s x)(x/x_\ell)^{-1} exp[-(x/x_\ell)] \tag{11a}$$

$$< u(x,y)u(x,o) > \sim \quad (a_t^2 a_\ell)^{-1}(y/y_t)^{-1} exp[-(y/y_t)] \tag{11b}$$

where $x_\ell = (a_\ell/\omega_s)$, $y_t = (a_t/\omega_s)$.

In the present example the transverse correlation length y_t is much greater than the longitudinal correlation length x_ℓ, since $a_t \gg a_\ell$; if as a transition approaches $\omega_s(T)$ decreases, both lengths grow. We are led naturally thereby to the picture of disc-like embryos in real space with correlated displacements whose normals are along q_s and within which the parallel planes have a shear modulation $\sim \cos(q_s x)$. The diffraction pattern from these embryos would be diffuse "rods" along the direction $\underline{q}_s$, and centered at approximately $\underline{q} \simeq \underline{q}_s$, which might be thought to be due to small inclusions of a long period second phase or defects, but within this model has nothing to do with either.

A more complete analysis would be worth doing. However, even the main features illustrated here go a long way toward explaining the extensively observed pretransitional features observed in many systems over the past few years, as discussed by Tanner at this meeting, and elsewhere.[2,3,4,16,17]

Summarizing: anomalous phonon dispersion behavior, at some $q \simeq q_s \neq 0$, and high anisotropy can lead directly to embryo-like regions[18] within which displacements are both correlated and modulated with a spatial period $\sim q_s^{-1}$ giving rise to intrinsic long period correlations. In this pedagogical discussion we have taken a single $\underline{q}_s$ (and its conjugate) as an example; in a real β-cubic system one could [14] have 8 equivalent (111) or 12 of the (110) equivalent $\underline{q}_s$ to combine to determine the range and structure of correlations.

III. NONLINEAR CONSIDERATIONS

The soft mode idea, which leads to second order transitions, has been called by de Gennes[19] an "Instability" transition, which would refer to $\omega^2(q)$ smoothly becoming negative, thus lattice instability i.e. the elastic susceptibility diverges.

However, we have seen at this workshop[20] that in the overwhelming majority of structural transformations, which show anomalous phonons that seem related to a resultant structure, the phonons soften only a small amount. Even in the case of significant but incomplete softening in ferroelectrics, through a long series of careful magnetic resonance studies Müller[21] concludes that the transition actually behaves like an order-disorder type involving sizeable correlated segments of titanium chains. Since in these many examples the phonon susceptibility does not diverge, and there seems to be evidence for arrays of phase and antiphase regions, we are led to consider first order transitions of the "Nucleation" type, in de Gennes' terminology. In this case local order parameters can change by large amounts, but due to spatial alternating "antiphase" regions the transition is macroscopically continuous. We show in this section how an anomaly in the phonon dispersion, such as that discussed in the previous section and ref. 2, can be incorporated into a phenomenological theory of this type of transition. Moreover, refs. 5, 6, 7 provide a quantitative basis for the present proposal for some transition metals.

A Landau Model for First Order Transitions: We suppose that the potential energy for distortion of a lattice by an amount $u(\underline{q}_s)$ (i.e. in the pattern of the "$\underline{q}_s$-mode") is carried to nonlinear fourth order terms

$$F = A\frac{u^2}{2} - \frac{u^3}{3} + \frac{u^4}{4} \quad . \tag{12}$$

(In general this function has three parameters, but the amplitude and energy scales can be chosen to leave only the shape variable $A(T)$, temperature renormalized, for study.)

If the cubic term were absent, one must have $A \to 0$ then < 0 to have a second order "instability-type" transition, i.e. $A = m\omega_s^2 = a(T - T_c)$ must soften completely.

Now, however, refer to Fig. 4, which shows $F(u(q_s))$ for a series of A-values. Remarkably, the inflection at $u = (1/2)$ first appears for $A = (2/8)$, then with only a small additional change to $A = (2/9)$ there is now the condition for a first order transition $u(q_s) = 0$ or $(2/3)$; moreover, the small amplitude motion about $u = 0$ is only lowered in frequency to $(8/9)\frac{1}{2} = 0.942$ of its value for the simple stability condition $U_{equilib} = 0$, $A \geq (1/4)$. Put another way, if $A = A(T) = m\omega_s^2$, hardly any "phonon softening" is needed to allow the first order transition, as is so often observed. Of course $A < 0$ corresponds to unconditional instability at $u = 0$.

Beyond the Quasiharmonic Concept: We recap at this point: (1) competing interactions can produce structural instabilities and anomalies in phonon dispersion curves, (2) dispersion anomalies and strong anisotropy will naturally produce "embryo"-like regions of internally modulated correlated structures, (3) if the displacement energies in mode-like distortions are computed to higher order of nonlinearity in the displacements, first order transitions can be controlled by very small changes with temperature in the quadratic (i.e. quasiharmonic frequency) coefficient.

We have surmised for some time[22] that the quasi-harmonic phonon anomaly above a transition temperature was a messenger telling us that for larger amplitude distortions (generally highly non-sinusoidal e.g. soliton-like) transitions into closely related patterns would take place. What is required for a more specific visualization is shown in Fig. 5 where, with potential forms for each q value like (12), we show an effective frequency as a function of both $u(q)$ and q. The two sets of curves are supposed to show that for $T > T_c$ the only stable minimum is at $u(q) = 0$ for all q, while for $T \simeq T_c$ modes not near q_s are still stable but around q_s there is now a minimim of potential energy at finite $u(0)$ i.e. the onset of a first order (Nucleation type) transition. Dynamically, this would show up as a strong order-disorder type of motion, and neutron scattering would be very broadened in both q space and energy. It would be impossible to describe this motion in quasi-harmonic language and simulations, similar to those recently reported by Kerr and Bishop[23] and in similar studies, are needed to quantify this phenomenological proposal.

Recent Quantitative Verification: Only recently has the author become aware of the calculations of refs. 5, 6, 7, which show quantitatively just the features proposed phenomenologically above. Figure 6 is taken from Figure 3 of ref. 5; the reader must refer to that paper for an excellent review and report of their work. What we reproduce here are the displacement energies for the $q_s = \frac{2}{3}[1,1,1]$ longitudinal mode, for Nb, MO, and Zr. For Zr, which shows the ω-phase instability, the behavior is exactly as surmised in our Fig. 5, and discussed in the previous section. Of course, in the work of Harmon et al. using the frozen phonon method, the temperature is effectively zero, so on our Fig. 5 it would correspond to $T \ll T_c$, for which we did not draw a curve. For higher temperature the authors showed[7] that anharmonic phonon perturbation theory could in fact stabilize a similar N-point instability; put in another way, at sufficiently high temperature the lower minimum of the thermally renormalized energy versus displacement would be at $u = 0$, corresponding to the stable high temperature phase. We are impressed by these calculations and the physical understanding which has guided them.

CONCLUDING REMARKS

Work remains to be done to refine an overall formalism, which incorporates our concepts. However, we believe that we have achieved an understanding of why simple soft mode theory fails in most applications while, at the same time, the wave number of the phonon anomaly is experimentally related to transformed structures.

In another paper we expect to present an extension within the present framework of the analyses of Cowley and Bruce, using a three-component order parameter for [110] instabilities, and one component for the (2/3) [111] ω-phase case. We set up a Landau formalism and determine the phase diagram, extending ideas in ref. 14 and 24.

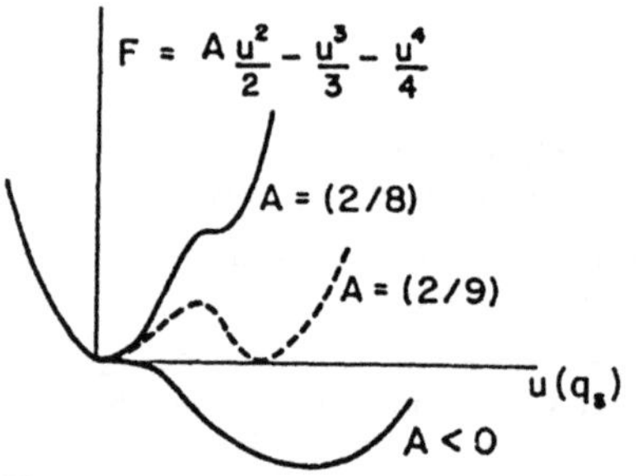

Fig. 4

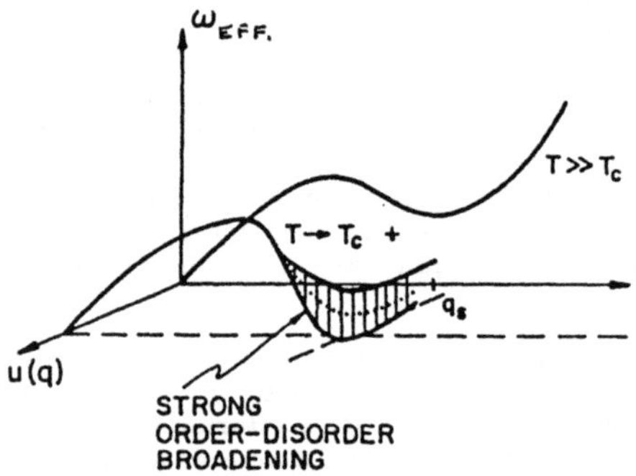

Fig. 5

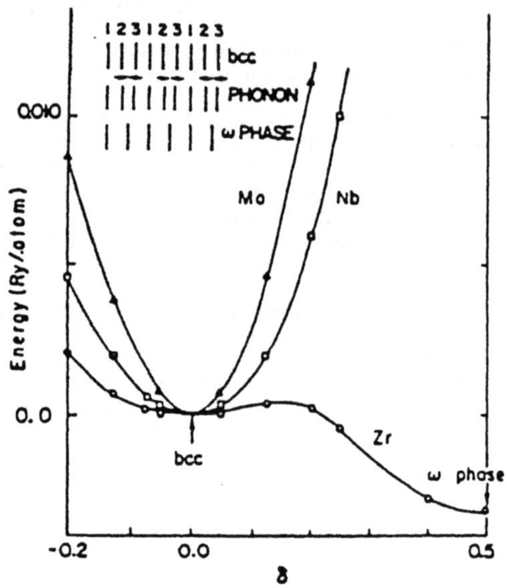

Fig. 6

4. Model Landau free energy function for first order transition, versus amplitude of anomalous mode at q_s. From Eq. (12) the inflection first appears at $A = (2/8)$, and the transition requires only small additional softening to $A = (2/9)$.

5. Augmented phonon dispersion surface: ω^2 versus q and amplitude $u(q)$.

6. Anharmonically unstable (2/3) [111] mode according to Harmon et al.[5]

On the experimental side we suggest that more complete studies of the precursor dynamics and the accompanying diffuse structure, for various directions around a q_s phonon anomaly, particularly versus temperature, are needed. In fact, it may be that the phenomena we have discussed here are responsible for the alkali metal instabilities, rather than proposed electronic effects i.e. charge density waves.

It is a pleasure to acknowledge discussions with many colleagues: Gerhard Barsch, Alan Bishop, Phil Clapp, Baruch Horovitz, Bill Kerr, Si Moss, Alec Müller, Steve Shapiro, T. Suzuki, Lee Tanner, Dieter Wohlleben, Manfred Wuttig, and Y. Yamada.

* Director's Fellow, Los Alamos National Laboratory; Conference support from the Los Alamos Center for Materials Science. Partial support from U.S. Dept. of Energy, Contract DE-FG02-85ER45214.

REFERENCES

1. C. Zener, Phys. Rev. **71**, 846 (1947), and references therein.

2. S.M. Shapiro, Y. Nada, Y. Fujii, and Y. Yamada, Phys. Rev. B **30**, 4314 (1984), and references therein.

3. Y. Morii and M. Iizumi, J. Phys. Soc. Japan **54**, 2984 (1985).

4. R.A. Robinson, G.L. Squires, and R. Pynn, J. Phys. F: Met. Phys. **14**, 1061 (1984).

5. K.-M. Ho, C.-L. Fu, and B.N. Harmon, Phys. Rev. B **29**, 1575 (1984).

6. Y. Chen, C.-L. Fu, K.-M. Ho, and B.N. Harmon, Phys. Rev. B **31**, 6775 (1985).

7. Y.-Y. Ye, Y. Chen, K.-M. Ho, B.N. Harmon, and P.-A. Lindgård, Phys. Rev. Lett. **58**, 1769 (1987).

8. C.M. Varma and W. Weber, Phys. Rev. Lett. **39**, 1094 (1977); A.L. Simons and C.M. Varma, Solid State Comm. **35**, 317 (1980); W. Weber, "Superconductivity in d- and f-Banned Metals", Academic Press, N.Y. (1980).

9. See reviews by K. Yosida, Prog. Low Temp. Physics **4**, 265 (1964) and by R.J. Elliott in Rado and Suhl "Magnetism" IIA, 385 (1965); also p. 486 of Introduction to Solid State Physics, C. Kittel, 3rd edition, John Wiley (New York).

10. V. Heine, this workshop; also Springer Series in Solid State Sciences **47**, 98 (1983), Springer-Verlag.

11. "Phase Transformations in K_2SeO_4 and Structurally Related Insulators", in "Incommensurate Phases in Dielectrics, 2", Ed. R. Blinc and A.P. Levanyuk, Elsevier (B.V. 1986).

12. M. Iizumi, J.D. Axe, Gi Shirane, and K. Shimaoka, Phys. Rev. B **15**, 4392 (1977); M.S. Haque and J.S. Hardy, Phys. Rev. B **21**, 245 (1980).

13. R.J. Elliott, J.A. Krumhansl, P.A. Leath, Reviews of Modern Physics **46**, 465 (1974), cf. Sec II.

14. R.A. Cowley and A.D. Bruce, J. Phys. C: Solid State Phys. **11**, 3577-3608 (1978).

15. Y. Morii and M. Iizumi, J. Phys. Soc. Japan **54**, 2948 (1985); M. Iizumi, J. Phys. Soc. Japan **52**, 549 (1983).

16. S.M. Shapiro, J.Z. Lorese, Y. Noda, S.C. Moss, and L.E. Tanner, Phys. Rev. Lett. **57**, 3199 (1986).

17. I.M. Robertson and C.M. Wayman, Phil. Mag. A **48**, 421 (1983); ibid. p. 443; ibid. p. 629.

18. Y. Noda and Y. Yamada, J. Phys. Soc. Japan **48**, 1288 (1980): have studied a number of special models which could lead to embryos; but all of these seem to require the existence of a defect, whereas the correlations we discuss here are intrinsic to the lattice response when certain phonon anomalies with strong anisotropy are present, as they are for example in TiNi(Fe), the subject of Ref. 2.

19. P.G. de Gennes, in "Fluctuations, Instabilities, and Phase Transitions", Geilo, Norway, 1973, Ed. T. Riste (Plenum, NY 1973).

20. S.M. Shapiro, "Neutron and X-ray Scattering Studies of Pre-Martensitic Phenomena", this workshop.

21. K.A. Müller, Springer Series in Solid State Sciences **69**, 234 (1986).

22. J.A. Krumhansl, Springer Series in Solid State Sciences **69**, 255 (1986).

23. W.C. Kerr and A.R. Bishop, Phys. Rev. B **34**, 6295 (1986).

24. B. Horovitz, J.L. Murray, and J.A. Krumhansl, Phys. Rev. B **18**, 3549 (1978).

Part II

Statics

Competing Interactions and the Origins of Polytypism

G.D. Price[1] *and J.M. Yeomans*[2]

[1]Department of Geological Sciences,
Universtity College London, Gower Street, London WC1E 6BT, UK
[2]Department of Theoretical Physics,
University of Oxford, 1 Keble Road, Oxford OX1 3NP, UK

1. INTRODUCTION

Crystal structures can be analysed at many levels. At the most fundamental level, they are described in terms of the relative distribution of their constituent atoms, or of the coordination polyhedra of the component cations and anions. It is becoming increasingly apparent, however, that many families of structures can be usefully described in terms of larger basic structural units or modules. If such an approach to the description of crystal structures is adopted, many complex solids may be systematised in terms of series of stacking variants of the simple subunits; these phases are known as polytypes. This relatively broad definition of polytypism closely follows that of THOMPSON [1] and is discussed at length by ANGEL [2]. The definition removes any chemical constraints upon stacking variants, and allows more than one type of module to be present in a given polytype family. Polytypism is however a special form of polymorphism, and although there is no constraint upon the chemistry or structure of the modules involved, to be considered polymorphic the various modes of module stacking should not affect the composition of the phase as a whole. It should also be noted that this definition of polytypism allows for the existence of 2- and 3-dimensional polytypic structures, comprised of prismatic (rod-like) and block modules respectively, as well as the classically more familiar layer modules of the 1-dimensional polytypes.

Polytypism is a common phenomenon, and examples of stoichiometric, non-metallic materials which exhibit polytypism include the mica minerals, and the classic polytypes with MX or MX_2 stoichiometry, characterised by SiC, ZnS, CdI_2 and MoS_2. Many metal systems, such as the binary alloys $TiAl_3$ and Cu_3Pd, also exhibit polytypism. Moreover included within the wider definition of polytypism, being composed of structurally compatible modules, are a range of minerals which include the pyroxenes, pyroxenoids, perovskites, spinelloids, zoisites, chlorites, sappharine and MnO_2 oxides. Polytypic series or families characteristically contain a large number of structures, which exhibit a variety of module stacking sequences. Simple, shorter period, commensurate modifications are most commonly observed, although polytypes may have very long period (in some cases in excess of 1000Å) or incommensurate repeats. Because the polytypes of a given compound are composed of virtually identical structural units, the free energy differences between them are small. In addition, the activation energy for transformation between them is often high. As a result, the transformation kinetics between polytypes is usually slow, and it is difficult to establish thermodynamic equilibrium. Undoubtedly, many metastable phases have been synthesised in such polytypic systems. However recent careful experimentation upon metallic and inorganic systems [e.g. 3,4] has established that at least some polytypes have thermodynamically definable fields of stability.

A variety of theories have been advanced to explain why some polytypic sequences appear to be more stable or occur more frequently than others [5]. Theories based upon growth or kinetic considerations may well explain how certain metastable sequences are formed. FRANK's [6] theory of polytypism assumes that the crystal initially grows from a screw dislocation in the stacking planes, so that the period of the polytype is determined by the step height of the growth spiral. Using this mechanism, it is possible to generate almost any polytype from shorter period structures, especially if the theory is modified to include growth resulting from interweaving spirals with different step heights [7]. However, these mechanisms are not universally applicable, as many examples of long period polytypes show no evidence of growth spirals. Moreover, the theory is unable to predict which polytypes will occur. Thermodynamically based models are required to explain the observed equilibrium phases found in systems such as the binary alloys, SiC and the spinelloids. Among the first of such theories was that of JAGODZINSKI [8], who argued that long period polytypes are stabilised by vibrational entropy effects. More recent workers [e.g. 9,10,11] however, have developed the idea that the relative stability of polytypic structures may be determined by considering effective interactions between the component structural units. To explain why long-period or incommensurate polytypic phases are found and to model the nature of the observed phase transformations between polytypes, it is necessary to appeal to interactions between structural units which are effective over a distance greater than that between first nearest-neighbour modules. In fact, it appears that competing interactions between first and further neighbour units are a prerequisite for the development of long period structures.

In this paper we will outline the major theoretical models which can be used to describe the behaviour of polytypes and related modulated structures. We will then briefly describe the structural characteristics of the spinelloids and other major polytypic families, and discuss their behaviour in terms of the theoretical models developed.

2. DEFINITIONS AND A SIMPLE MODEL

2.1. Introduction

Our aim in this section is to summarise theoretical work on model systems which have or could be used to describe polytypism or related modulated order in materials. At this stage we give few physical examples: these appear in section 4, where the properties of various compounds are described and their behaviour is related to the theories elucidated here. Moreover, little detail of the theoretical techniques used to derive the results is presented - references are given for those interested. Instead our aim is to compare and contrast the phase diagrams of various models of modulated phases and to emphasize the features necessary to generate long period structures. In this way we hope to clarify one of the main problems in understanding modulated phases in materials: a better understanding of which mechanism is responsible for the stabilisation of complicated structures in each of the various compounds considered.

In section 2.2 we define terms useful throughout the article. Then, in section 2.3, the Frenkel-Kontorova model, one of the first and simplest examples of the commensurate - incommensurate phase transition will be described. This model is important because it emphasizes the role of competition and of domain walls in the physics of modulated structures.

In section 3 we demonstrate how many of the same features appear in spin models with competing interactions. We emphasize that, in a system with short-range competing interactions, long period structures can be stabilised either by long-range interactions at zero temperature or by the effects of entropy at finite temperatures. In section 5 we consider a class of model with a continuous, rather than a discrete variable, on each lattice site. A conclusion attempts to summarise the similarities of and differences between the various models considered.

2.2. Definitions

A commensurate structure has average periodicity, a, which is a rational fraction of the period, b, of an underlying lattice. If a is not a rational fraction of b the structure is said to be incommensurate. Experimentally if the ratio a/b is a complicated rational fraction it may be impossible to determine whether the structure is commensurate or incommensurate.

In both experimental systems and theoretical models the period of the modulated structure changes as a function of, for example, temperature or the strength of the pinning potential. If the wavevector is close to a commensurate value it may lock-in to that value over a range of the controlling parameter. It has proven useful to give (rather colourful) names to three different types of behaviour [12,13]:

(i) incomplete devil's staircase: in which the wavevector locks-in to all possible commensurate values but incommensurate regions remain between each commensurate phase.

(ii) complete devil's staircase: in which the wavevector locks-in to all commensurate values. The commensurate phases fill parameter space with no intervening incommensurate regions.

(iii) harmless devil's staircase: in which the wavevector locks-in to a (possibly infinite) subset of commensurate values. Different phases are separated by first order phase transitions.

To some extent the distinction is useful only to theorists, as high order commensurate phases are stable only over a range of the control parameter that would be impossible to resolve experimentally, and hence a continuous variation of the wavevector, with lock-ins to the low order commensurate phases, would be observed in an experiment. Perhaps the most useful experimental distinction is that in cases (i) and (ii) increasing the resolution would show more lock-ins, whereas in (iii) this need not be the case.

2.3 The Frenkel-Kontorova Model

a) Continuum model

One of the earliest theories of the commensurate-incommensurate phase transition was due to FRENKEL and KONTOROVA [14] and FRANK and VAN DER MERWE [15]. This theory is important because it emphasizes two vital features in the theory of modulated structures - the role of competing terms in the Hamiltonian and the existence of domain walls. We first consider the continuum theory, where the effects of the discrete lattice are ignored, and then describe the results of more recent work [12] which includes lattice effects.

The Frenkel-Kontorova model can be thought of as an array of atoms, coupled by harmonic springs of natural length, a, lying in a potential, V, of periodicity, b. It is defined by the Hamiltonian:

$$H = \sum_n \{(x_{n+1} - x_n - a)^2/2b^2 + V(1 - \cos(2\pi x_n/b))\} \quad (1),$$

where x_n is the position of the nth atom. The physics of this model is determined by the competition between the elastic energy which favours an incommensurate separation between atoms, a, and the tendency of the atoms to sit at the bottom of the potential wells leading to a commensurate structure.

It is helpful to introduce a phase P_n, which measures the position of the atoms relative to the minima of the potential, defined by:

$$x_n = nb + bP_n/2\pi \quad (2).$$

The model was originally solved in the continuum limit:

$$P_n - P_{n-1} = dP/dn \qquad (3),$$

when (1) transforms to:

$$H = \int \{(dP/dn - d)^2/2 + V(1 - \cos P)\}dn \qquad (4),$$

where $d = 2\pi(a-b)/b$ measures the misfit between the two competing length scales. Minimising the energy, H, is equivalent to finding a solution of:

$$d^2P/dn^2 = V\sin P \qquad (5),$$

which is the one-dimensional sine-Gordon equation.

The solution comprises a regular lattice of domain walls, where the phase changes by 2π, with spacing l and width $l_o = l/\sqrt{V}$. Hence the structure remains commensurate to the greatest extent possible, with the misfit being taken up in local regions of phase slip which become narrower as the pinning potential increases. This is a very general feature of incommensurate phases. The domain walls repel exponentially and hence are, in the continuum limit, equispaced.

Comparing the stability of this phase to the commensurate phase one finds, as expected, that walls first appear when V becomes small enough or d large enough that the wall formation energy becomes negative. Because the walls repel each other their number increases smoothly from zero as the threshold, d_c, is passed and the commensurate-incommensurate transition is continuous with:

$$l = 1/\ln(d_c - d) \qquad (6).$$

One can show by expanding about each commensurate phase in turn that there is lock-in at an infinite number of commensurate values as d is varied. However, for V small, where the continuum approximation is expected to be correct, lock-in only occurs over a fraction $O(\sqrt{V})$ of the parameter space. Hence this is an example of an incomplete devil's staircase.

b) Discrete model

Solution of the Frenkel-Kontorova model in the discrete case where lattice effects are not ignored by making the approximation (3) is much more complicated and has only been carried out relatively recently [12]. As the potential V increases the continuum solution (where one expands about each commensurate phase in turn and assumes the solutions can be superimposed) breaks down and there is a crossover from an incomplete to a complete devil's staircase, where commensurate phases fill the parameter space with every possible commensurate phase being stable for some range of d.

3. SPIN MODELS

Recently considerable insight into the behaviour of polytypic compounds has been obtained by mapping them onto spin models. This is based on the idea that if, as is the case for most polytypes, the basic building blocks of the crystal structure can lie in one of two positions they can be mapped onto a two-state, Ising spin variable, $s_{ij} = \pm 1$. The difference in energy between different orientations, parallel or antiparallel, of the blocks is then expressed through the coupling constants in the Hamiltonian.

In preparation for a more detailed account of this mapping, given in section 4, we summarise work on Ising spin models in which competing interactions result in modulated structures.

3.1. Short-Range Competing Interactions at Zero Temperature

We shall consider first the ground state (T = 0) behaviour of the Ising model on a simple cubic lattice defined by the Hamiltonian:

$$H = -1/2\, J_0 \Sigma_{ijj'} s_{ij} s_{ij'} - J_1 \Sigma_{ij} s_{ij} s_{i+1j} - J_2 \Sigma_{ij} s_{ij} s_{i+2j} - J_3 \Sigma_{ij} s_{ij} s_{i+3j} \qquad (7)$$

where i labels the layers perpendicular to a chosen axial direction and j, j' label sites within a layer. The coupling within the layers, $J_0 > 0$, will be assumed ferromagnetic but we shall consider both signs of the axial couplings, J_1, J_2, and J_3.

The ground state is found by minimising the energy, H. The first term ensures that spins within layers will always be ferromagnetically aligned, but along the axial direction the ordering will depend on the values of J_1, J_2 and J_3 as shown in Fig. 1. Note that only a small number of short-period structures appear as ground state phases. This is a result of the limited range of the interaction.

It will be helpful here, and in particular when we start dealing with longer period structures, to introduce a notation to describe the axial ordering. This was first used for spin models by FISHER and SELKE [16] but turns out to be almost identical to that introduced by ZDHANOV and MINERVINA [17] to categorise classical polytypes. First we define a band to be a sequence of consecutive layers with the same spin value terminated by layers of opposite spin. For example, a state where the repeating spin sequence is:

$$....++--++---.... \qquad (8)$$

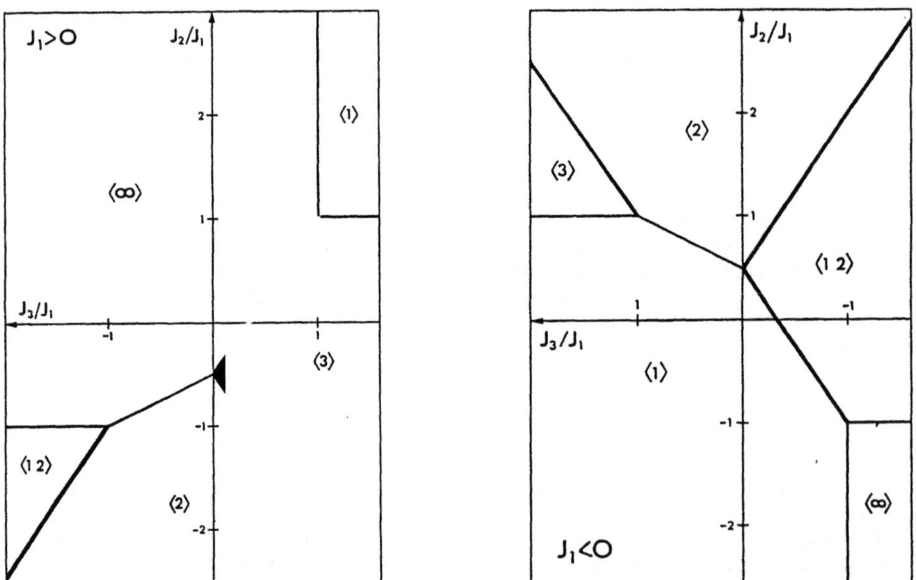

Figure 1. The ground state of the three-dimensional Ising model defined by (7).

comprises three 2-bands followed by one 3-band. The notation $\langle n_1, n_2 \ldots n_m \rangle$ is then used to describe the state where a wavelength (or half a wavelength if there are an odd number of bands) contains m bands of length $n_1, n_2 \ldots n_m$. Hence (8) is represented by ⟨2223⟩ or $\langle 2^3 3 \rangle$.

Consider now the phase boundaries between the different ground states in Fig. 1. On six of the boundaries, identified by bold lines in the figure, the ground state is infinitely degenerate, with an infinite subset of structures, both periodic and random, having the same energy. In the vicinity of these boundaries it seems reasonable that small perturbations can stabilise certain of the degenerate structures, which may be of arbitrarily long periods at the expense of others. Two possible perturbations are longer range interactions and entropic effects at finite temperatures. Much discussion is currently in progress as to which is most relevant in experimental systems: the answer is likely to be compound dependent. We discuss each possibility in turn.

3.2. Long-Range Interactions

One way of introducing long period phases into spin models is to increase the range of the interactions, while remaining at zero temperature. Ground state phases have periods up to twice the interaction range. Obviously, as the number of parameters, J_i, increases it becomes impossible to elucidate exactly the ground state of the model. However, various special cases have been considered which allow us to understand the behaviour in the limit that the interaction range becomes infinite.

BAK and BRUINSMA [18] have considered the one-dimensional Ising model with long-range, convex, antiferromagnetic interactions. This would include power law decays and exponentially decaying Coulomb interactions. They were able to prove that, as a function of an applied magnetic field, there is a complete devil's staircase. SHINJO and SASADA [19] have shown that if the interaction is of finite range the staircase becomes harmless. Several other authors [13,20] have treated one-dimensional Ising models with long-range oscillatory interactions. Elastic interactions between defect planes in lattices come into this category [21]. In this case there are still, in general, a large number of commensurate phases, but the devil's staircase is harmless.

Although these results have been obtained for zero temperature, it is expected that the addition of a ferromagnetic in-plane interaction, J_0, preserves the general phase structure at finite temperatures, although individual phases may appear or disappear.

3.3. Finite Temperatures

If long-range interactions are not present another peturbation is needed to break the degeneracy of the multiphase lines. When the behaviour of the Hamiltonian (7) at finite temperatures is considered, the stable phase corresponds to that with minimum free energy. The entropic contribution to the free energy is structure dependent and hence stabilises a subset of the degenerate ground state phases.

We consider first the case $J_3 = 0$ which is simpler but contains all the essential physics in the problem. This is the so-called ANNNI, or axial next nearest neighbour Ising model [22], which has received a great deal of attention in the literature as one of the simplest spin models where modulated structures are stable [16,23,24]. To obtain long period phases we must consider $J_2 < 0$ so that the first and second neighbour axial interactions compete. The ground state can then be read off from Fig. 1 to be:

$$\langle\infty\rangle,\ J_1/|J_2| > 2;\ \langle 2\rangle,\ 2 > J_1/|J_2| > -2;\ \langle 1\rangle,\ -2 > J_1/|J_2| \tag{9}$$

$J_1/|J_2| = 2$ and -2 are multiphase points where all phases comprising bands of length > 2 and all phases containing only 1- and 2-bands respectively are degenerate.

The phase diagram as a function of the interaction parameters and the temperature is shown in Fig. 2(a) for $J_1 > 0$ and Fig. 2(b) for $J_1 < 0$. At finite temperatures an infinite number of modulated structures of arbitrarily long periods are stable. The region of phase space over which a given structure locks-in rapidly becomes very narrow with increasing wavelength. Hence in Fig. 2 it is only possible to show a representative sample of the shorter period phases with many widths exaggerated for clarity. Note that, for $J_1 > 0$, all the stable phases consist of bands of length > 2 whereas for $J_1 < 0$ only 1- and 2- bands appear. This is of course a consequence of the particular degeneracies of the multiphase points. As will be seen, the predominance of bands of certain lengths is very reminiscent of polytypic behaviour, for example in silicon carbide.

Not all possible commensurate phases are stabilised, but only particular sequences. For example, in the limit that the temperature tends to zero, only $\langle 2^{n-1}3 \rangle$ and $\langle 12^n \rangle$, n = 1, 2, 3 are stable for $J_1 > 0$ and $J_1 < 0$ respectively. As the temperature is increased new phases appear in a regular pattern: the first phase to become stable between $\langle n_1 \rangle$ and $\langle n_2 \rangle$ is $\langle n_1 n_2 \rangle$, for example $\langle 232^2 3 \rangle$ appears between $\langle 23 \rangle$ and $\langle 2^2 3 \rangle$. Further increases in temperature lead to the appearance of more complicated combinations of structures through repeated application of the same algorithm. This pattern is a general feature of all models with competing interactions that have been studied so far. At T > Tc/2, where Tc

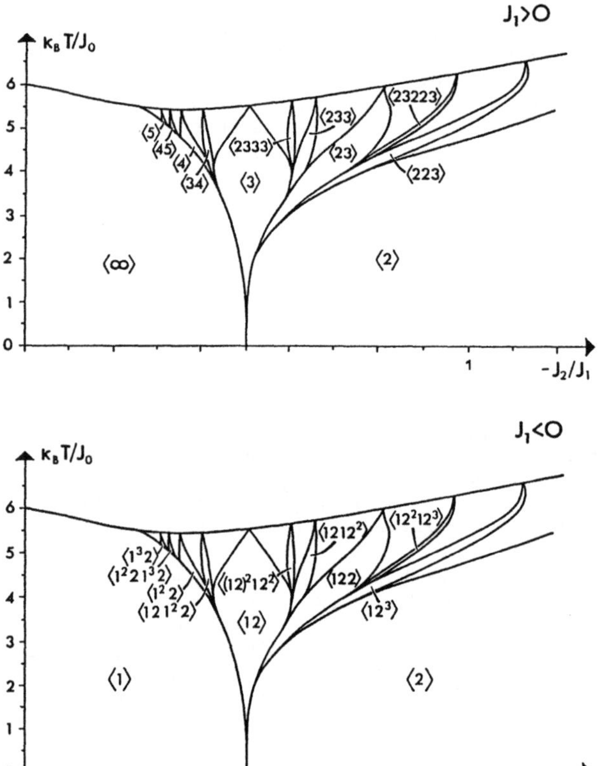

Figure 2. The ANNNI model phase diagrams as a function of temperature.

is the transition temperature to the disordered phase, incommensurate structures are also thought to be stable.

These results were obtained using mean-field theory, which is believed to give a good representation of the phase diagram of this and similar models in three dimensions [25] and low-temperature series [17]. Details of the theoretical techniques are given in the papers cited and several reviews [26,27].

It is important to comment on the physical mechanism by which the long period phases are stabilised. To do this it is perhaps easiest to think of the modulated structures in terms of regular arrays of domain walls. The exact definition of a wall can be chosen for convenience, for example in a phase such as <6> it is most natural to think of a wall as a boundary between up- and down-bands, whereas in the $\langle 2^n 3 \rangle$ sequence the 3-bands can be considered as regularly spaced walls.

At finite temperatures the domain walls fluctuate due to local excitations or spin-flips which contribute to the entropic term in the free energy. Fluctuations will cause the domain walls to collide and interact. It is these interactions (together with the intrinsic wall energy) which determine the spacing of the walls and hence the period of the associated modulated structure. The effective wall-wall interactions are long range and hence can stabilise phases of arbitrarily long wavelength. They fall off very quickly (exponentially) with distance: this is why the long period phases become narrow so rapidly with increasing period. Multiwall interactions are instrumental in stabilising combination phases such as $\langle 232^2 3 \rangle$.

We may summarise by stating that the ANNNI-mechanism for stabilising long period modulated structures is entropic fluctuations of the domain walls. Hence the long period phases only exist at finite temperatures.

A very similar phase diagram is seen when the third neighbour interactions in (7) are taken into account [28]. Near each of the multiphase lines an infinite subset of the ground state phases appears as stable phases at finite temperatures. More phases appear according to the usual combination rules as the temperature is increased.

3.4. An Isotropic Model

In the ANNNI model the axial direction is picked out as special by the definition of the Hamiltonian and the resulting modulated structure is thereby forced to be one-dimensional. This is a consequence of theorists trying to simplify the problem, but in many crystal structures the competing interactions are expected to be isotropic.

Little work has been done on models with isotropic competing interactions because many of the standard theoretical techniques break down when dealing with three-dimensional modulated phases. However, some insight into their behaviour can be obtained by just considering the ground state of a model with first and second neighbour competing interactions along each cube edge, together with interactions between spins across each face diagonal. The interaction parameters are defined in Fig. 3a and the interesting region of the ground state phase diagram shown is in Fig. 3b. The notation used to describe the stable structures is an obvious extension of that introduced in section 3.3. For example $\langle 2,2,\infty \rangle$ comprises 2-bands in two directions and ferromagnetic ordering in the third.

The most striking feature of this phase diagram is that modulation can occur in one, two or three directions as typified by the phases $\langle 2,\infty,\infty \rangle$, $\langle 2,2,\infty \rangle$ and $\langle 2,2,2 \rangle$ respectively. This can be explained by considering the crossing energy of sets of domain walls at right angles to each other: if this is positive, domain walls do not like to cross and hence there will be regions of the phase diagram where one-dimensional modulation is favoured even though the initial Hamiltonian is

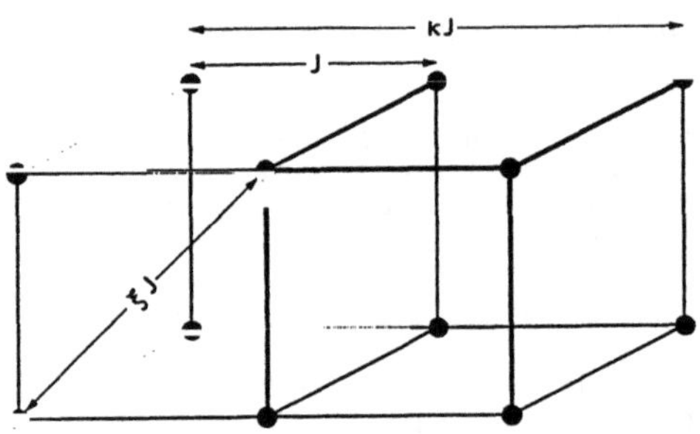

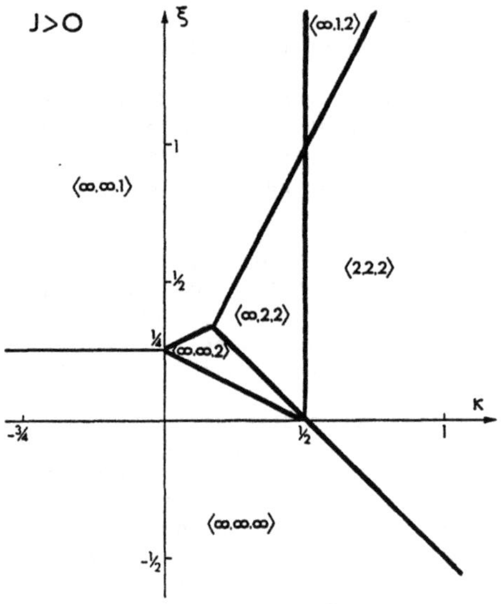

Figure 3. (a) The interactions of the isotropic model. (b) The ground state diagram for the isotropic model [38].

isotropic. The ground state phase boundaries shown in bolder type in Fig. 3b, are multiphase lines and longer period phases, similar to those observed in the ANNNI model are stabilised close to them at finite temperatures [38].

4. POLYTYPE STRUCTURES AND BEHAVIOUR

Polytypic behaviour is epitomized by the spinelloids [10]. The family takes its name from its most commonly occurring member, the spinel structure, from which the other structures in the family can be derived. Like spinel, spinelloids generally have an AB_2O_4 stoichiometry, and have structures which are based upon a nearly cubic close-packed arrangement of oxygen ions, within which cations occupy both tetrahedrally and octahedrally coordinated sites. The cations define a 'basic structural unit' within the oxygen framework (Fig. 4a) from which all spinelloids can be constructed. The arrangement of the basic structural units is invariant in two orthogonal directions, and generates an infinite sheet, part of which is shown in Fig. 4b. Variations in the packing of this sheet in the third dimension give rise to the observed range of spinelloid structures. In the spinel structure the component spinelloid sheets are packed parallel to (110). Adjacent sheets are related by a glide operator of the type 1/4[112](110) which, when regularly repeated, generates the spinel structure shown in Fig. 4c. If the basic structural unit shown in Fig. 4a is represented by the arrow illustrated, the stacking sequence along [110] of spinel can be described by the code ..↑↓↑↓.. or <1>. In addition to the glide operator, successive idealised spinelloid sheets may also be related by a mirror or twin operator. For example in the idealised beta-phase polymorph of Mg_2SiO_4, wadsleyite (Fig. 4d), basic spinelloid structural units are alternately related by mirror and glide operators and can be represented by the stacking code ..↑↑↓↓.. or <2>.

There are an infinite number of possible stacking sequences which can be generated by combining mirror and glide operators. However only six of the possible sequences have been reported as forming crystal structures. These six structures are the spinel, <1>, and beta-phase, <2>, described above, and structures which correspond to the codes <12>, <122>, <3>, and <13>. The phase relationships between spinelloid polytypes in the Ni_2SiO_4-$NiAl_2O_4$ system have been studied in detail by AKAOGI et al. [29]. They found that compositions close to $Ni_3Al_2SiO_8$ can adopt one of three spinelloid structures. At pressures, P, less than 5GPa a

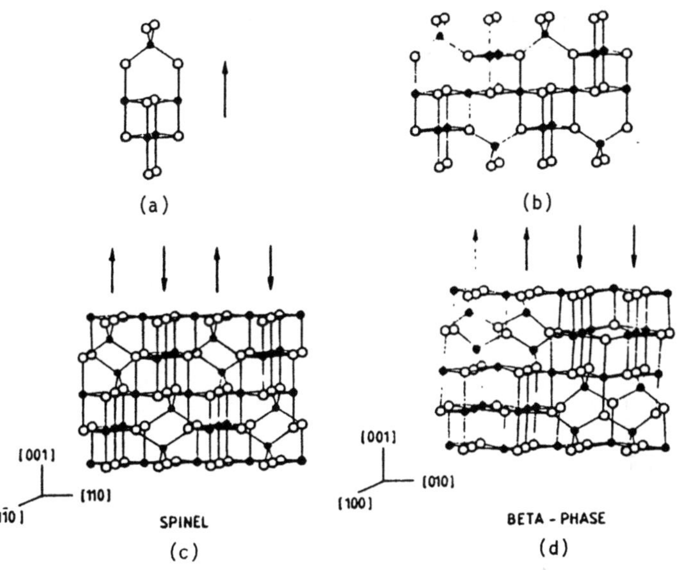

Figure 4. (a) The spinelloid basic structural unit, (b) the constituent spinelloid sheet, (c) the spinel structure, (d) the beta-phase structure.

spinelloid with the beta-phase, <2>, structure is encountered. However, with 5<P<7GPa, a spinelloid with the <122> structure is stable, while with P>7GPa the <12> structure is adopted. These transformations are virtually independent of temperature, and are reported to be first order in nature (see Fig. 5).

It is possible to interpret the behaviour of polytypic materials, such as spinelloids, in terms of an Ising spin model, by mapping the basic polytypic structural unit onto a magnetic spin variable [9,10]. The interactions between the units are then represented by a Hamiltonian, such as (7), with competing first and further neighbour interactions. For given J_n and k_BT/J_o, the stable polytype will correspond to the spin phase with minimum free energy. Transformations between polytypes occur because of changes in temperature, or changes in the effective interaction energies J_n as a function of pressure, temperature and chemical environment. As the conditions to which the polytype is subjected vary, the point defined by the coordinates J_n and k_BT/J_o may describe a trajectory which passes through many different phases. The exact form and extent of this path, and hence the sequence of stable phases, will critically depend on the exact relationship between J_n and the external conditions. However, PRICE and YEOMANS [10,11,30] qualitatively accounted for the observed sequence of transformations in the spinelloid system (Fig. 5) by mapping its behaviour onto an axial Ising model. They contended that variations in J_1 and J_2 as a function of increasing pressure defines a trajectory in interaction energy space (Fig. 1) which starts in the <3> phase, passes through the <2> and <12> phases, and ends in the <1> phase. The <122> phase is stabilised between <12> and <2> by, for example, longer range interactions. The failure to observe any longer-period structures (e.g. <322>, $<12^3>$, etc.) in this system, may be accounted for by the very small region of stability predicted for the longer-period phases.

Similarly, axial Ising models have been used to describe polytypic behaviour of other inorganic and metallic systems, such as $MgSiO_3$ pyroxenes, which occur as one of three polymorphic structures, known as the minerals clinoenstatite, orthoenstatite and protoenstatite. These three phases can be considered as polytypes with packing sequences <∞>, <2> and <1> respectively. In this system, the role of temperature appears to be most important in determining the relative magnitudes of J_1 and J_2. At temperatures above 980°C, protoenstatite is the stable phase, but on cooling it is thought to convert first to orthoenstatite, <2>, and then below 650°C to clinoenstatite. The mechanism and kinetics of this

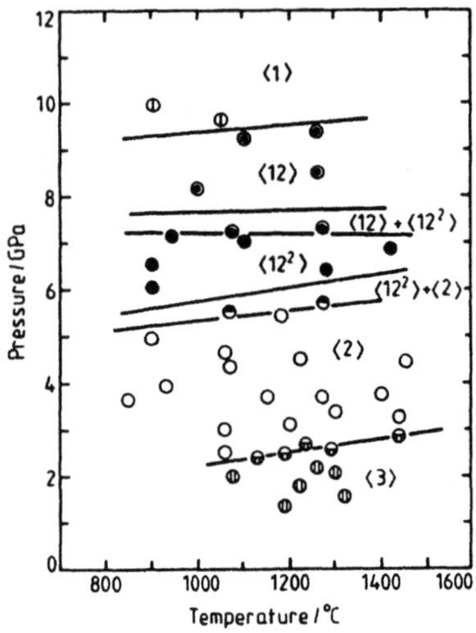

Figure 5. Phase diagram for the Ni-Al-Si spinelloid system [29].

inversion, however, make it difficult to define transformation temperatures exactly. The inferred sequence of polytypic phase transformations, <1>-<2>-<∞>, is compatible with a simple trajectory in the J_1, J_2 space of Figs. 1. No intermediate phases (e.g. <12> or <3>) have been found in this system although high-resolution transmission electron microscopy has revealed that considerable stacking disorder is developed in some partially inverted enstatites.

In addition to the mineral polytypic behaviour described above, axial Ising models can been used to explain the wurtzite, <1>, to zinc-blende, <∞>, transformations which occur in the classic ZnS and AgI polytypes, transformations between hexagonal perovskite polytypes, and all the recently observed reversible transformations in the SiC system [9]. JEPPS and PAGE [3] have reported three sets of reversible SiC transformations (all determined under slightly different physical or chemical conditions) which correspond to <1> ↔ <∞>, <∞> ↔ <3> and <3> ↔ <2>. The occurrence of these SiC polytypic inversions is particularly striking given the topological distribution of these phases on the ANNNI phase diagram (Fig. 2). It is also interesting to note that the large majority of stable polytypes observed in SiC are comprised of 2- and 3-bands, suggesting a ferromagnetic first neighbour interaction. For CdI_2, however, 1- and 2-bands predominate, indicating that the effective J_1 is antiferromagnetic. Similar behaviour has also been reported in the binary alloys (e.g. $TiAl_3$, Ag_3Mg), and again the behaviour in these systems can be qualitatively discussed in terms of axial Ising models [27,31].

Although axial Ising models have undoubtedly been very useful in helping to understand the principles of polytypism, they have a number of limitations. For example, in some metallic systems, such as Cu_3Pd, two-dimensional modulated or polytypic behaviour has been reported [32]. In this case it would appear that it would be more appropriate to describe the polytypism in terms of an isotropic rather than an axial spin model (see section 3.4).

Another serious problem with the ANNNI model is revealed when the crystallographic analogue of the spin-flips, which stabilize the high temperature, long-period phases characteristic of model (Fig. 2), are considered. For metallic systems, the spin-flip simply maps onto an exchange of unlike atoms, and is likely to be structurally quite feasible. Indeed, high-resolution transmission electron microscopy (HRTEM) [4] reveals significant disorder at domain boundaries in binary alloy phases, as would be expected if they were entropy stabilised. However, similar HRTEM on phases such as the spinelloids and SiC, shows them generally to have straight domain walls. In addition, 'spin-flips' in materials such as SiC would produce physically unrealistic, and energetically unfavourable atomic configurations. In this analysis, therefore, it may be more appropriate to discuss polytypism in covalently bonded materials, such as SiC, as resulting from long-range interactions (section 3.2), rather than being entropy stabilised. Indeed, calculations by PRICE et al. [33] on the energetics of the spinelloid system, have indicated that J_3 is non-zero, and that long-range interactions must be included in a full description of polytypism in such systems. However, before we can conclude our discussion of the origins of polytypism, we must consider one further class of model, that provides another way of understanding entropy stabilisation and which may be more widely applicable.

5. MODELS WITH CONTINUOUS VARIABLES

Given that one of the main criticisms of using the ANNNI model to represent polytypic compounds is that the elementary spin excitations needed to stabilise the long period phases correspond to unphysical rearrangements of atoms, it is important to consider models in which the spin variable on each lattice site can vary continuously. One of the most detailed studies of a model of this sort with competing interactions has been carried out by AXEL and AUBRY [20]. We shall summarise their work and then discuss briefly similar systems which have been considered by other authors.

The Axel-Aubry model is defined by the Hamiltonian:

$$H= \Sigma_j \{c_1(u_{j+1} - u_j)^2/2 - c_2(u_{j+2} - u_j)^2/2 + V(u_j)\} \qquad (10),$$

where $V(u_j)$ is a double-well potential which is approximated by the superposition of two parabolas:

$$V(u_j) = (u_j - z_j)^2/2, \; z_j = \text{sign}(u_j) = +/-1 \qquad (11).$$

One interpretation of u_j is as the displacement of atom j from the jth site of the lattice.

The phase diagram of (10) is very similar to others we have considered in this paper. This follows because the Axel-Aubry model can be mapped onto a one-dimensional Ising model with spins s_j at lattice site j which interact through long-range oscillatory interactions. It was claimed in section 3.2 that such a system exhibits a harmless devil's staircase. Moreover the Axel-Aubry model is a low-temperature mean-field approximation to the ANNNI model with u_j representing the deviation of the magnetisation from saturation in layer j. Hence one finds in its phase diagram the ANNNI sequences of phases, although the topology is affected by the natural choice of variables for this problem.

Although (10) is a one-dimensional Hamiltonian, addition of ferromagnetic in-plane interactions is not expected to affect the ground state phase diagram. Moreover, consideration of the effect of temperature on the Axel-Aubry model shows that new commensurate phases appear through the usual combination rules as the temperature is increased, and that incommensurate phases become stable as the paramagnetic transition is approached.

A similar system which exhibits comparable behaviour has been studied by JANSSEN and TJON [34,35,36]. They consider the usual linear chain of atoms which interact with harmonic second and third neighbour interactions but an anharmonic first neighbour coupling. The ground state phase diagram for this system is very similar to that of the Axel-Aubry model except for the appearance of incommensurate phases which fill an increasing portion of the phase diagram as the depth of the pinning potential decreases. It has recently been pointed out by BENKERT et al. [37] that a model of this sort would provide a useful way of explaining why some compounds form incommensurate structures whereas others prefer a locked-in wavevector.

6. DISCUSSION

Perhaps the overwhelming impression resulting from this analysis is that the phase diagrams of all the different models considered are very similar. In each case defining a Hamiltonian with competing interactions has led to modulated structures. To obtain long period structures three different mechanisms have been proposed. Firstly, long-range repulsive or oscillatory interactions as described in section 3.2. These could be Coulombic or elastic in origin. Secondly entropic effects at finite temperatures, as in the ANNNI model treated in section 3.3. This mechanism would be identified experimentally by fluctuating domain walls and the fact that the long period phases disappear as the temperature tends to zero. Thirdly long period phases are also produced by Hamiltonians with continuous spin variables interacting through competing interactions as in the Axel-Aubry and Janssen-Tjon models described in section 5.

There are two possible physical interpretations of this latter model, which may be useful in applications to polytypes. Firstly, recall that the $V(u_j)$ in (10) is a double well potential which allows the atom on each lattice site to lie in one of two potential minima. u_j measures the atomic deviations from the minima and the two different interpretations depend on what causes this deviation. Most obviously one can think of the terms c_1 and c_2 in (10) as being competing elastic energies. In this case the long period phases will exist at zero temperature and each plane of atoms will be displaced through the same amount. However the u_j can also be thought of as the average displacement of an atom in layer j because of thermal fluctuations. In this case the long period phases would only exist at finite temperatures when the u_j are non-zero and one has, as in the ANNNI-model, an entropic mechanism for stabilising the modulated structures but without the ANNNI problem of discrete spin-flips that are unphysical in many polytypic compounds.

Details of the phase diagram - whether one obtains a complete, incomplete or harmless Devil's staircase - depend on the finer details of the Hamiltonian. In general it will be difficult to distinguish between the different cases experimentally unless the harmless staircase was very harmless! Perhaps it is worth stressing that the same mechanism can lead to both incommensurate and locked-in modulated structures [37].

Finally, it is important to stress that, in the case of a harmless staircase where the phase transitions are first order, many phases appear as metastable states over considerable regions of the phase diagram. Hence there can be an infinite number of metastable states which correspond, in general, to very small differences in free energy. This leads to severe experimental difficulties, and the understanding of polytypic compounds will require a study of the metastability effects and kinetics which may mask the equilibrium phase diagrams.

The models described in this paper have helped to elucidate possible mechanisms for the stabilisation of modulated structures as equilibrium phase. The outstanding problem in many cases is now to ascertain which is the most relevant physics for a particular polytypic compound.

7. ACKNOWLEDGEMENTS

It is a pleasure to thank Volker Heine for many stimulating discussions. GDP gratefully acknowledges the receipt of a 1983 University Research Fellowship from the Royal Society.

8. REFERENCES

1. Thompson, J.B. (1981). In Structure and bonding in crystals II, eds M. O.Keeffe and A. Navrotsky (Academic Press, New York) p 167.
2. Angel, R. J. (1987) Z. Krist. 176, 193.
3. Jepps, N. W. and Page, T. F. (1984) J. Crystal Growth 7, 259.
4. Loiseau, A., van Tenderloo, G., Portier, R. and Ducastelle, F. (1985) J. de Physique 46, 595.
5. Pandey, D. and Krishna, P. (1984) J. Crystal Growth 7, 213.
6. Frank, F.C. (1951) Phil. Mag. 42, 1014.
7. Trigunayat, G. C. and Chadha, G. K. (1971) Phys. Status. Solidi A4, 9.
8. Jagodzinski, H. (1954) Neues Jahrb. Mineral. Monatsh. 3,49.
9. Smith, J., Yeomans, J. and Heine, V. (1984). In Proceedings of NATO advanced studies institute on modulated structure materials, edited by T. Tsakalakos. (Dortrecht, Nijhoff) p23.
10. Price, G. D. and Yeomans, J. M. (1984) Acta Cryst B40, 448.
11. Yeomans, J. M. and Price G. D. (1986) Bull. Min. 109, 3.
12. Aubry,S. (1978) In Solitons in Condensed Matter Physics eds. A. R. Bishop and T. Schneider (Springer, Berlin) p. 264.
13. Villain, J. and Gordon, M. (1980) J. Phys. C. Solid St. Phys. 13, 3117.
14. Frenkel, J. and Kontorova, T. (1938) Phys. Z Sowjet 13, 1.
15. Frank, F. C. and van der Merwe, J. H. (1949) Proc. Roy. Soc. (London). A198, 205.
16. Fisher, M. E. and Selke, W. (1980) Phys. Rev. Lett. 44, 1502.
17. Zdhanov, G. S. and Minervina, Z. (1945) J. Phys. (Moscow) 9, 151.
18. Bak, P. and Bruinsma, R. (1982) Phys. Rev. Lett. 49, 249.
19. Shinjo, K. and Sasada, T. (1985) J. Phys. C. Solid St. Phys. 18, L261.
20. Axel, F. and Aubry, S. (1981) J Phys. C Solid St Phys. 14, 5433.
21. Stoneham, A. M. and Durham, P. J. (1973) J. Phys. Chem. Solids 34, 2127.
22. Elliott, R. J. (1961) Phys. Rev. 124, 346.
23. Bak, P. and von Boehm, J. (1980) Phys. Rev. B21, 5297.
24. Selke, W. and Duxbury, P. M. (1984) Z. Phys. B57, 49.
25. Szpilka, A. and Fisher, M. E. (1986) Phys. Rev. Lett. 57, 1044.
26. Bak, P. (1982) Rep. Prog. Phys. 45, 587.
27. Yeomans, J. M. (1987) Solid State Physics, edited by H. Ehrenreich, F. Seitz and D. Turnbull (Academic Press) In press.
28. Barreto, M. N. and Yeomans, J. M. (1985) Physica 134A, 84.
29. Akaogi, M., Akimoto, S., Horioka, K., Takahashi, K., and Horiuchi, H. (1982) J. Solid State Chem. 44, 257.
30. Price, G. D. (1983) Phys. Chem. Minerals, 10, 77.
31. de Fontaine, D. and Kulik, J. (1985) Acta Metall. 33, 145.
32. Broddin, D., van Tendeloo G., van Landuyt, J., Amelinckx, S., Portier, R., Guymont, M. and Loiseau, A. (1987) Phil. Mag. 54 395.
33. Price, G.D., Parker, S.C. and Yeomans, J.M. (1985) Acta Cryst. B41, 231.
34. Janssen, T. and Tjon, J. A. (1981) Phys. Rev. B24 2245.
35. Janssen, T. and Tjon, J. A. (1982) Phys. Rev. B25 3767.
36. Janssen, T. and Tjon, J. A. (1983) J. Phys. C. Solid St Phys. 16 4789.
37. Benkert, C., Heine, V. and Simmons, E. H. (1987) Europhysics Letts. 3, 833.
38. Upton, P. and Yeomans, J. (1987) Europhysics Letts. In Press.

On the Systematics of Phase Transformations in Metallic Alloys

L.E. Tanner

Chemistry & Materials Science Department,
Lawrence Livermore National Laboratory,
Livermore, CA 94550, USA

Phase transformations in metallic alloys proceed by diffusional replacive or diffusionless displacive mechanisms. These are generally viewed as independent processes without significant interactive coupling. However, new studies of displacive transformations, particularly regarding pretransformation phenomena, suggest that such coupling may indeed occur. If so, it should have a significant effect on the mode, morphology and mechanisms of the replacive transformations. In this paper we examine these possibilities in light of the most recent observations and thinking in this field.

1. INTRODUCTION

Metallurgists are continually searching for new alloy formulations and new and unique methods for materials processing for the purpose of enhancing and optimizing technologically useful properties in metallic solids [1-3]. The control and manipulation of phase transformations are particularly important in accomplishing these ends through the development of complex single or multiphase microstructural assemblages, where the constituent phases may be stable, metastable or their combination [2,3]. Under appropriate conditions, a phase transformation becomes possible when a crystalline phase is brought into an unstable or metastable state; for example, by using a familiar processing procedure when a phase is only stable at high temperatures, to cool it below its single-phase transus. In the traditional view [2-4], the transformation will proceed by either a <u>replacive</u> or a <u>displacive</u> mechanism. The former involves thermally activated diffusional atomic rearrangements to produce one or more new phases, either with or without compositional change. Typical replacive processes include precipitation, spinodal decomposition, ordering, eutectoid decomposition, etc. [3]. The latter proceeds by athermal diffusionless atomic displacements (i.e., cooperative lattice shears and/or shuffles) forming a new single phase of lower symmetry without a compositional change [4]. Displacive processes in metallic systems are martensitic [4] or omega-type [5]. As a general rule, displacive transformations are alternative (metastable) structural changes that the high-temperature stable phase <u>may</u> undergo, usually initiating at rather low temperatures, when cooling is sufficiently rapid to avoid replacive transformations at intermediate temperatures.

As implied above, for the most part these two processes have been dealt with independently with only moderate concern for the possibility of any type of interactive coupling. Overall, displacive transformations are less well understood and this has undoubtedly limited their full development and utilization. In particular, there have been difficulties in elucidating the primary origins of the displacive behavior [6], and in establishing suitable models for nucleation [7,8]. Recently, however, significant advances in both experiment and theory related to comparable transformations in non-metallic materials [9,10] have begun making an impact on metallurgical studies [8,11,12]. As a consequence, a new view of displacive behavior is emerging that is based on the highly anisotropic properties of the parent phase leading to influential structural changes well before the transformation temperature is reached [13]. Specifically, phonon spectra have been observed to exhibit temperature-dependent anomolies (dips) along just those branches related to the displacements involved in the eventual phase change [6, 9,10,14]. The phonon wavelengths at which these anomalies appear are invariably those of intrastructural modulations which evolve on cooling [13-16]. Thus, the parent phase effectively "prepares itself" via the development of periodic displacement patterns (usually incommensurate) that mimic the new product phase structure [12]. This is a very powerful concept that is currently being confirmed and expanded experimentally [11-16] and has already led to new theoretical models and explanations [17-20]. KRUMHANSL [6,14] has recently critically reviewed some of the differing theoretical ideas and suggests that the most promising approach follows a proposal put forth by ZENER [21] many years ago; namely, that structural instabilities and anomalous phonon behavior stem from competition between short-range (unstable) ionic or bonding forces and longer-range or electron gas non-bonding forces.

Looking further, if these pretransformational effects extend to well above room temperature, we are struck by the likelihood that they may indeed couple with thermally activated processes and thus perhaps significantly influence the mode, morphology and kinetics of replacive transformations at intermediate temperatures. For example, the strain-mediated predisplacive structures may provide heterogeneous or "pseudo-homogeneous" nucleation sites where the incipient symmetry change induced by these distortions could likely bias the nature of the transformation process. Indeed, there have been unusual, unexplained microstructures observed [22-25] that could be effectively analyzed from this viewpoint. In addition, there are also numerous reports of the anomalous enhancement of diffusion in high-temperature phases as displacive transformation temperatures are approached [26-31]. This too may be a manifestation of pretransformation and, furthermore, would certainly be expected to affect replacive transformation kinetics if this behavior extends into the undercooled metastable state. Our purpose here is to review certain of these observations in the context of phase transformation systematics modified by the new view of displacive processes.

2. DISCUSSION

2.1 Thermodynamics and Kinetics

Starting from the traditional viewpoint [2-4], the systematics of phase transformations are described using a series of interrelated diagrams (Figs. 1-4) which illustrate the influence of thermodynamic and kinetic factors in a prototypical alloy system A-B. The partial phase diagram in Fig. 1 shows that solute B destabilizes the low-temperature α phase in favor of high-temperature β and the curve marked M_s indicates the start temperature for a displacive martensitic transformation $\beta\rightarrow\alpha'$ [32]. The equilibrium phase boundaries are determined by graphical thermodynamics using the common tangent construction to hypothetical Gibbs free energy (G) vs. composition (C) curves [3], as displayed for several temperatures in Fig. 2. Crossover of the curves indicates the composition where $G^\beta = G^\alpha$ at each temperature and its locus is given by the dashed curve $T_0^{\beta\rightarrow\alpha}$ in Fig. 1. It will become evident below that T_0 represents a critical demarcation for the ensuing transformations once β is undercooled.

Considering an alloy of composition C_0, we allow it to cool from above the $\beta/\beta+\alpha$ transus. In the range $T>T_0$ (Fig. 2a), β can only transform via replacive phase separation ($\beta\rightarrow\beta+\alpha$) as is shown for the slowest cooling rate R_1 in the time-temperature-transformation (TTT) diagram in Fig. 3. The TTT plot is derived from the nucleation, growth and diffusional mobility characteristics of system A-B [2,3]. Cooling into the range where $T<T_0$ makes it thermodynamically possible for a composition-invariant metastable transformation to proceed as well. However, the martensitic $\beta\rightarrow\alpha'$ transformation requires a critical undercooling $\Delta T^-(= T_0-M_s)$ [2-4,33]; hence, it will

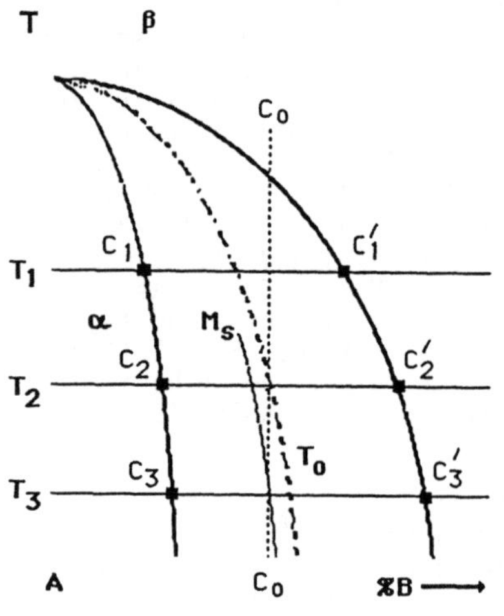

Figure 1

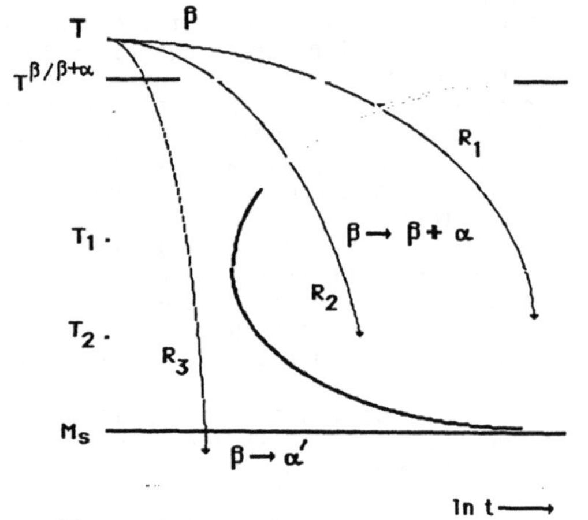

Figure 3

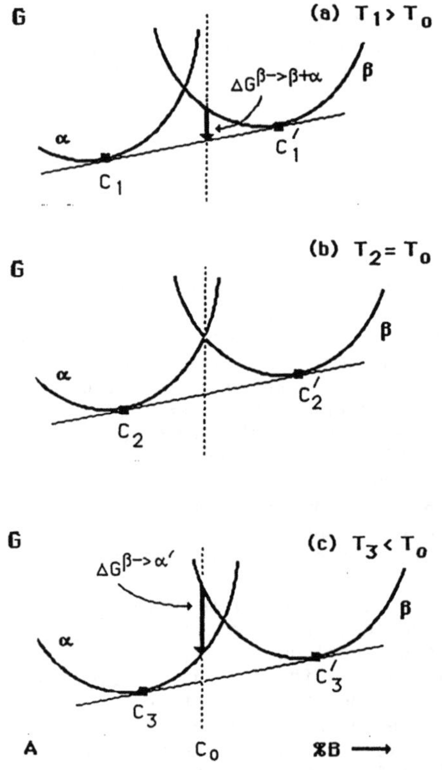

Figure 2

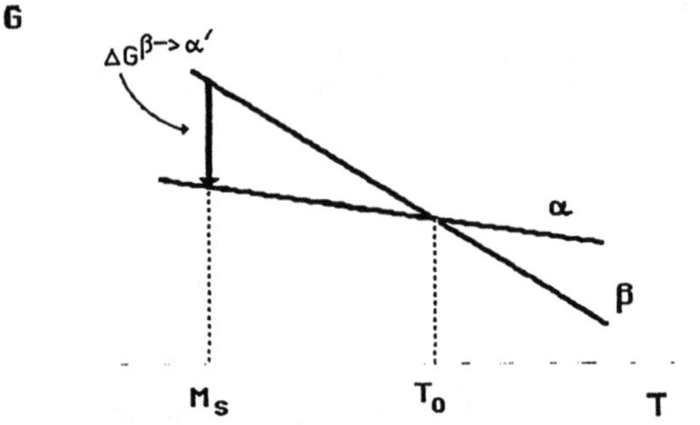

Figure 4

Figure captions on the facing page.

not start until M_s is reached as indicated in Figs. 1 through 4. Furthermore, it should be obvious that cooling must be sufficiently rapid in order to bypass replacive transformation at intervening temperatures $T>M_s$, viz., that $R_3>>R_2>R_1$ as indicated in the TTT plot (Fig.3).

The foregoing is based on the assumption that the martensitic transformation is from one fully commensurate (i.e., homogeneous) phase to another (viz., $\beta\to\alpha'$ at M_s, see Fig. 4) [33]. Modifying this approach to incorporate the pretransformation behavior described in Section 1, we introduce a modulated (incommensurate) transitional state β' in which an intermediary to the product phase α' develops continuously within the parent β phase as M_s is approached [13]. This is depicted in Fig. 5 as a free energy path that lies below those for the homogeneous commensurate α and β phases and bridges them between $T^{\beta'}$ and T_o', respectively. We also alter the TTT diagram (Fig. 3) by adding a graded shading in the region to the left of the "C" curve as is shown in Fig. 6. The increasing density of shading with decreasing temperature schematically represents the evolution of β' modulations with cooling. This, of course, implies that the replacive transformations to the right of the "C" curve are perhaps more properly identified as $\beta'\to\beta+\alpha$.

2.2 Pretransformation Microstructures

Observations of premartensitic modulations have been made in a wide variety of alloy systems using x-ray [12,34], neutron [15] and electron scattering [35,36] and by transmission electron microscopy (TEM) [35,36]. The effects are seen on two levels: (a) a generic long-wavelength strain response to the perturbation of the normal lattice periodicity, viz., elastic (Huang) diffuse scattering (HDS) emanating from Bragg peaks and a quasi-periodic striated strain contrast ("tweed") in TEM images; and (b) the direct effects of the specific incommensurate modulation, e.g., satellites flanking Bragg peaks and an atomic level image of the associated periodic microstructure. Examples of TEM and electron diffraction observations for $Ni_{62.5}Al_{37.5}$ from the work of TANNER et al. [37] are given in Fig. 7. Fig. 7a is a two-beam amplitude contrast image showing the tweed striations parallel to {110} traces of the cubic B2 parent phase. Fig. 7c is a high-resolution many-beam phase contrast image

Fig. 1. Phase diagrams for hypothetical A-B alloy system; M_s is the start temperature for the <u>metastable</u> displacive (martensitic) transformation (see text for details)

Fig. 2. Schematic Gibbs free energy vs. composition curves for A-B alloys at temperatures below $T^{\beta/\beta+\alpha}$, viz., $T_1>T_2>T_3$; at C_o:
(a) $\Delta G^{\beta\to\beta+\alpha}$ is the free energy decrease for the equilibrium replacive transformation, and (c) $\Delta G^{\beta\to\alpha'}$ is the free energy decrease for the metastable displacive (martensitic) transformation

Fig. 3. Time-Temperature-Transformation (TTT) diagram for the A-B alloy of composition C_o; trajectories indicate continuous cooling at rates $R_3>>R_2>R_1$; "C" curve delineates start of equilibrium $\beta\to\beta+\alpha$ replacive transformation

Fig. 4. Free energy vs. temperature curves for the A-B alloy of composition C_o; $\Delta G^{\beta\to\alpha'}$ is the free energy decrease for the metastable martensitic transformation

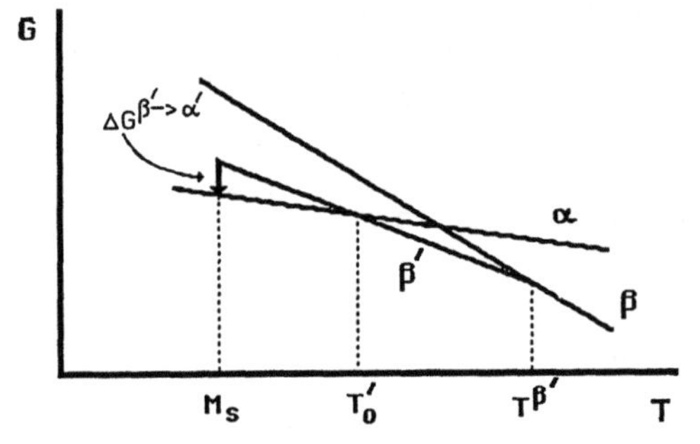

Fig. 5. Modification of Fig. 4 to show that β passes through a structurally modulated transitional state β' on cooling to M_s; $\Delta G^{\beta' \to \alpha'}$ is the free energy decrease for the <u>metastable</u> martensitic transformation

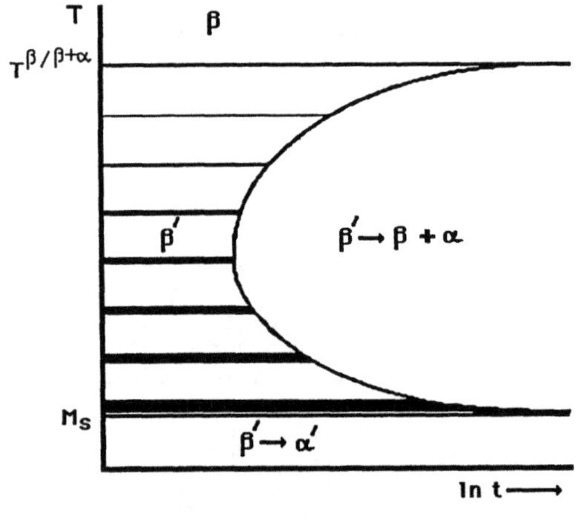

Fig. 6. Modification of Fig. 3 to show evolution of the β' transitional state with respect to the replacive transformation temperature

revealing the underlying fine-scale (~1.3 nm spacing) incommensurate periodic shear modulations of {110}<1$\bar{1}$0> type, which are the precursor to a 7R (5$\bar{2}$ stacking-type) martensitic structure [4,38]. Diffraction shows diffuse streaking (HDS) and satellites along <ζζ0> (Fig. 7b); the position of the satellites is ζ ~ ±0.16, which is inversely proportional to the modulation spacing described above. The temperature dependence of the satellites is shown in Fig. 7d. These same effects have been documented with considerable accuracy by neutron examinations of the same crystal by SHAPIRO et al. [15]. The latest results [16] confirm the static nature of the modulations and show that they are directly related to an anomalous temperature-dependent dip in the transverse acoustic <110> phonon branch at ζ ~ 0.16. Further details are discussed in the contribution by SHAPIRO [39] in these proceedings.

Besides the premartensitic modulations that the above Ni-Al and other <u>beta phase</u> (B2-type) alloys exhibit, there is also a second form of displacive modulation present (Fig. 8). These are omega phase-type distortions of <111> longitudinal-type related to an anomalous dip at ζ ~ 2/3 in the longitudinal acoustic <ζζζ> phonon branch [40]. The diffuse scattering for this case is evident when examining a <110>* diffraction zone [25] as shown in Fig. 8b. Dark field imaging (Fig. 8c) with this diffuse intensity reveals an indistinct dispersed contrast typical of a distortional pattern, but different from the premartensitic shear modulations described earlier.

2.3 Replacive Transformations of Modulated Microstructures

There have been numerous observations on the heat treatment of the pretransformation modulated state, but few systematic studies have attempted to elucidate

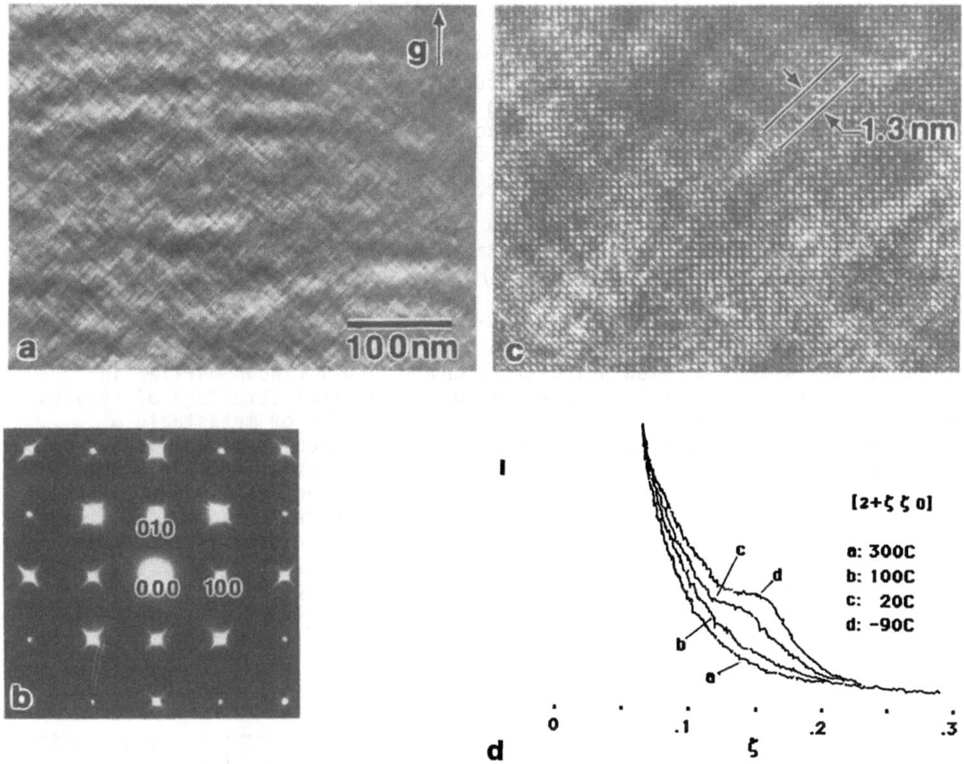

Fig. 7. TEM observations of $Ni_{62.5}Al_{37.5}$ at room temperature, (001) orientation: (a) two-beam bright-field amplitude contrast image of {110}<1$\bar{1}$0> tweed; (b) multi-beam high-resolution phase cntrast image of 7R-type premartensitic modulations; (c)zone axis ED from (b); (d) temperature dependence of diffuse scattering and satellites in ED (see Refs. [15, 16, 37] for further details)

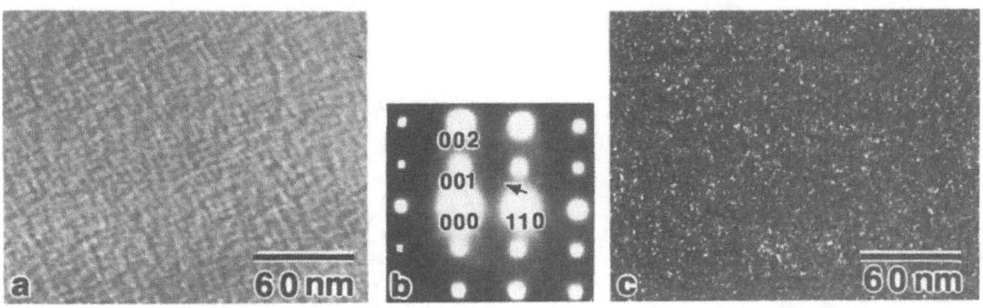

Fig. 8. TEM observations of $Ni_{62.5}Al_{37.5}$ at room temperature, (110) orientation: (a) two-beam bright-field amplitude contrast image; (b) SAED from (a); (c) dark-field amplitude contrast image using "athermal ω" diffuse scattering indicated by the arrow in (b)

microstructural changes in terms of coupled replacive and displacive mechanisms. KUBO and WAYMAN [22] aged Cu-Zn β(B2) alloys in thin-foil form at low temperatures and reported that a {100} spinodal decomposition structure developed. However, ROBERTSON and WAYMAN [23] subsequently determined the transformation habit to actually be {110}, thus, showing that the premartensitic distortional modulations do indeed evolve directly into long-wavelength compositional modulations. They found the same behavior for a Cu-Ni-Al β(DO_3) phase [23], as well as for the Ni-Al β(B2) phase [24] that we described in Section 2.2. Unfortunately, the aging treatments in all cases were not sufficiently long in duration to allow full development of new stable or metastable phases. Accordingly, further work is needed to enable complete evaluation of these processes. However, the results are a very firm indication of transformation coupling behavior.

Another illustrative example involves Ni-Al from the work of REYNAUD [25] who found that the ω-like longitudinal distortions in β(B2) (shown earlier in Fig. 8) transform in-situ with thermal aging to the Ni_2Al structure of related symmetry. This is analogous to the well-known formation of metastable ω precipitates in Ti- and Zr-based alloys [41]. The quench-retained disordered bcc solid solutions of these alloys invariably contain "athermal ω", i.e., the modulated microstructure, as shown in Fig. 9b for as-quenched Ti-15Mo. Subsequent aging results in the sharpening of the diffuse ω reflections (c.f., Figs. 9a and 9c) and marked changes in the microstructure. The once-irresolvable patches of "athermal ω" strain contrast (Fig. 9b) become discrete second-phase ellipsodal precipitates of metastable "isothermal ω" (Fig. 9d) with lower composition than the starting β. This is part of a low-temperature sequence in these systems in which aging first promotes rejection of solute (Mo) from the

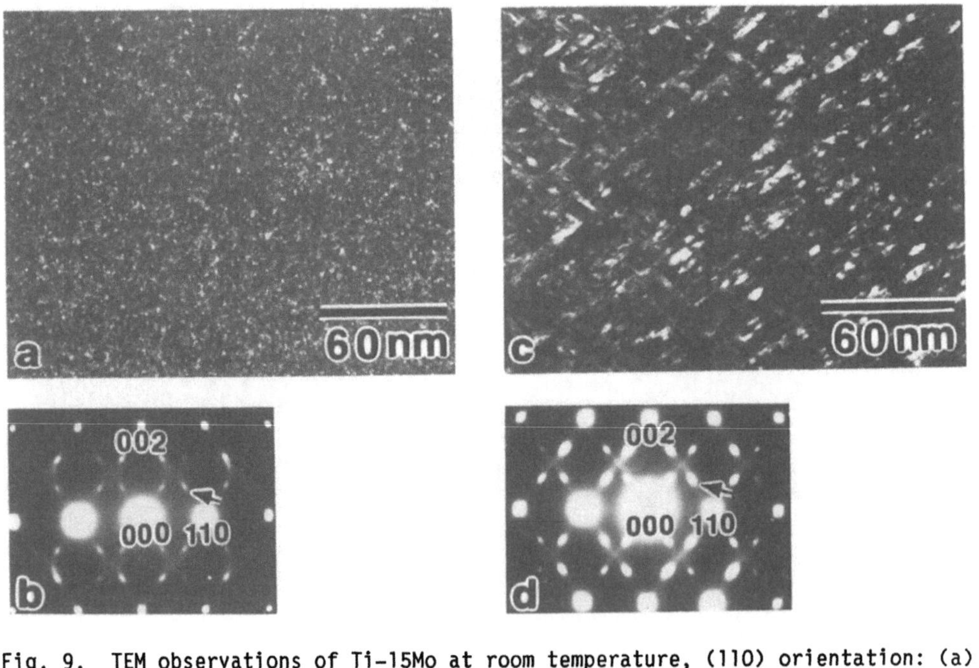

Fig. 9. TEM observations of Ti-15Mo at room temperature, (110) orientation: (a) as-quenched; dark-field amplitude contrast image using "athermal ω" diffuse scattering indicated by the arrow in (b); (b) SAED from (a); (c) aged; dark-field amplitude contrast image using "isothermal ω" reflection indicated by the arrow in (d); (d) SAED from (c)

ω-distorted regions to create the metastable phase dispersed in a solute-enriched β matrix. With continued aging, the ω is replaced by stable hcp α of still lower Mo concentration, which further enriches the β. Thus, the alloy approaches its equilibrium two-phase state via sequential free energy reducing steps where, once again, the coupling of replacive and displacive processes appears to play a significant role in the overall transformation. A further rather dramatic example of this analogous type was found in the course of the decomposition of a rapidly solidified Zr-Al alloy by BANERJEE and CAHN [42]. In this case, as-solidified bcc β subsequently phase-separated leading to ω distortions in solute-lean and solute-rich β; these structures, in turn, gave rise to the formation of disordered "isothermal ω" and ordered Zr_2Al, respectively.

2.4 Anomalous Diffusion

AARONSON and SHEWMON [26] were the earliest investigators to link the unusual enhanced diffusion of certain metals and alloy solid solutions with the proximity of phase transformations, and to associate the behavior with ZENER's [21] ideas on the incipient displacive instabilities of their lattices, viz., those due to anomalous TA<110> effects. Figure 10 shows a typical plot of diffusion constant (D) against reciprocal temperature (1/T) indicating anomalous non-Arrhenius behavior as $T^{\beta/\beta+\alpha}$ is approached. SANCHEZ and DE FONTAINE [27] and HERZIG and KOHLER [28] dealt with certain bcc Ti- and Zr-based alloys and proposed that the "athermal ω" distortions (LA<111>-type) provided more effective models, though LE GALL et al. [29], studying Ti-Hf alloys, preferred the TA<110> effects related to the bcc→hcp transformation. The ZENER [21] approach is the only appropriate one for explaining enhanced diffusion in displacively transforming fcc phases, as indicated by CHAPMAN et al. [30] for C in Fe-30Ni and by BOKSHTEYN et al. [31] for Co-rich alloys. The models are all basically rather similar, incorporating the concept that the temperature-dependent "softening" of the lattice in one manner or another effectively enlarges and geometrically enhances the activated complex for diffusion, thus allowing easier atom/vacancy interchanges.

The diffusion measurements are made above the single-phase transus, but we know that pretransformation effects continue to evolve as temperature falls into the undercooled (metastable) regime. Accordingly, we can expect enhanced atomic transport to do the same and therefore, to have a significant effect on speeding up replacive transformations. Furthermore, this may help to explain thermally activated transformations at temperatures in many alloys lower than one might normally predict.

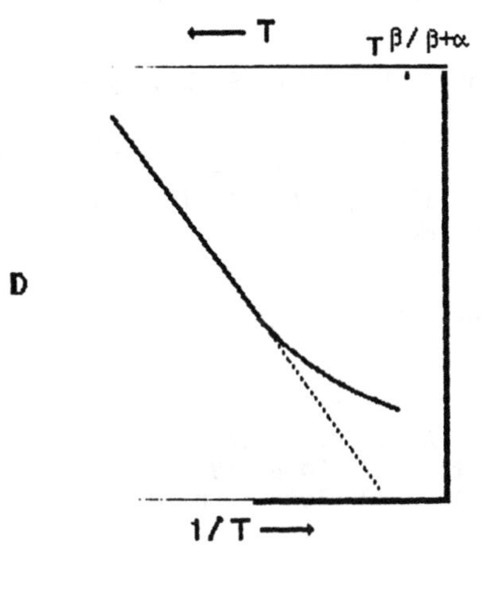

Fig. 10 Typical anomalous temperature dependence of diffusion in those alloy systems where a high-temperature → low-temperature phase transformation is approached from above (see Refs. [26-31])

3. CONCLUDING REMARKS

As the area of displacive transformations in metallic alloys advances to new and more definitive understandings, it is clear from what we have touched on here that the entire field of phase transformations is due for critical re-examination. Certain systems can be used as ideal models and be studied in a systematic manner with the latest high-resolution experimental probes. Besides redirecting and fine-tuning our knowledge, such research efforts are quite likely to reveal still better ways to optimize the structure/property relationships of many known alloys as well as new ones.

It is a pleasure to acknowledge the stimulating interactions with many of my colleagues: T. Barbee, P. C. Clapp, U. Dahmen, D. DeFontaine, R. Gronsky, J. A. Krumhansl, S. C. Moss, D. Schryvers, S. M. Shapiro and M. Wuttig. Thanks are also due to R. LeSar and K. Sims for encouragement and support. Work was performed under the auspices of the U.S. Department of Energy, Contract No. W-7405-Eng-48.

References

1. See for example: (a) The Mechanism of Phase Transformations in Metals, Monograph and Report Series No. 18 (Inst. of Metals, London,1956); (b) The Mechanism of Phase Transformations in Crystalline Solids (Inst. of Metals, London, 1969); (c) Phase Transformations, Amer. Soc. for Metals, Metals Park, OH, 1970; (d) Solid-to-Solid Phase Transformations, ed. by H. I. Aaronson, D. E. Laughlin, R. F. Sekurka and C. M. Wayman (The Met. Soc. AIME, Warrendale, PA, 1982); (d) Phase Transformations in Solids, ed. by T. Tsakalakos, Mat. Res. Soc. Symp. Proc., Vol. 21 (North-Holland, New York, 1984)
2. J. W. Christian: The Theory of Phase Transformations in Metals and Alloys (Pergamon Press, Oxford, 1965)
3. D. A. Porter, K. E. Easterling: Phase Transformations in Metals and Alloys (Van Nostrand and Reinhold, New York, 1981)
4. Z. Nishiyama: Martensitic Transformation (Academic Press, New York, 1978)
5. S. K. Sikka, Y. K. Vohra, R. Chidambaram: in Prog. Mat. Sci., 27 (Pergamon, London, 1982) p. 245
6. J. A. Krumhansl: in Nonlinearity in Condensed Matter, ed. by A. R. Bishop, D. K. Campbell, P. Kumar and S. E. Trullinger, Springer Ser. Sol.-St. Sci., Vol. 69 (Springer, Berlin, 1987) p. 255
7. A. H. Heuer, M. Ruhle: Acta Met., 33, 2105 (1985)
8. See Proc. Int. Conf. on Martensite Transformations (ICOMAT), held in: (a) Cambridge, USA, 1979 (MIT Press, Cambrige, MA, 1979); (b) Leuven, Belgium, J. de Phys., 43, Colloq. C4 (1982); (c) Nara, Japan, 1986 (Jpn. Inst. Met., Tokyo, 1986)
9. S. M. Shapiro: Met. Trans. A, 12A, 567 (1971)
10. Incommensurate Phases in Dielectrics, 2, ed. by R. Blinc, A. P. Levanyuk (Elsevier, Amsterdam, 1986)
11. Proc. Symp. on Pretansformation Behavior Related to Displacive Transformations, ed. by L. E. Tanner, W. A. Soffa, Met. Trans. A, 19A (1988) in press
12. V. V. Kondrat'yev, V. G. Pushin: Phys. Met. Metall., 60, No. 4, 1(1985)
13. G. R. Barsch, J. A. Krumhansl, L. E. Tanner, M. Wuttig: Scripta Met., 21, 1257 (1987)
14. J. A. Krumhansl: in these proceedings
15. S. M. Shapiro, J. Z. Larese, Y. Noda, S. C. Moss, L. E. Tanner: Phys. Rev. Lett., 57, 3199 (1986)
16. S. M. Shapiro, J. Z. Larese, S. C. Moss and L. E. Tanner: to be published
17. G. Guenin, P. F. Gobin: Met. Trans. A, 13A, 1127 (1982)
18. G. R. Barsch, J. A. Krumhansl: Met. Trans. A, 19A (1988) in press
19. P.-A. Lindgard, O. G. Mouritsen: Phys. Rev. Lett., 57, 2458 (1986)
20. (a) Y. Morii, M. Iizumi: J. Phys. Soc. Jpn., 54, 2948 (1985); (b) M. Iizumi: J. Phys. Soc. Jpn., 52, 549 (1983)

21. C. Zener: Phys. Rev., 71, 846 (1947)
22. (a) H. Kubo, C. M. Wayman: Met. Trans. A, 10A, 633 (1979); (b) H. Kubo, I. Cornelis, C. M. Wayman: Acta Met., 28, 405 (1980)
23. I. M. Robertson, C. M. Wayman: Met. Trans. A, 15A, 269 (1984)
24. I. M. Robertson, C. M. Wayman: Met. Trans. A, 15A, 1355 (1984)
25. F. Reynaud: J. Appl. Cryst., 9, 263 (1976)
26. H. I. Aaronson, P. G. Shewmon: Acta Met., 15, 385 (1967)
27. (a) J. M. Sanchez, D. de Fontaine: Acta Met., 26, 1083 (1978); (b) J. M. Sanchez: Phil. Mag., 43, 1407 (1980)
28. C. Herzig, U. Kohler: Acta Met., 35, 1831 (1987)
29. G. Le Gall, D. Ansel, J. Debuigne: Acta Met., 35, 2297 (1987)
30. L. R. Chapman, D. A. Powers, M. Wuttig: Scripta Met., 16, 437 (1982)
31. S. Z. Bokshteyn, B. S. Bokshteyn, S. T. Kishkin, L. M. Klinger, L. M. Mirskiy, N. G. Orekhov, I. M. Razumovskiy: Phys. Met. Metall., 46, No. 2, 136 (1978)
32. This is a rather simple, but typical phase diagram quite effective for illustrative purposes here. Examples of actual systems with this configuration and appropriate phase transformations are Fe-Ni and Ti-Mo [3].
33. L. Kaufman, M. Cohen: in Prog. Met. Phys., 7 (Pergamon, London, 1958) p. 16534.
34. Y. Tsunoda, N. Kunitomi: in Proc. ICOMAT-86/Nara (Jpn. Inst. Met., Tokyo, 1986) p.349
35. L. E. Tanner, A. R. Pelton, R. Gronsky: J. de Phys., 43, Colloq. C4, C4-169 (1982)
36. I. M. Robertson, C. M. Wayman, Phil. Mag. A, A48, 421, 443 and 629 (1983)
37. L. E. Tanner, A. R. Pelton, G. Van Tendeloo, D. Schryvers, M. Wall, submitted to Scripta Met. (1987)
38. V. V. Martynov, K. Enami, L. G. Khandros, S. Nenno,A. V. Tkachenko: Phys. Met. Metall., 55, No. 5,136 (1983)
39. S. M. Shapiro: in these proceedings
40. R. A. Robinson, G. L. Squires, R. Pynn: J. Phys. F: Metal Phys., 14, 1061 (1984); (b) R. Pynn: J. Phys. F: Metal Phys., 8, 1 (1978); (c) S. C. Moss, D. T. Keating, J. D. Axe: in Phase Transitions (Pergamon Press, New York, 1973) p. 179; (d) C. Stassis, J. Zaretsky, N. Wakabayashi: Phys. Rev. Lett., 41, 1726 (1978).
41. (a) E. W. Collings: in Applied Superconductivity (Plenum, New York, 1986) p.38; (b) in The Physical Metallurgy of Titanium Alloys (Amer. Soc. for Metals, Metals Park, OH)
42. S. Banerjee, R. W. Cahn: Acta Met., 31, 1721 (1983)

Neutron and X-Ray Scattering Studies of Premartensitic Phenomena

S.M. Shapiro

Brookhaven National Lab., Upton, NY 11973, USA, and
Laboratoire Léon Brillouin (commun CEA-CNRS), CEN-Saclay,
F-91191 Gif-sur-Yvette, Cedex, France

1. Introduction

Competing interactions play an essential role in all types of phase transitions. In the much studied spin glasses it is well known that the competing ferro- and antiferromagnetic interactions lead to a spin's frustration resulting in the frozen spin glass state at low temperatures /1/. In ferroelectrics, the competition is between short-range interatomic forces and longer range electric dipolar forces which leads to the structural instability associated with ferroelectric ordering /2/. In metals, where martensitic phase transitions occur, the competing forces are not as well defined, but are likely due to the interatomic forces and the longer range electronic interactions. Depending upon the electronic band structure and the phonon dispersion curves of the solid, a variety of structures and microstructures can appear over a wide range of temperatures and compositions.

In this paper I shall discuss neutron and X-ray investigations of some metallic alloys which are known to exhibit martensitic transformations. It will be shown that precursor effects are usually present in the diffuse scattering and in the phonon dispersion curves, but the transition cannot be described in terms of the soft mode picture used in the Landau and Devonshire theory to describe structural phase transitions /2,3/. In addition, it will be seen that it is inappropriate to look at these microstructures as incommensurate systems, but more correctly as a coherent coexistence of two phases.

Martensitic transitions have been studied by metallurgists for nearly one hundred years /4/. The term martensitic was first used to describe the plate like structure observed in quenched hardened steels. Nowadays, the term martensitic is also used to describe transitions in non-ferrous metals and even oxides. In a martensitic transformation the atoms exhibit an anomalous shear-like displacive motion. In the low temperature phase the atoms are only slightly displaced from their more symmetric positions of the high temperature phase. This contrast with order disorder transitions where atoms must shift by unit cell lengths in order to achieve the ordered state. Martensitic transitions are usually first order, exhibit hysteresis, but frequently look continuous with temperature. Defects, nucleation, and growth play an important role in the transitions /5/.

In order to obtain microscopic information about the nature of the martensitic transitions, metallurgists have performed extensive electron and X-ray studies on many systems /6/. Usually observed in the diffraction photographs are diffuse spots, rods or planes of scattering (referred as REL spots, rods or walls ; REL = REciprocal Lattice) corresponding to 3, 2 or 1 dimensional correlations. The real space image of this diffuse scattering provides important information about its origin. Ultrasonic studies are usually performed to probe the elastic behavior of the solid particularly the expected shear anomaly. Although these measurements have great sensitivity to small changes, they are restricted to the long wavelength or small momentum response of the system. The diffraction effects observed suggest that anomalies extend out into the Brillouin zone, away from the zone center.

The electron and X-ray observations cannot establish whether the features observed are dynamic or static in origin since no energy analysis of the diffuse scattering is possible. This is because the energy resolution required to separate phonon effects is $\delta E/E_i \sim 10^{-8}$ since the incident energy (E_i) of the photon is of the order of keV and phonon energies are in the range of 10 meV. Presently this resolution is impossible to achieve using X-ray generators, but new instruments situated at synchrotron radiation sources are approaching the required resolution /7/. Inelastic neutron scattering, because of the much lower incicent energy, requires a modest resolution of $\delta E/E \sim 0.1$, which is easily achievable with existing triple axis neutron instruments situated at the various reactors in the world. Thus neutron scattering studies of martensitic materials are indispensible in probing the dynamical behavior of the phase transformation.

Most of the earlier neutron studies of structural phase transitions have been performed on perfect or nearly perfect solids /2,3/. In these systems soft modes, i.e. lattice vibrational modes whose frequencies go to zero, or nearly zero, as the transition is approached, are present. Only relatively few systems exhibiting martensitic phase transitions have been studied and some of them are listed in Table I. These are classified by a qualitative estimate of the amount of mode softening which is present. Only few systems exhibit nearly complete softening. The superconducting A15 materials Nb_3Sn and V_3Si exhibit almost complete softening in the $C' = 1/2(C_{11}-C_{12})$ elastic constant /8/ and nearly the entire [110]-TA_2 branch shows a large decrease in energy as the martensitic transition is approached /9/. The alloy In-Tl also exhibits complete softening of the elastic constant C' as measured by ultrasonic techniques /10/. Neutron scattering studies, on the other hand, show no anomalous temperature dependence of any acoustic modes /11/implying that the lattice softening occurs only for long wavelength excitations ($\lambda > 250$ Å). The omega (ω) phase alloy is also listed to emphasize the similarity between martensitic systems and those BCC based systems exhibiting ω phases. In the latter, the mode expected to be responsible for the ω transition is the $[\zeta\zeta\zeta]$ -LA, $\zeta = 2/3\ \zeta_{ZB}$ where ζ_{ZB} is the zone boundary wavevector. Earlier studies revealed a very broad inelastic response which was temperature dependent, but no clear soft mode was observed /12/. Studies on ω phase materials are limited largely because no known alloy exhibits a reversable ω transition. Most of the systems listed exhibit only a partial softening where the fractional change of phonon energy is $\Delta E/E \sim 10\text{-}20\%$. In addition, this anomalous behaviour is not restricted to one q value as seen in many structural phase transitions (either commensurate or incommensurate), but occurs over a broad region of q space.

In the remainder of this paper, I shall review the neutron and X-ray studies of 2 compounds that have recently been studied : Ni-Ti(Fe) /13,14/ and Ni-Al /15/ alloys. I believe the behavior exhibited in these systems is typical of many of the compounds listed in Table I.

2. Ni_xTi_{1-x}(Fe) alloy

The Ni_xTi_{1-x} alloy exhibits the phenomenal property of shape memory /16/. For x = 50, it is an ordered alloy with a simple cubic, CsCl or β-phase structure. The martensitic phase transition occurs near room temperature with considerable hysteresis upon heating and cooling. It was realized that premartensitic (PM) effects were present but the structure of the PM and martensite phase remained unresolved for many years. A breakthrough come when it was realized that by alloying with Fe one can depress the martensitic transition temperature revealing more clearly the PM phase /17/. For the composition $Ni_{46.8}Ti_{50}Fe_{3.2}$ the martensite phase is completely supressed and two well-defined phase transitions occur /14,18/. Below T_I = 232 K, superlattice reflections appear and at T_{II} = 224K a first order rhombohedral distortion occurs.

Ultrasonic studies in the high-temperature cubic phase revealed a small softening of the elastic constant C' with decreasing temperature /19/. Inelastic neutron scattering measurements showed that the major softening occurs, not at

Table I

Martensitic Systems Exhibiting Phonon Anomalies

Alloy	**Transition**	**References**
No Mode Softening :		
$Zr_{80}Nb_{20}$	BCC	/12/
Nearly complete Softening		
Nb_3Sn	Sc → Tet	/9/
Ni-T_1(Fe)	SC → Rhombo → Mono	/13/,/14/
InTl	FCC → FCT	/10/,/11/
Partial Mode Softening :		
Li	BCC → 9R	/27/
Na	BCC →HCP	O. Blaschko et al: Phys. Rev.B **30**, 1667 (1984)
Au-Cu-ZN	BCC → Rhombo	M. Mori et al: Solid St. Commun. **17**, 127 (1975)
Cu-Al-Ni	FCC → Rhombo	S. Hoshino et al: Jpn J. Appl. Phys. **14**, 1233 (1975)
Cu-Al-Zn	FCC → 2H FCC → 9R	G. Guenin et al: J. Physique **C4**, C4-597 (1982)
Fe-Pd	FCC → FCT	M. Sato et al: Phys. Rev. Lett. **12**, 2117 (1982).
Fe-Pt	FCC → BCC	K. Tajima et al: Phys. Rev. Lett. **37**, 519 (1976)
Fe-Ni	FCC → BCC	Y. Endoh : unpublished
Tl	BCC → HCP	M. Iizumi: J. Phys. Soc. Jpn **52**, 549 (1983)
Ni-Al	SC → 7R	/15/

q=0, in the elastic regime, but near q = 1/3[110] /13/. This is shown in Fig.1 where neutron intensity contours of the [ζζ0]-TA_2 phonon branch are plotted at two different temperatures. The atomic displacements associated with this branch are along the ζζ0 direction and the limiting slope for q → 0 corresponds to the elastic constant $C' = 1/2(C_{11}-C_{12})$. At high temperatures $T_I < T = 350$ K, the phonon energy widths for $\zeta > 0.2$ become very broad but a clear minimum is seen near $\zeta = 1/3$. As the temperature is lowered to T = 240 K, there is a dramatic build up of inelastic scattering localized near $\zeta = 1/3$ as shown in the bottom portion of Fig.1. This is consistent with a major softening of the branch near $\zeta = 1/3$. Anomalous inelastic scattering at low energies is also observed in the [ζζζ] -LA branch near $\zeta = 1/3$, but this is temperature independent.

Elastic neutron and X-ray diffraction scans along the [ζζ0] direction revealed critical scattering about $q_I = 1/3(1,1,0)$ which develops into a sharp peak at T_I. Additional satellites, with the same temperature dependence, also develop near $q_{II} = \frac{1}{3}(1,1,1)$. The intensity of the satellites increase smoothly with

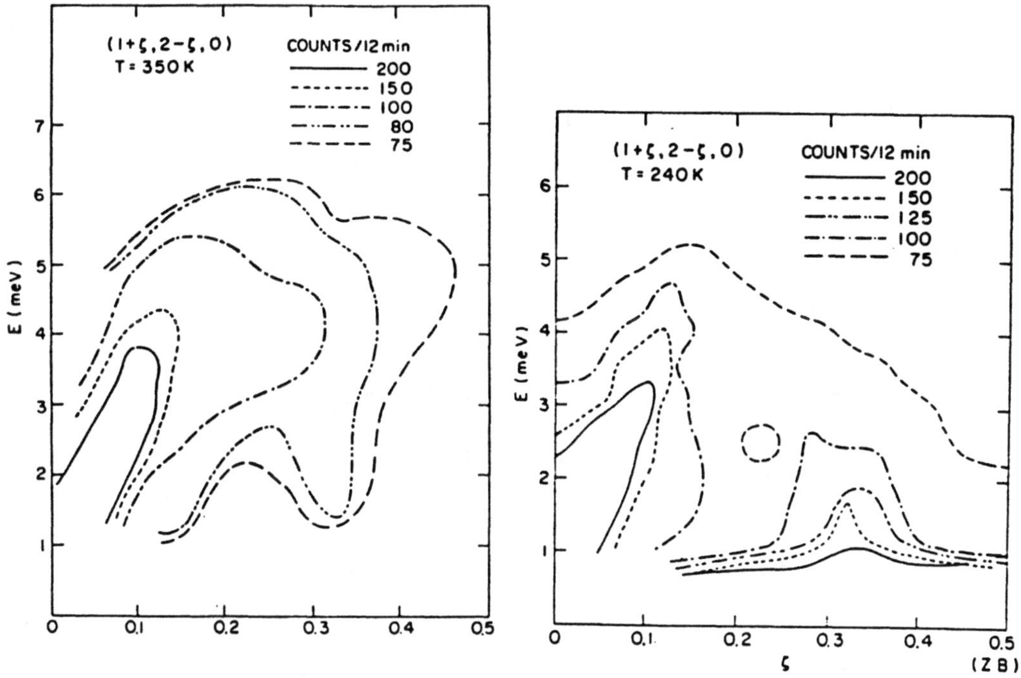

Figure 1. Inelastic intensity contours for the [ζζ0]-TA_2 phonon branch in Ni-Ti-(Fe). The background is 55 cts/12 min. (from ref. 13)
Left : T = 350 K
Right : T = 240 K.

decreasing temperature without being affected by the rhombohedral distortion of the cubic lattice which occurs at T_{II} = 224 K /18/.

An X-ray examination of the q_I and q_{II} type satellites measured in many different Brillouin zones revealed an intensity pattern that could be explained by displacements arising from a linear combination of three phonon modes with wavevectors $q^{(1)}$ = 1/3(1,1,0), $q^{(2)}$ = 1/3(0,1,1) and $q^{(3)}$ = 1/3(1,0,1) with atomic displacements along $[1\bar{1}0]$, $[01\bar{1}]$ and $[\bar{1}01]$, respectively.

A closer inspection of the satellites indicates that the are not located precisely at the commensurate 1/3 positions. Because of the metallic nature of the system it is tempting to describe the system as being a charge density wave (CDW) transition with the incommensurability related to the Fermi wavevector. Figure 2 shows that this cannot be a correct interpretation. In Fig.2, the intensity contours of two q_{II} type satellites measured at T = 226 K are shown.

The left portion shows that the satellite measured in the (1,1,1) Brillouin zone is slightly incommensurate with the cubic phase lattice and the incommensurate wavevector, δ, is directed along the 111 direction. The satellite shown on the right was measured in the (1,1,2) Brillouin zone and the incommensurability is seen to be larger and oriented in a different direction. Similar effects were observed for other q_{II} satellites as well as the q_I = 1/3(1,1,0) type satellites. A schematic representation of these measurements is shown in Fig.3 for the (001) and (011) scattering planes. In this figure, the lengths of the arrows is not significant but the directions indicate the

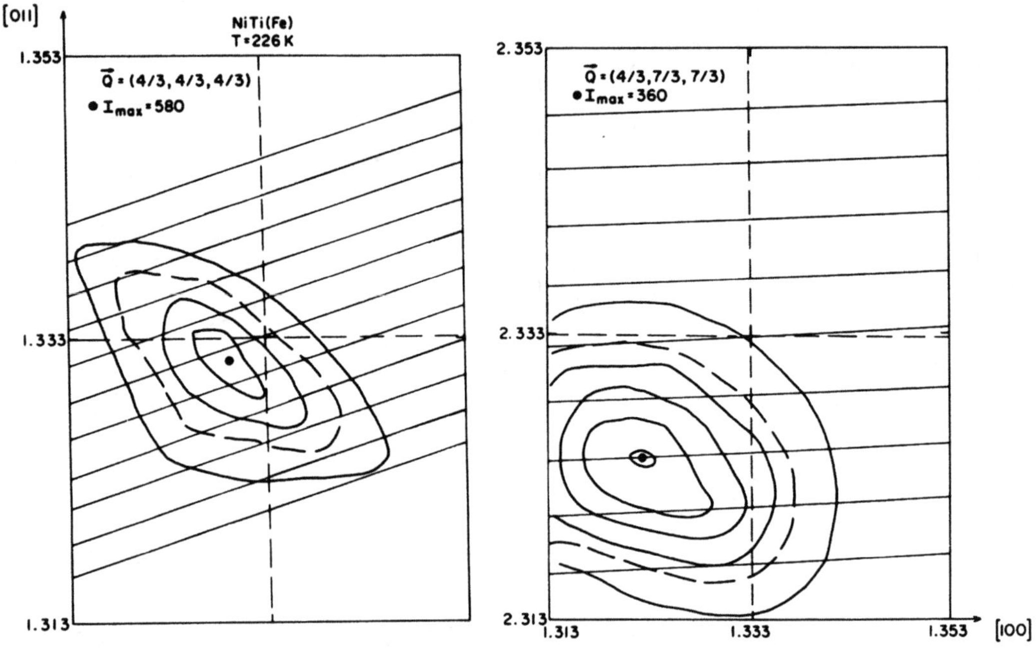

Figure 2. X-ray equi-intensity contours for two Q_{II} type satellites. The parallel lines corresponds to different settings of the position-sensitive detector. The dashed lines represent half the maximum intensity (from ref. 14).

orientation of $\vec{\delta}$. In general, the further from the origin the larger the quantity $|\delta|$. Also there is a distinct "swirl" like pattern for the q_I satellites.

The scattering vector for each satellite can be defined as

$$\vec{Q} = \vec{\tau} + \vec{q}_c + \vec{\delta}$$

where $\vec{\tau}$ is a reciprocal lattice vector and $\vec{q}_c$ is a reduced wave vector measured from $\vec{\tau}$ to the 1/3 position. We demonstrated above that the incommensurate wave vector $\vec{\delta}$ depended upon $\vec{\tau}$. If one had a CDW or a mass density wave, δ would be a constant, independent of which Brillouin zone it is being measured. This is clearly not the case in Ni-Ti and an alternative explanation of the origin of the satellite is needed. Before discussing this, let us look at another martensitic system exhibiting some similar features.

3. Ni_xAl_{1-x} alloy

In the Ni_xAl_{1-x} alloy, the CsCl structure can be maintained at room temperature for $45\% < x < 63\%$ /20/. Below room temperature a martensitic phase transition occurs whose nature and low-temperature structure is not fully understood. $Ni_{63}Al_{37}$ has been extensively studied at room temperature by electron and X-ray diffraction /21,22/. Diffuse streaks have been observed in the electron diffraction patterns and their real space image have shown that a "tweed" microstructure is present /23,24,25/. In addition, ultrasonic studies have shown that there is a small softening of the elastic constant C' prior to the phase transition /26/. These observations have revealed the importance of precursor effects in the martensitic phase transformations.

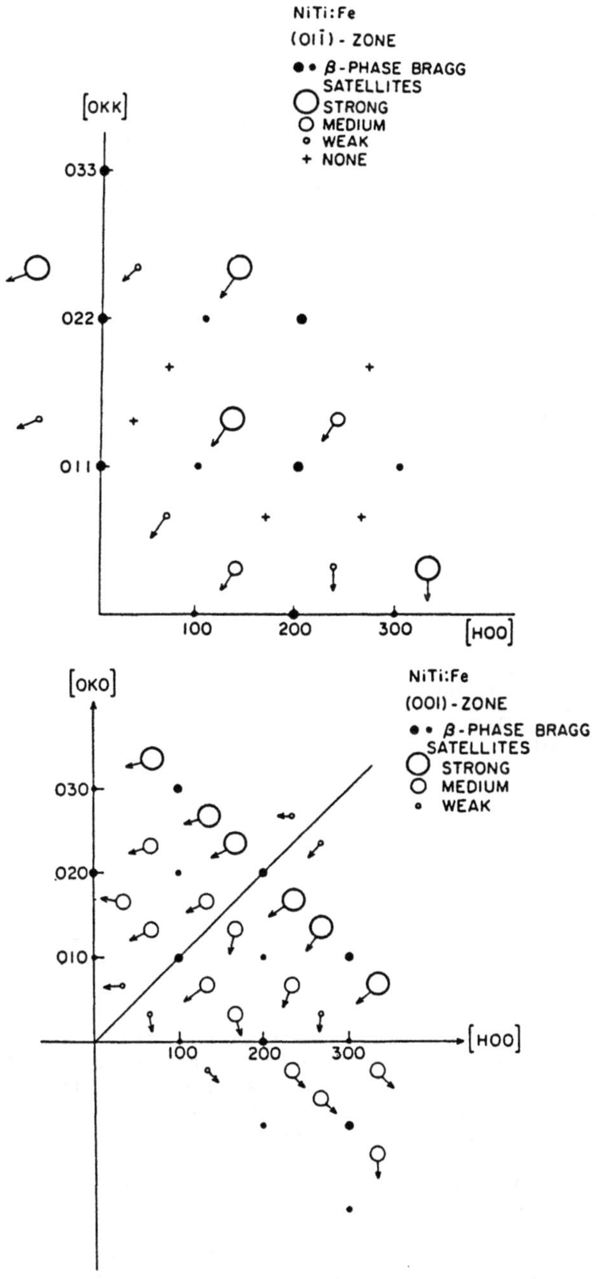

Figure 3. Schematic representation of observed satellites in Ni-Ti(Fe) at T = 226 K. The arrows indicate the direction of the incommensurate wavevector, but their length is not significant (from ref. 14).
Top : q_{II} satellites measured in (011) zone.
Bottom : q_I satellites measured in (001) zone.

Neutron scattering experiments have shown that the major anomaly in the phonon dispersion curve is not in the long wavelength regime probed by ultrasonic studies but occurs at finite q values as shown in Fig.4 /15/. Here is shown the dispersion curve of the $[\zeta\zeta 0]$-TA_2 phonon branch (atomic displacements parallel to $[\bar{1}10]$) measured on Ni_xAl_{1-x} for x = 63, 58 and 50 atomic percent. The composition dependence of the anomalous region is shown in Fig.4b by subtracting the measured phonon energy E from that determined by a harmonic sine wave dispersion : $E_s \sim v_s$

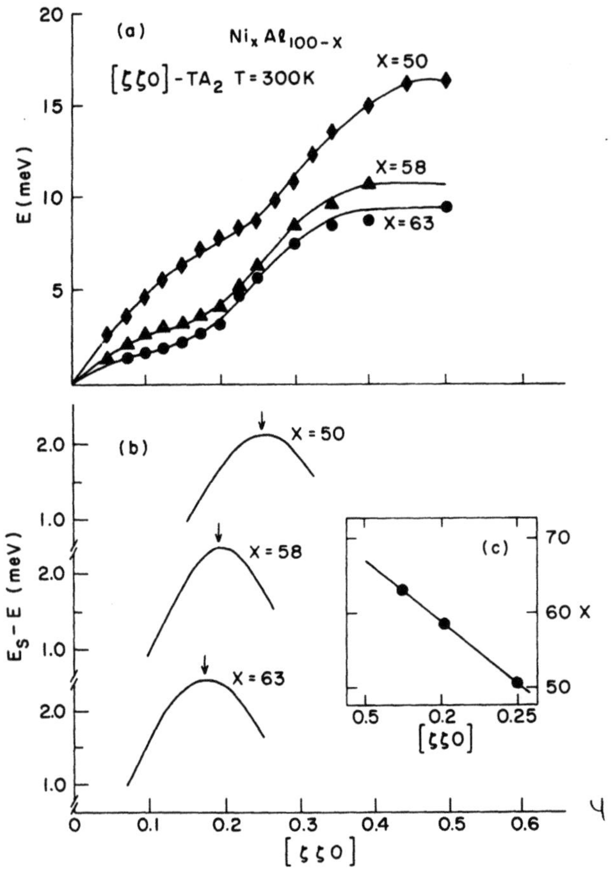

Figure 4. a) Dispersion curves of [110]-TA_2 (displacements along [110]) branch of Ni_xAl_{1-x} for x = 63,58,50 at %. b) Deviation of measured phonon frequency E from that determined using a sine wave dispersion, E_s. c) variation of the maxima of $E-E_s$ with concentration X (From ref.15).

$\sin \pi\zeta$ where v_s is the velocity of sound. E_s - E is plotted for each x and the maximum corresponds to the largest deviation. It is seen in Fig.4c that the position of the maximum anomaly varies linearly with x.

The temperature dependence of the dispersion curve was measured in the x = 58 and 63 atomic per cent samples and showed similar behavior. There is a modest amount of softening (10% - 20%) of this branch with decreasing temperature but restricted mostly to the q region where the anomalous behavior occurs. This amount of softening is typical of most martensitic phase transformations /27/.

All three samples studied exhibit diffuse elastic scattering measured along the same [ζζ 0] direction, but the temperature dependence is very different. For x = 58 no phase transition occurs and the diffuse scattering is nearly structureless and only weakly temperature dependent. In x = 63, however, there is structure in the diffuse scattering and it is strongly temperature dependent as shown as Fig.5. At high temperature broad satellites are present on either side of the (200) Bragg peaks at ζ = 0.13. These are localized along the [110] ridge of diffuse scattering. The intensity of this peak increases continuously on cooling, but shows a distinct thermal hysteresis (Fig.5b) indicating a first order phase transition. The peak also sharpens on cooling to low temperatures.

As in NiTi, the phonon softening appears to induce the elastic diffuse scattering since they both occur in the same wavevector region. Examination of different Brillouin zones show that the satellites cannot be described by a simple

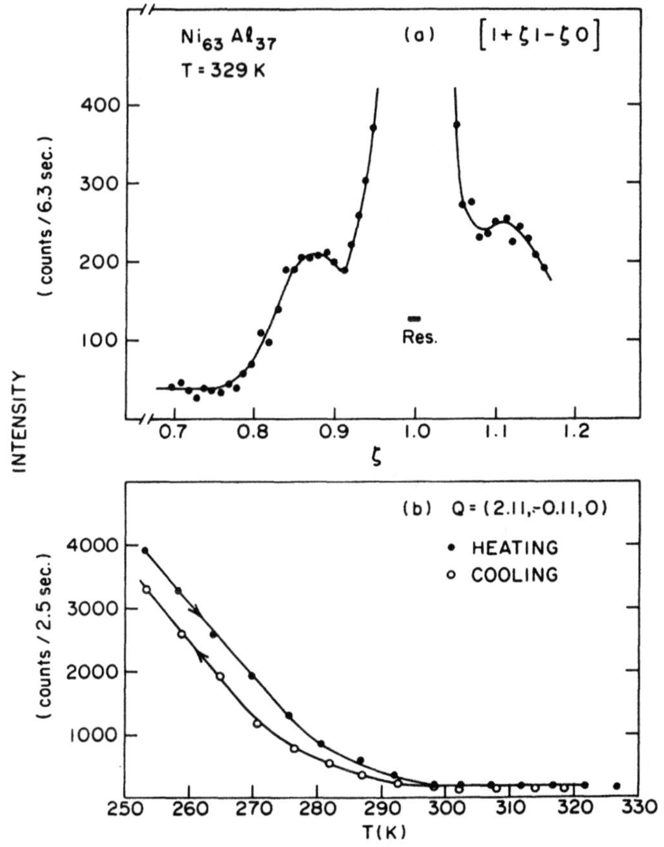

Figure 5. a) Elastic neutron scattering spectrum measured along [ζζ0] direction about 200 Bragg peak. b) Temperature dependence of the intensity measured at Q = (2.11 - 0.11,0) measured on heating and cooling (From ref. /15/).

density wave since the satellite positions vary from zone to zone as shown in Fig.6. Figure 6a is a scan along [ζζ0], through the (2,0,0) Bragg peak; the same scan as in Fig.5a only at lower temperature. It is seen that the intensity of the satellites on either side of the Bragg peak is equal or greater than the cubic Bragg peak intensity. Figure 6b, is a parallel scan only through the (1,0,0) Bragg peak. It is clearly seen that the satellite position, $\vec{\delta}$ depends strongly upon the Brillouin Zone being measured. In this case, δ appears to be proportional to Q.

A clearer picture of what is occurring is obtained by looking at the superposition of the cubic reciprocal lattice and those of the measured satellites as shown on the top of Fig.6. It appears that we have a coexistence of the cubic phase with two variants or twins of the low-temperature orthorhombic phase. Thus, it is improper to view the peaks appearing in Fig.5 as satellites generated by a density wave. A more correct description is that the new low temperature, the product phase, is growing coherently within the high-temperature parent phase.

4. Discussion

In the traditional soft mode picture of structural phase transitions the displacements associated with a particular phonon are just those required to generate the low-temperature structure. As the transition temperature is approached, the frequency of the mode decreases and the system becomes unstable and distorts into the new low-temperature phase. New Bragg peaks develop and the symmetry of the low-temperature phase is a subgroup of the symmetry of the

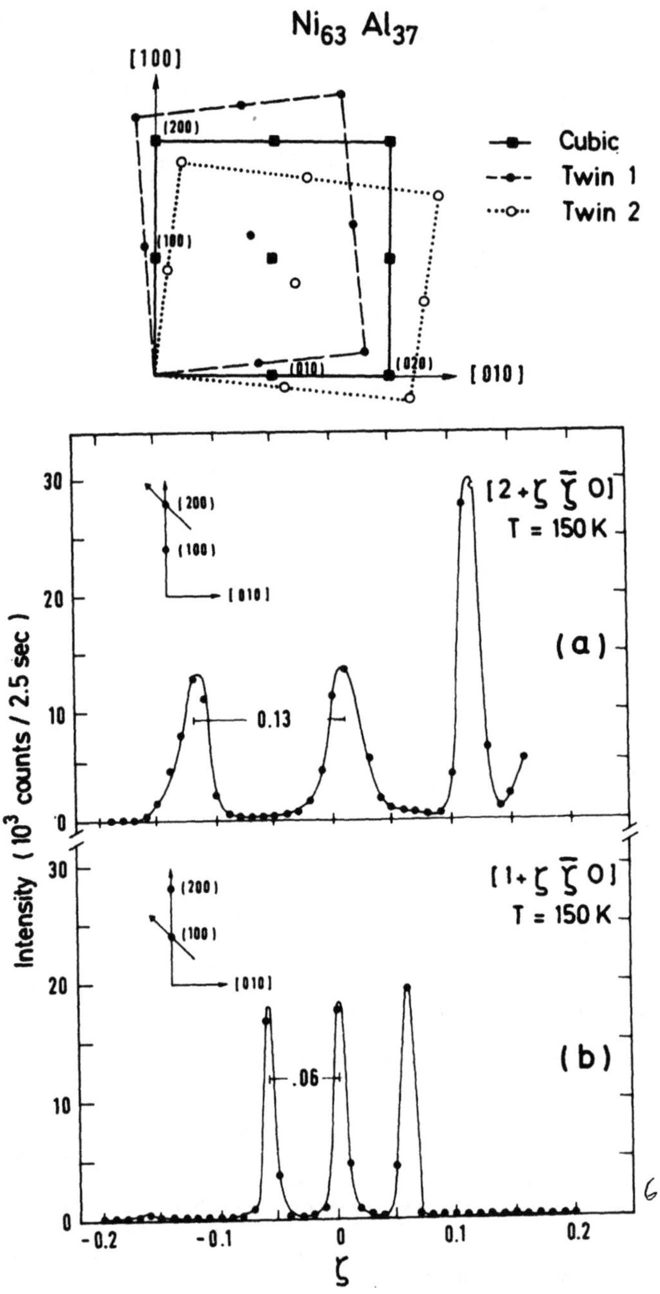

Figure 6. a) Same as Fig.5a only at T = 150 K. b) Same as Fig.6a only about (100) Bragg peak. Top: Representation of observed features in reciprocal space.

high-temperature phase. There is a direct symmetry relationship between the soft mode and the new structure.

In the martensitic transformations, the relationship between the phonon softening and the low-temperature structure is not clear. Most of the systems listed in Table I show only a small (<20%) softening, compared to nearly complete softening in "traditional" structural phase transitions occurring in $SrTiO_3$ and

KH_2PO_4 (KDP) /2,3/. Recently LINDGARD and MOURITSEN /28/ showed that it is possible to have a martensitic transformation with only a small change in a mode frequency due to the anharmonic coupling of two modes. The phonon anomalies and their small temperature dependence observed are consistent with their ideas, but the origin of the anomalies is not clear. Since these are good metals it is reasonable to expect that the electronic structure will play an important role via the electron phonon coupling /29,30/.

The new Bragg peaks appearing at low temperatures cannot be described as a simple condensation of a density wave because their wavevector $\vec{\delta}$ varies with Q. Several explanations have been proposed. YAMADA et al /31/ proposed a "Modulation Lattice Relaxation" model which is based on the cubic parent phase having a particular type of phonon dispersion curve plus embryos of the low temperature product phase. Due to the interactions between the product and the parent phase, modulated lattice relaxations occur in the parent phase. This model was successful to explain the "swirl" pattern of $\vec{\delta}$ observed in NiTi and given in Fig.3b. KRUMHANSL /32/ used a model based on stacking solitons proposed by HOROVITZ et al /33/ to explain the similar peak shifts observed in ω phase Zr-Nb alloys, a system which has much in common with the martensitic systems described here. Finally SALAMON et al /18/ proposed that for NiTi the peaks can be indexed as commensurate reflections on a rhombohedral "ghost" lattice, but with the main Bragg peaks at the original cubic position. This picture is similar to that indicated in NiAl on the top of Fig.6.

Features presented above may occur in many of the systems mentioned in Table I but the studies of the elastic scattering in most experiments are incomplete. It is important to measure the diffuse scattering and new peaks in several Brillouin zones. Most diffraction studies are performed only in one or two Brillouin zones. A recent X-ray diffraction study on the Au-Cd alloy covering many Brillouin zones revealed behavior very similar to Fig.3 observed in NiTi /34/.

The picture of the martensitic systems emerging from these studies is that at high temperatures there exist small regions of the product phase distributed throughout the parent phase. These embryonic regions can grow in size and become more numerous as the temperature is lowered. Thus, above any transition temperature there is a premartensitic phase where there is a coexistence of the parent and product phase. Diffuse scattering may appear when the product embryos are very small. Structure in the diffuse scattering will develop as the embryos grow in size. This nucleation process must depend upon the microscopic forces of the parent phase since phonon anomalies are present in the systems mentioned in Table I. The relationship between the phonon anomalies and low-temperature structure needs more clarification.

Indeed, more work needs to be done with the microscopic techniques discussed here to characterize other martensitic transitions before a complete picture of martensitic phase transitions emerges.

Acknowledgements : It was a pleasure to have as collaborators in this work : Y. Fujii, J.Z. Larese, M. Meichele, S.C. Moss, Y. Noda, M.B. Salamon, S.K. Satija, L.E. Tanner and Y. Yamada. The stimulation provided by discussions with J.A. Krumhansl is also gratefully acknowledged. Work at Brookhaven is supported by the Division of Material Sciences, U.S. Dept. of Energy under contract DE-AC02-76CH00016.

References

1. D. Shearington, S. Kirkpatrick : Phys. Rev. Lett. **35**, 1792 (1975)
2. A.D. Bruce, R.A. Cowley : **Structural Phase Transitions** (Taylor and Francis, Ldt, London 1981)
3. G. Shirane : Rev. Mod. Phys. **46**, 437 (1977)

4. F. Osmund : Compt. Rend. **118**, 532 (1894)
5. N. Nakanishi : Prog. in Mat. Science **24**, 143 (1980)
6. See papers in ICOMAT-82 : International Conference on Martensitic Transformations, J; Physique **C4** (1982)
7. A.J. Burkel, J. Peisel, B. Dorner : Europhys. Lett. **3**, 957 (1987).
8. N.G. Pace, G. Saunders : Proc. Roy. Soc. **A326**, 521 (1972)
9. J.D. Axe, G. Shirane : Phys. Rev. **B8**, 1965 (1973)
10. D.J. Gunton, G.A. Saunders : Solid State Comm. **14**, 865 (1974)
11. T.R. Finlaxson, M. Mostoller, W. Reichardt, H.G. Smith, Solid Stat. Commun. **53**, 461 (1985)
12. S.C. Moss, D.T. Keating, J.D. Axe ; **Phase Tranformations,** L.E. Cross, Ed. (Pergamon Press, N.Y. (1973)) p.179.
13. S.K. Satja, S.M. Shapiro, M.B. Salomon, C.M. Wayman : Phys. Rev. B **29**, 6031 (1984)
14. S.M. Shapiro, Y. Noda, Y. Fujii, Y. Yamada: Phys. Rev. B **30**, 4314 (1984).
15. S.M. Shapiro, J.Z. Larese, Y. Noda, S.C. Moss, L.E. Tanner: Phys. Rev. Lett. **57**, 3199 (1986)
16. W.J. Buehler, J.V.G. Frich, C. Wiley : J. Appl. Phys. **34**, 1475 (1963)
17. M. Matsumoto, T. Honma : Trans. Jpn. Met. Suppl. **17**, 199 (1976)
18. M.B. Salamon, M.E. Meichele, C.M. Wayman, Phys. Rev. B **31**, 7306 (1985)
19. N.G. Pace, G. Saunders : Phil. Mag. **20**, 73 (1970)
20. A.J. Bradley, A. Taylor, Proc. Roy. Soc. A**159**, 232 (1939)..
21. I.M. Robertson, C.M. Wayman : Philos. Mag. A **48**, 421,443,629 (1983)
22. P.Georgopoulos, J.B. Cohen : Acta Met. **29**, 1535 (1981)
23. A. Lasalmonie : Scripta Met. **11**, 577 (1977)
24. L.E. Tanner, A.R. Pelton, R. Gronsky, J. Physique (Paris) **C4**, 169 (1982)
25. L.E. Tanner : these proceedings
26. K. Enami, J. Hasunuma, A. Nagasawa, S. Nenno : Scripta Met. **10**, 879 (1976)
27. See for example : G. Ernst, C. Artner, O. Blaschko, C. Krexner : Phys. Rev. B **33**, 6465 (1986)
28. P.A. Lindgard, O.G. Mouritsen : Phys. Rev. Lett. **57**, 2458 (1987)
29. S.K. Sinha : **Dynamical Properties of Solids,** edited by G.K. Horton and A.A. Marradudin (North Holland, N.Y. 1980) p1.
30. M.A. Krivoglaz : Zh. Eksp. Teor. Fiz. **84**, 355 (1983) Sov. Phys. JETP **57**, 295 (1983)
31. Y. Yamada, Y. Noda, M. Takemoto : Solid. Stat. Comm. **55**, 1003 (1985)
32. J.A. Krumhansl : unpublished
33. B Horovitz, J.L. Murray, J.A. Krumhansl, Phys. Rev. B **18**, 3549 (1978)
34. Y. Noda, M. Takimoto, T. Nakagawa, Y. Yamada : to be published.

Statics and Dynamics of Twin Boundaries in Martensites

B. Horovitz[1;2;3], *G.R. Barsch*[1], *and J.A. Krumhansl*[2]

[1]Materials Research Laboratory and Department of Physics, Pennsylvania State University, University Park, PA 16802, USA

[2]Laboratory of Atomic and Solid State Physics, Cornell University, Ithaca, NY 14853, USA

[3]On leave from the Department of Physics, Ben-Gurion University, Beer-Sheva, Israel

The formation of a coherent array of twins is studied for a tetragonal to orthorhombic displacive transition. This structure is stabilized by an interface strain, coupling the twin band and the parent phase. The interface restoring forces for twin boundary oscillations result in unusually low frequencies and a limiting $\sim$ (wavevector)$^{1/2}$ dispersion.

1. Introduction

The formation of twin boundaries and twin bands are a common occurrence in structural and martensitic phase transitions /1,2/. The symmetry of the parent (untransformed) phase allows for forming a few distinct variants or "twins" of the product (transformed) phase. The coexistence of two of these twins results in a localized twin boundary. The microscopic orgin of the structural transition is of interest by itself and was recently summarized by one of us /3/. Here we show that twin boundaries are stabilized by parent-product elastic forces. These forces determine the spacing between twin boundaries and their normal mode frequencies.

We study in particular a tetragonal to orthorhombic (T-O) transition. In addition to well known examples such as In-Pb, Mn-Fe, Mn-Ni /4/ this transition was also recently observed in the Copper-Oxide high T_c superconductors /5-8/; furthermore, twin boundaries were also observed by electron microscopy /6-8/. We show elsewhere /9/ how the interactions between electrons and twin boundaries can be responsible for the enhancement of T_c.

2. Statics

It was recently shown /10-13/ that a static solution for a twin boundary or for a periodic array of twin boundaries can be produced entirely by displacive distortions of a high symmetry phase without any need for dislocations. Explicit solutions were given in a continuum theory by allowing for both nonlinear elasticity and for nonlocal strains (i.e. strain gradients). The significance of this description, as will emerge below, is that large scale motion at low frequency is allowed by the coherent twin boundary. This is very different from the dynamics of dislocations whose motion involves discontinuities in the strain field and hence drag and damping.

The T-O transition is essentially a two-dimensional square to rectangular transition since there is no symmetry change in the z direction. There are two ways to deform a square ($a \times a$) to a rectangle ($a \times b$ or $b \times a$), hence the T-O transition has two twins related to each other by a reflection in the (110) plane. Geometrical considerations then show that the twins can be smoothly joined if the twin boundary is the (110) or ($1\bar{1}0$) plane /2,14/.

Consider r, s as coordinates and $u(r,s)$, $v(r,s)$ displacement fields in the (11) and ($\bar{1}1$) directions respectively; the order parameter for the T-O transition is then

$$e_2 = (\frac{\partial u}{\partial s} + \frac{\partial v}{\partial r})/\sqrt{2}\,. \tag{1}$$

Since under reflection in the (11) plane $e_2 \to -e_2$ the free energy $F(e_2)$ must be symmetric in e_2. The two twins of the orthorhombic phase correspond to the two degenerate minima of $F(e_2)$ at $e_2 = \pm\epsilon/\sqrt{2}$.

A twin boundary is essentially a topological soliton which interpolates (in space) between the two degenerate minima. Explicit solutions for a twin boundary and for a twin boundary lattice (TBL) were found by using a Landau-Ginzburg expansion for $F(e_2)$, keeping nonlinear terms and gradients of e_2 /11,13/. A TBL is a periodic solution whose periodicity 2ℓ varies in some finite range. Since the absolute ground state is a single variant one needs an additional force in the system to stabilize the TBL and also determine its periodicity.

The additional required force is provided by an interface between a parent phase and a twinned product phase, a configuration well known in martensitic transitions /1,2,14/. The geometry, as illustrated in Fig. 1, defines a habit plane interface which intersects the twin boundaries. The intersection angle is determined such that the strain for parent-product matching (without dislocations) does not diverge. Twinning on the product side is then essential to allow for an equal average lattice constant on both sides of the habit plane, i.e. it is an invariant plane strain. In this scenario the system is in a thermodynamically metastable state, the TBL being stabilized by a two-phase interface. Another type of interface is that between two perpendicular TBL's, as seen in some cases /6-8/. We expect the analysis below to be valid also for this case.

We proceed to evaluate the strain energies associated with the habit plane interface. We define a collective coordinate S_n for the position of the $n-th$ twin boundary. The collective coordinate concept is well defined if its eigenfrequencies are well below the continuum of the acoustic phonons. As found below, this is indeed the case for wavevectors $k \gtrsim 1/L_2$.

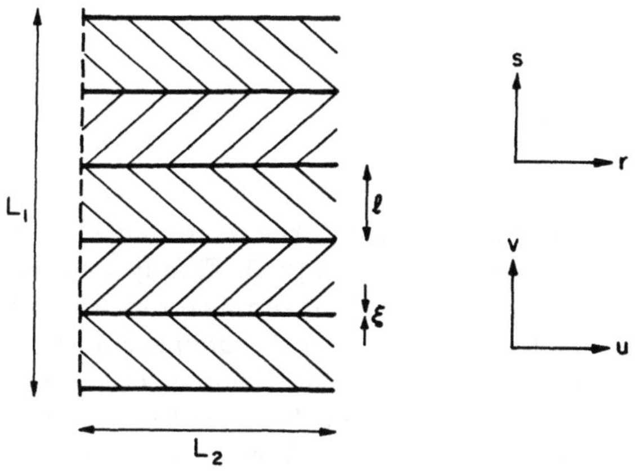

Fig. 1 The habit plane (dashed line) separates a twinned produce phase (on right) from the parent phase (on left). L_1, L_2, and L_3 are dimensions of the product phase (L_3 is in the z direction, perpendicular to r and s). The separation between twin boundaries is ℓ and their width is ξ. The [110] and $[1\bar{1}0]$ directions are r and s, respectively, and the ion displacement fields are $u(r,s)$ and $v(r,s)$.

We assume the case of sharp twin boundaries, $\ell \gg \xi$. The displacement field has then the one-dimensional form

$$
\begin{aligned}
&u(s) = \epsilon(s - S_0) + \bar{u} && S_0 \leq s \leq S_1 \\
&u(s) = -(-)^n \epsilon(s - S_{n-1}) - \sum_{j=1}^{n-1} (-)^j \epsilon(S_j - S_{j-1}) + \bar{u} \\
& && S_{n-1} \leq s \leq S_n (n = 2, 3, \cdots, M)
\end{aligned} \tag{2}
$$

Define a small displacement coordinate δ_n from the expected ground state position $n\ell$, i.e.

$$S_n = n\ell + \delta_n \tag{3}$$

We assume an even number M of twin boundaries with boundary conditions $S_M = S_o + M\ell$ and $u(S_M) = u(S_o)$ so that $\delta_o = \delta_M$ and $\sum_{n=1}^{M} (-)^n \delta_n = 0$. The values of ℓ, $\bar{u}$ and δ_n are left as variational parameters. Thus we focus on the essential "soft" coordinate part of the continuum solution without the need for its details or for the details of the free energy functional.

We next need to determine the habit plane by finding the invariant plane strain. We find that the orientation of the habit plane is a sensitive function of ϵ and the volume ratio of the two twins. For simplicity we make one allowable choice that the habit plane is perpendicular to the twin boundaries; we have examined also other cases and the results below do not change qualitatively.

The next step is to determine the fringing elastic field for the displacements $u(r,s)$, $v(r,s)$ in the parent phase. In the s direction we can assume periodic boundary conditions and define the Fourier components $u(r,k)$, $v(r,k)$ with real k, i.e.

$$u(r,s)=\sum_k u(k)e^{qr}e^{iks}$$
$$v(r,s)=\sum_k v(k)e^{qr}e^{iks}\ . \tag{4}$$

For $r\to-\infty$ we can use linear elasticity theory which yields the relation $q(k)$. We find two solutions which vanish as $r\to-\infty$, i.e. $Req=\gamma\mid k\mid>0$. Finally we need to match the solutions (4) and (2) across the habit plane. For simplicity we assume that $\epsilon\ell$ is smaller than a lattice constant so that epitaxy type discommensurations can be avoided. The parent-product matching is possible in principle, as guaranteed by the construction of the habit plane. We therefore extend (4) to $r=0$ and continuity with (2) yields $u(k)\sim u_o(k)$, where $u_o(k)$ is the Fourier transform of (2). While this is not an exact procedure for all k we expect it to be accurate for low k components; the reason is that the resulting fringing field is of long range - expanding $u_o(k)$ with k and substituting in (4) yields $u(r,s)$, $v(r,s)\sim 1/r$. This long range field should not be sensitive to the details of the matching near the habit plane.

The interface elastic energy is now obtained by substituting (4) in the elastic energy and integrating on $-\infty<r<0$. The integral on $exp(\gamma\mid k\mid r)$ then yields $\sim\mid k\mid$ terms instead of the usual $\sim k^2$ energies. The interface energy is thus of the form

$$E_{IN}=\alpha L_1L_3\sum_k\mid k\mid\mid u_o(k)\mid^2 \tag{5}$$

where α is of the order of the elastic constants.

The elastic energy of the product phase involves the creation energy E_o of a twin boundary per unit area

$$E_{TB}=E_oL_2L_3\cdot L_1/\ell[1+0(e^{-\ell/\xi})]\ . \tag{6}$$

Since the twin boundaries are exponentially localized in a width ξ /11,13/ the interaction between them is $\sim exp(-\ell/\xi)$. This interaction can be neglected relative to (5) if
$\mid k\mid L_2exp(-\ell/\xi)\ll 1$; for $\ell/\xi\simeq 20-100$ /5-7/ this is a safe assumption even for a macroscopic L_2. Substituting (2) in (5) yields (for $\delta_n=0$ and $\bar{u}=-\ell\epsilon/2)E_{IN}=0.27\alpha\epsilon^2L_1L_3\ell$ which combined with (6) has a minimum at

$$\ell=[E_oL_2/(0.27\alpha\epsilon^2)]^{1/2}\ . \tag{7}$$

We have analyzed data on twin bands in $In-T\ell$ /15/ which result from a cubic-tetragonal transition. We find that $\ell\sim(L_2)^\sigma$ where $\sigma=0.5-0.5$, consistent with $\ell\sim\sqrt{L_2}$ of Eq. (7); the data on L_2 is in the range $2.7-10.5\mu m$.

3. Dynamics

The motion of a twin boundary results in a surprisingly non-local effect, as illustrated in Fig. 2. Since the strain e_2 in the twins is fixed at $\pm\epsilon$ the result is a displacement $u(s) \sim \int_{S_m}^{s} ds' e_2(s')$ which affects the whole stack of twins with $s > S_m$. Thus an apparently simple local motion of a boundary results in a coherent macroscopic motion.

To present this idea more precisely, consider the kinetic energy of the product phase (is $\partial/\partial t$ and ρ is the mass density)

$$E_k = \tfrac{1}{2}\rho L_2 L_3 \int_{S_0}^{S_M} \dot{u}^2(s)ds = \tfrac{1}{2}\rho L_2 L_3 \epsilon^2 \sum_{n=1}^{M} \dot{\eta}_n^2 \tag{8}$$

where the normal mode $\eta_n (n = 1, 2, \cdots, M)$ is

$$\eta_n = 2\sum_{j=1}^{n-1} (-)^j \delta_j + \delta_o - \delta u \tag{9}$$

and $\delta u = \ell/2 + \bar{u}/\epsilon$. The nonlocal transformation from δ_n to η_n implies that $E_k \sim \sum_k \mid \dot{\eta}(k) \mid^2 \sim \sum_k \mid \dot{\delta}(k) \mid^2 / sin^2(\frac{1}{2}k\ell)$. Thus the *effective kinetic mass* of the twin boundary motion diverges as $k \to 0$. This divergence is a most significant aspect of our results. It implies that possible damping due to coupling of localized lattice imperfections to δ_n vanishes as $k \to 0$.

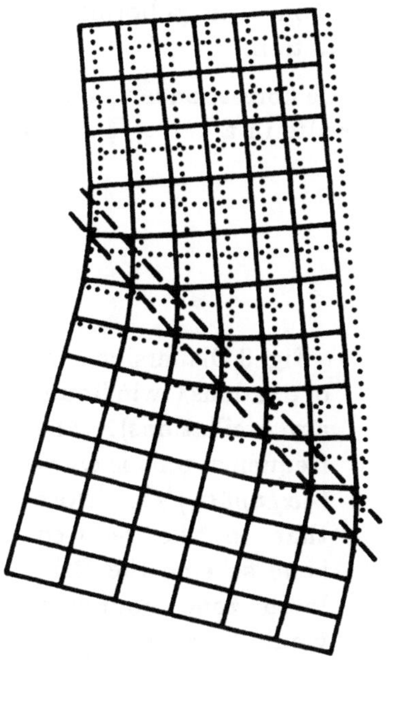

Fig. 2 The lower dashed line is a twin boundary separating the lattices with full lines. Moving the twin boundary to the upper dashed line requires the dotted lattice. Lattice sites then shift perpendicular to the twin boundary motion.

The parent phase has a kinetic energy which by using (4) is smaller than (8) by a factor of $| k | L_2$. Thus for $| k | \gg 1/L_2$ the parent phase follows adiabatically the dynamics as dictated by (8). The elastic energy is then dominated by the parent-product interface (due to the degeneracy of the twin boundary translation) while the kinetic energy is dominated by the bulk product phase.

This scenario has an analog in a linear elasticity problem; Love waves /16/ describe localized waves in a layer of material A with thickness L_2 which is attached to a bulk material B. When the elastic constant of material A is vanishing, to mimic the above scenario, the dispersion of Love waves is $\omega \sim \sqrt{k}$ for $| k | \gtrsim 1/L_2$. This is also obvious by comparing Eqs. (5,8) for the $u(s)$ field.

In our case Eq. (5) is valid only for describing the twin boundary motion. The cancellation between nonlinear and nonlocal energies which leads to the TBL also produces the vanishing $\sim exp(-\ell/\xi)$ of the product phase bulk contribution to the restoring force. In terms of $\eta(k)$, the Fourier transform of η_n, we find from Eqs. (5,8) the dispersion

$$\omega^2(k) = \frac{4\alpha}{\pi\rho\ell L_2} sin^2(\tfrac{1}{2}k\ell) \sum_{p=-\infty}^{\infty} [\frac{1}{| \frac{k\ell}{2\pi} - p |} - \frac{1}{| \frac{1}{2} - p |}] \,. \tag{10}$$

This dispersion is shown in Fig. 3.

For $k \to 0$ $\omega \sim \sqrt{k}$, representing the long range elastic force mediated through the parent phase. (Recall however, that for $| k | L_2 \lesssim 1$, ω should become linear in $| k |$.) This mode is represented by anti-parallel motion of neighboring twin boundaries (Fig. 4) which results in a low k strain modulation.

Equation (10) has a zero mode also at $k = \pi/\ell$; this mode corresponds to a rigid translation of the whole twin boundary lattice which results in a $k = \pi/\ell$ modu-

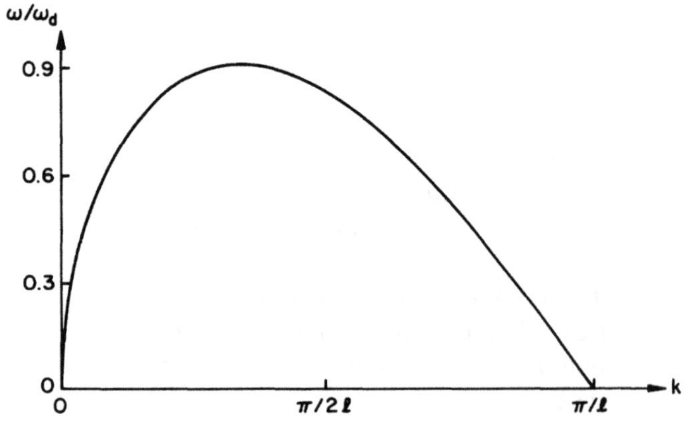

Fig. 3 Dispersion curve of dyadons (twin boundary oscillations). The frequency unit is $\omega_d = (4\alpha/\pi\rho\ell L_2)^{1/2}$. Note that in the reduced zone scheme the Brillouin zone boundary would be at $q = \pi/2\ell$.

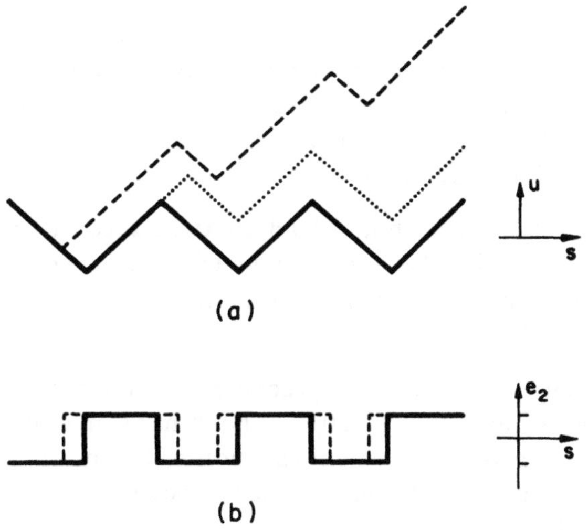

Fig. 4 a) The displacement field $u(s)$ of a TBL (full line), the effect of the motion of a single twin boundary (dotted line), and the effect of a $q_1 \to 0$ oscillation (dashed line). b) The strain e_2 of a TBL (full lines) and the effect of a $q_1 \to 0$ oscillation (dashed lines).

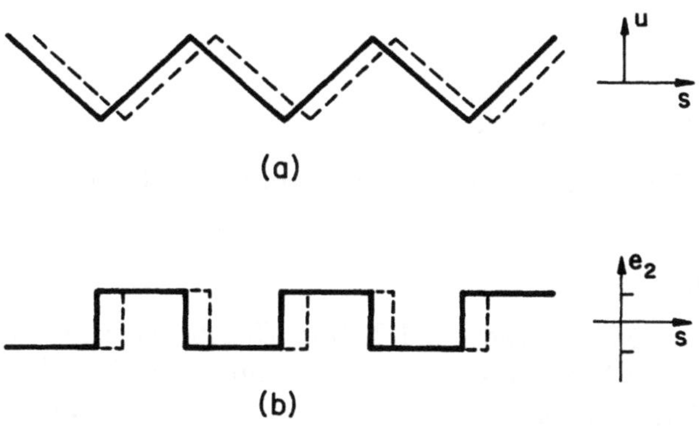

Fig. 5 a) The displacement field $u(s)$ of a TBL (full line) and the effect of a $q_1 = \pi/\ell$ oscillation (dashed line). b) The strain e_2 of a TBL (full line) and the effect of a $q_1 = \pi/\ell$ oscillation.

lation of the strain (Fig. 5). This translation mode may be pinned by impurities or discreteness effects and then $\omega(\pi/\ell)$ is finite.

The predicted frequencies are $\omega \lesssim \omega_{ac}\sqrt{\ell/L_2}$ where ω_{ac} is an acoustic phonon with $q \simeq \pi/\ell$. For values of $\ell \simeq 100 - 1000$Å and $L_2 \simeq 10^4$Å /6-8/ we estimate $\omega \lesssim 10^{10}\ sec^{-1}$. In units of temperature this corresponds to $\sim 1°K$; we then expect an unusual contribution to the specific heat at these low temperatures /9/.

We can define the collective coordinates S_n also to include transverse wavevectors corresponding to a periodic bending of each wall. This bending involves mainly the elastic constants of the orthorhombic phase and we expect that the frequency is linear with the transverse wavevector. We estimate the room temperature thermal fluctuations of a twin boundary position as $\sim 1\%$ of the lattice constant.

4. Conclusions

We have studied the statics and dynamics of coherent, reversible twin boundary arrays. Since dislocations are avoided, the resulting dynamics corresponds to coherent low frequency elementary excitations which involve a quasi-macroscopic motion; we propose to call these excitations "dyadons".

The unusual low frequency collective dynamics should lead to anomalies in the specific heat at temperatures of order $\sim 1°K$. These low frequencies can account for anomalies in elastic constants, which imply a much lower sound velocity than that inferred from neutron scattering /17,18/.

It is also known that by applying external stresses a large scale reversible motion of habit planes and twin boundaries is possible /19,20/. The "shape memory effect" /21/ which has been studied extensively is a manifestation of the motions discussed here; they are remarkable in that events on the atomic scale lead to a coherent, reversible, macroscopic effect.

We are pleased to acknowledge support from the U.S. Department of Energy, grant No. DE-FG-02-85ER-45214.

References

1. J.W. Christian, Met. Trans. 13A, 509 (1982).
2. A.G. Khachaturyan, Theory of Structural Transformations in Solids, (J. Wiley, N.Y., 1983).
3. J.A. Krumhansl, in this volume.
4. L. Delaey, M. Chandrasekharan, M. Andrade and J. Van Humbeck, in Solid → Solid Phase Transformations, H.I. Aaronson, D.E. Laughlin, R.S. Sekerka, and C.M. Wayman, Eds., (Met. Soc. of AIME, Warrendale, P.A., 1982) p. 1429.
5. I.K. Schuller, D.G. Hinks, M.A. Beno, D.W. Capone, L. Soderholm, J.P. Locquet, Y. Bruynseraede, C.U. Segre and K. Zhang, Solid State Commun. 63, 385 (1987).
6. G. Van Tendeloo, H.W. Zandbergen and S. Amelinckx, Solid State Commun. 63, 389 (1987).
7. C.H. Chen, D.J. Werder, S.H. Liou, J.R. Kwo and M. Hong, Phys. Rev. B 35, 8767 (1987).
8. R. Beyers, G. Lim, E.M. Engler, R.J. Savoy, T.M. Shaw, T.R. Dinger, J.W. Gallagher and R.L. Sandstrom, Appl. Phys. Lett. 50, 1918 (1987).
9. B. Horovitz, G.R. Barsch and J.A. Krumhansl, Phys. Rev. B 36, 8895 (1987).
10. G.R. Barsch and J.A. Krumhansl, Phys. Rev. Lett. 53, 1069 (1984).
11. G.R. Barsch and J.A. Krumhansl, Metal. Trans. A (to be published).

12. F. Falk, Z. Phys. B 51, 177 (1983).

13. A.E. Jacobs, Phys. Rev. B 31, 5984 (1985).

14. M.S. Wechsler, D.S. Lieberman and T.A. Read, Trans. AIME 197, 1503 (1953).

15. A. Moore, J. Graham, G.K. Williamson and G.R. Raynor, Acta. Met. 3, 579 (1955).

16. B.A. Auld, Acoustic Fields and Waves in Solids, (J. Wiley, N.Y., 1973) Vol. II p. 94-102.

17. T.R. Finlayson, M. Mostoller, W. Reichardt and H.G. Smith, Sol. State. Comm. 53, 461 (1985).

18. L.R. Testardi, in "Physical Acoustics," Ed. W.P. Mason and R.N. Thurston, 10, 193 (Academic Press, N.Y., 1973).

19. H.P. Weber, B.C. Tofield, and P.F. Liao, Phys. Rev. B 11, 1152 (1975).

20. S.W. Meeks and B.A. Auld, Appl. Phys. Lett. 47, 102 (1985).

21. See e.g. Shape Memory Effects in Alloys, J. Perkins, ed. (Plenum, N.Y., 1975).

The Frenkel-Kontorova Model with Nonconvex Interparticle Interactions

S. Marianer, A.R. Bishop, and J. Pouget

Los Alamos National Laboratory,
Los Alamos, NM 87545, USA †, and
ICTP, P.O. Box 586, I-34100 Trieste, Italy

We present an analytical and numerical study of a chain of atoms moving in a periodic potential with nonlinear, nonconvex interparticle interactions, described by the Hamiltonian

$$H = \sum_n \frac{1}{2} \dot{u}_n^2 + A(u_{n+1} - u_n)^4 - B(u_{n+1} - u_n)^2 - \cos(u_n) \ .$$

The ground state is shown to be homogeneous for B < 1/8 and dimerized for B > 1/8. The nonconvexity is shown to also play an important role when excitations are considered. In the dimerized phase, we define the staggered order parameter $v_n = (-1)^n u_n$ and map the model to the on-site ϕ^4 problem. In particular we find a localized kink solution, v = tgh(x/d), with a width d varying from infinity at B = 1/8 to zero at B = 3/16, where the interparticle interactions in the ground state cross over from the nonconvex region to the convex one. We also show that at this point the kinks are pinned to the lattice. These results are verified by a direct numerical simulation of the discrete model. Finite temperature effects are discussed in terms of a displacive phase transition at B = 1/8 becoming order disorder transition at B = 3/16. When coupling between the springs is introduced by adding a strain gradient term $G(u_{n-1} - 2u_n + u_{n+1})^2$ to the Hamiltonian, we observe a crossover from an infinite kink width at B = 1/8 to a finite width for B > 3/16, determined by the competition between the effective double well and the interspring coupling strengths.

Physical systems with competing interactions that include incommensurate length scales have become a subject of intense interest because they can lead both to modulated ground states and to unusual dynamics and excitations. Most of the theoretical studies to date have been limited to cases where the interparticle interactions are convex and where there are two competing length scales. For these cases it was shown [1] that the ground states are always periodic or quasi-periodic and that the phase transitions are continuous. These results are valid only for convex interparticle interactions and it was shown by Aubry, Fesser and Bishop [2] that first order phase transitions might occur when nonconvexity is introduced. See also Griffiths et al. [3] and Marchand et al. [4]. It is well known that nonconvex interparticle interactions can exist in condensed matter systems. Example are, the RKKY oscillatory exchange interactions between localized spins in metals, and oscillating indirect interactions mediated by elastic strains [5]. Nonconvex interparticle interactions are also present in a Ginzburg Landau energy functional for the strains in materials undergoing elastic phase transitions (e.g. Barsch and Krumhansl [6]).

† Permanent address.

To study the effects of nonconvexity we use here an extension of the familiar (e.g. Aubry and LeDaeron [1]) Frenkel-Kontorova model into which we introduce degenerate double well interparticle springs. The Hamiltonian for the system is given by:

$$H = \sum_n \frac{1}{2} \dot{u}_n^2 + A(u_{n+1} - u_n - a)^4 - B(u_{n+1} - u_n - a)^2 - cos(u_n) \quad . \tag{1}$$

This is in general a model with <u>three</u> competing lengths: $L_{1,2} = a \pm L_0$, the minima of the double well spring ($L^2_0 = B/2A$) and $L_3 = 2\pi$, determining the minima of the substrate potential. However, we limit ourselves here to the study of the homogeneous and dimerized phases by setting a = 0, i.e. <u>two</u> competing lengths. For this case the configuration where the particles are at the minima of the substrate potential, $u_n = 0$, is such that the interparticle interaction energies are at their maxima. It is then easy to see that the system can lower its energy (for a large enough value of B) by dimerization, and the ground state configuration is $u_n = \frac{1}{2}(-1)^n u_0$ where u_0 is determined by the competition between the substrate and interparticle energies: Substituting in (1) and minimizing with respect to u_0 we obtain

$$4Au_0^3 - 2Bu_0 + 1/2\, sin(u_0/2) = 0 \quad . \tag{2}$$

Eq. (2) has the solutions $u_0 = 0$ and $u_0 = \pm[(B\text{-}1/8)/(2A\text{-}1/192)]^{1/2}$ where we have approximated sin(u/2) by $u/2 - u^3/64$). Thus, for $B < 1/8$ the ground state will be homogeneous, $u_n = 0$, while for $B > 1/8$ it will be dimerized.

To study excitations of the system in the dimerized phase, we consider first the staggered order parameter $v_n = (-1)^n u_n$. In the ground state $v_n = 0$ (homogeneous phase) and $v_n = + (-)u_0/2$ dimerized long-short (short-long) springs, respectively. Substituting a continuum approximation $v_{n\pm1} = v_n \pm hv' + \frac{1}{2}h^2v''$, in the Hamiltonian and equations of motion we obtain:

$$H = \dot{v}^2/2 + v'^2/2\,(2B - 48Av^2) + 16Av^4 - 4Bv^2 - cos(v) \tag{3}$$

$$\ddot{v} = 8Bv - 64Av^3 - sin(v) + (2B - 48Av^2)v'' - 48Avv'^2 \quad . \tag{4}$$

We have obtained traveling wave solutions to equation (4) numerically (Fig. 1), a limit of which are the solitary wave solutions of Fig. 2. The physical meaning of the kink solutions in the context of our model is a change between two topologically inequivalent ground states from a "short-long" spring length

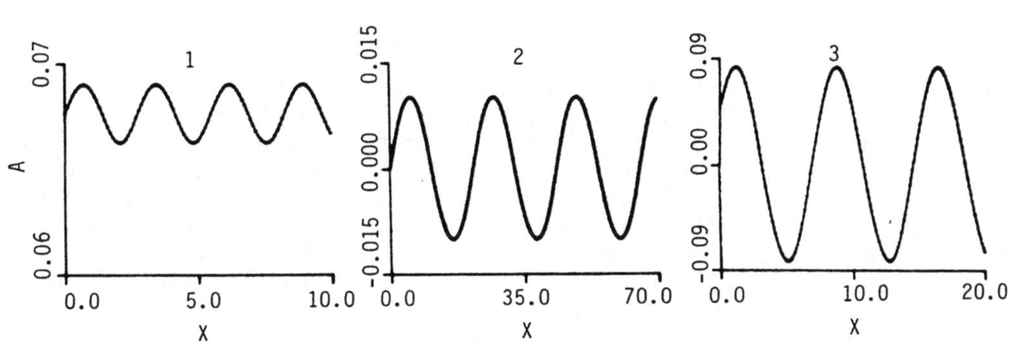

Fig. 1. Traveling wave solutions of eqn. (4). The numbers are from the ϕ^4 terminology [7].

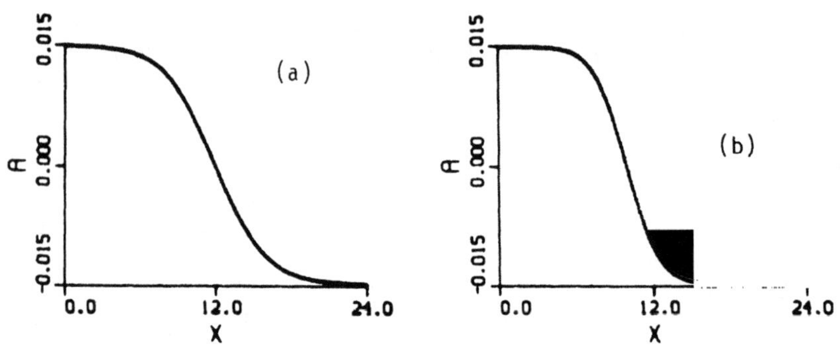

Fig. 2. The solitary kink solution obtained as an infinite period limit of solution No. 2 in Fig. 1. (a) $v_{kink}=0$. (b) $v_{kink}>0$.

configuration at one end of the chain, to a "long-short" one at the other end. The main properties of the kink solutions are: (i) their width d is a decreasing function of the velocity and of B; and (ii) strong pinning of the kink occurs for B = 3/16 where the static kink width vanishes and the continuum approximation breaks down. Since equations (3) and (4) were obtained in a continuum approximation, we have checked that the above properties of the kinks are valid for the discrete model (1) as well. To this end we numerically solved the equations of motion derived from (1), imposing, via boundary conditions, a single kink in the chain. To obtain the static defect configuration, we started from homogeneous initial positions and random velocities for the particles, and "cooled" the chain by adding a damping term εu_n to the equations of motion. After the velocities became sufficiently small, we turned off the damping and checked that the final configuration obtained is indeed a static solution of the equations of motion. For $B < 3/16$ (Fig. 3a) the change from $u_n = \frac{1}{2}(-1)^n u_0$ at the left of the chain to $u = \frac{1}{2}(-1)^{n+1} u_0$ at the right is indeed gradual and the continuum approximation is appropriate, while for $B > 3/16$ (Fig. 3b) the kink is pinned to one site having an abrupt "Ising wall" shape. All these properties can be analyzed by mapping our Hamiltonian to an on-site "ϕ^4" model, as we now show.

We assume v to be small and expand sin(v) and cos(v) in equations (3) and (4) to $O(v^4)$. Defining a dimensionless displacement $W = v/(u_0/2)$ we obtain:

$$H = (u_0/2)^2 (W^2/2 + B(1 - 3(u_0/L_0)^2 W^2)W'^2 + 2(B - 1/8)(W^2 - 1)^2 \quad . \tag{5}$$

This is the Hamiltonian for the on-site ϕ^4 problem [7] with an effective spring constant $C = 2B(1 - 3(u_0/L_0)^2W^2)$, and a depth $E = (u_0/2)^2(B - 1/8)$. We can therefore apply known results to the present problem. The main effect of the nonlinear interparticle interaction is to introduce a dependence on W in the effective spring constant C. This plays an important role, determining the stability limits of the different traveling wave solutions to (5), as will be discussed elsewhere. In this report we focus on the effects of the nonlinearity on the traveling kinks (fig. 2). The asymptotic values of W in these solutions are $W = +\frac{1}{2}u_0$ as $z \to +\infty$, where $z = x - ct$ and c is the kink velocity. the kink width d and its energy E will thus be given from the ϕ^4 theory [7] as:

$$d = \frac{u_0}{2}\left\{2B[1 - 3(u_0/L_0)^2]/[B - 1/8]\right\}^{1/2} \tag{6}$$

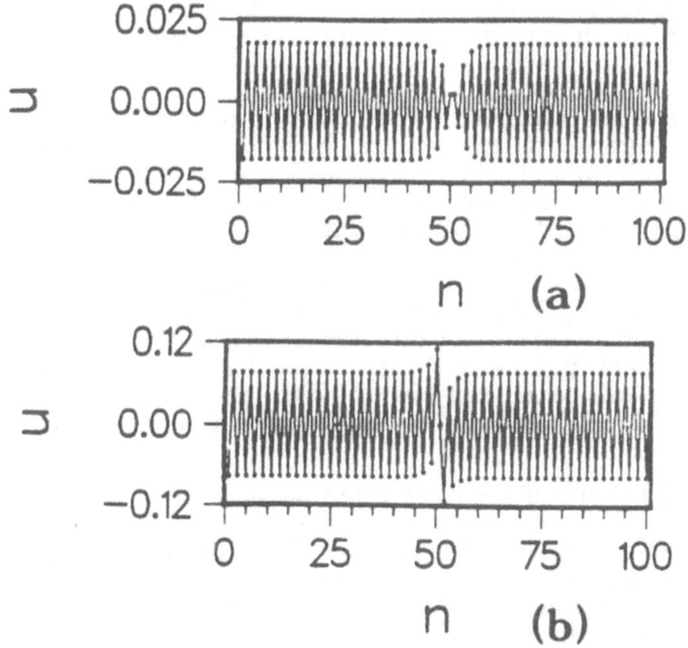

Fig. 3 The discrete solutions with one kink: (a) $1/8 < B < 3/16$; (b) $B > 3/16$.

$$E = \frac{1}{4} u_0^2 \left\{ 2B[1 - 3(u_0/L_0)^2][B - 1/8] \right\}^{1/2} \tag{7}$$

The kink's width is infinite at the phase transition ($B = 1/8$) and decreases to $d = 0$ when u reaches the value $L_0/\sqrt{3}$ (at $B = 3/16$). It has zero creation energy both at the phase transition, where its amplitude vanishes, and when $u_0 = L_0/\sqrt{3}$, where the whole chain except one atom is in its ground state. It is now easy to see that at the point $u_0 = L_0/\sqrt{3}$ the interparticle interaction changes from concave (for $u < L_0/\sqrt{3}$) to convex (for $u_0 > L_0/\sqrt{3}$). A similar effect was noted by Barsch and Krumhansl [6] in their study of solitons in ferroelastic materials. This point is seen more clearly when we consider the linearized phonons in the dimerized ground state. The dispersion relation is then:

$$\omega^2 = \cos(u_0/2) + 4B\,[3(u_0/L_0)^2 - 1]\,\sin^2(k/2)\ . \tag{8}$$

For $k = \pi$, this has a homogeneous component plus upward (downward) curvature when the ground state is in the concave (convex) region of interparticle interactions. Note that although the ground state is dimerized the linearized phonon spectrum consists of one branch only. Splitting into acoustic and optic branches and in particular opening a gap at $k = \pi$ occurs only if nonlinear effects are included. As will be shown elsewhere, these will become important only at large enough amplitudes A: viz.

$$A > \{2B[1 - 3(u_0/L_0)^2] + \cos(u_0/2)\}/\{3(u_0/L_0)\}\ . \tag{9}$$

The analogy with the ϕ^4 theory can be applied to moving kinks and to finite temperature effects as well. For traveling kinks the limiting sound velocity is c

$= 2B[1 - 3(u_0/L_0)^2]$, which vanishes when $u_0 = L_0/\sqrt{3}$. This explains both the decrease of the kink's width with velocity (Fig. 2) and the pinning of the kink when $u_0 = L_0/\sqrt{3}$.

When finite temperatures are considered, we can distinguish two regions [7]: (1) the displacive region near $B = 1/8$ with a characteristic "transition" temperature $k_BT \simeq (0.4)\ 2B[1-3(u_0/L_0)^2][B-1/8](u_0/2)$, which is the creation energy of the kink (eq. 7); and (2) the order-disorder region near $B = 3/16$ with a transition temperature $k_BT \simeq 0.2(B-1/8)(u_0/2)$, which is the barrier height of the effective double well. These have the following interpretation for the present model: At $B = 1/8$ the "melting" of the dimerized springs is characterized by large domains (of the order of the kink's width) of nearly undimerized portions of the chain separating dimerized phases. At $B = 3/16$, on the other hand, there exist many dimerized regions separated by Ising walls (i.e. a width of the order of a lattice constant) in which the chain's dimerization changes from a "long-short" to a "short-long" spring configuration.

Finally we return to the shape of the kink when $u_0 = L_0/\sqrt{3}$. The sharp jump in the value of W from $+u_0/2$ to $-u_0/2$ is smoothed out if we include higher derivatives in the continuum Hamiltonian, or equivalently longer range interparticle interactions in the discrete version. Such terms are present if we consider, for instance, an expansion of the energy in the elastic string. To see the effect of such a term we add $1/2\, G\, (W'')^2$ to the Hamiltonian (eq. 5). Substituting a solution $v = \mathrm{tgh}(z/d)$ in the equation of motion we obtain to leading order in 1/d:

$$d = \left\{ \left[\frac{1-3(u_0/L_0)^2}{8G}\right]^2 + \frac{B-1/8}{8G} - \frac{1-3(u_0/L_0)^2}{8G} \right\}^{-1/2} . \qquad (10)$$

Eqn. (10) shows a crossover from an infinite width at $B = 1/8$ to a finite nonzero one, $d = \sqrt{[8G/(B-1/8)]}$, at $u_0 = L_0/\sqrt{3}$. Note that for $G > 0$ we obtain well defined kink solutions for all values of B ($u_0 \to L_0$ as $B \to \infty$).

To summarize we have presented a study of the ground state and excitations of the Frenkel-Kontorova model with nonconvex interparticle interactions, emphasizing the special effects of the nonconvexity on the ground state and on the excitations. Our study here has been limited to nonconvexity with two competing length scales. As indicated earlier, a third length scale can be introduced by choosing a nonzero value for a in eqn. (1). This was done by Marchand, Hood and Caille [4] in their study of the ground states of (1), assuming small displacements u_n and thus replacing $\cos(u_n)$ by $1 - u^2_n/2$. The phase diagram obtained in this study consists of various modulated configurations with first and second order phase transitions between them. On the other hand Barsch et al. [6,8] have shown that the present model with a strain gradient term may be useful in the description of twin boundary dynamics in martensite materials. In this case the substrate potential models the parent phase and the other terms are the expansion of the free energy as a function of the strain and strain gradients. We are currently studying this model. We have obtained additional ground state configurations (as the strain gradient term is increased). These consist of configurations where $u_n = n\cdot a$ for $n = 1...N$ and $u_n = n\cdot b$ for $n = N+1...M$ (the dimerized phase is where $M = N = 1$, but different M, N can be obtained as the parameters are varied). The case

(N,N) with N $\gg$ 1 corresponds to the twin boundary lattice described by Barsch et al. [8]. A complete phase diagram is in preparation and will be published elsewhere.

We are grateful for valuable discussion with Baruch Horovitz and Philip Rosenau. Rosenau has recently discussed continuum approximation schemes for lattices with general interparticle interactions and substrate potentials [9,10]. This work was supported in part by the US DOE.

References

1. S. Aubry and P. Y. LeDaeron, 1983 Physica 8D 381.
2. S. Aubry, K. Fesser and A. R. Bishop, 1985, J. Phys. A 18 3157.
3. R. B. Griffiths and W. Chou, 1986, Phys. Rev. Lett. 56 1929.
4. M. Marchand, K. Hood, and A. Caille, 1987, Phys. Rev. Lett. 58 1660.
5. J. Villain and M. B. Gordon, 1980, J. Phys. C13, 3117.
6. G. R. Barsch and J. A. Krumhansl, 1984, Phys. Rev. Lett. 53 1069.
7. S. Aubry, 1975, J. Chem. Phys. 62 3217; 1976 ibid 64 3392.
8. G. R. Barsch, B. Horovitz and J. A. Krumhansl, 1987, Los Alamos Workshop on Competing Interactions. (These Proceedings).
9. P. Rosenau, Phys. Lett 118A 222 (1986).
10. P. Rosenau, Phys. Rev. B 36 (in press) (1987).

Cu_3Pd Observed by High-Voltage Electron Microscopy

J. Kulik, S. Takeda, and D. de Fontaine*

Materials and Chemical Sciences Division,
Lawrence Berkeley Laboratory, Berkeley, CA 94720, USA
*Permanent address: College of Education, Osaka University,
1-1 Machikaneyama-cho, Toyonaka-shi, Osaka 560, Japan

Abstract

Cu-Pd samples of compositions varying from 16 to 26 at.% Pd were irradiated *in situ* in a 1.5 MeV electron microscope at various temperatures. Low temperature (90 K) irradiation produced completely disordered solid solutions. Irradiation at room temperature up to as high as about 500 K produced steady state short range order (SRO) which, for specimens of 18% or more Pd, is characterized by diffuse intensity at $[1,\pm q,0]$ and equivalent positions in reciprocal space (modulated SRO). In general, q is a function of composition, temperature and irradiation dose. High temperature irradiation tended to produce the expected equilibrium long range order — either $L1_2$ or a long period superstructure depending on composition and temperature. The 18 and 20% samples irradiated at room temperature exhibited steady state *modulated* SRO even though the expected equilibrium structure is one of *unmodulated* order ($L1_2$). It is suggested that spinodal ordering is responsible for this latter effect. An f.c.c.-based Cu-Pd phase diagram is proposed incorporating ordering stability loci and a metastable Lifshitz point.

1. Introduction

It is well known that, below about 800 K, Cu-Pd f.c.c. solid solutions undergo ordering reactions in the range of about 10 to 30 at.% Pd. Evidence has come from x-ray diffraction [1-5], and electron diffraction and microscopy [5-9]. From these and other data, SUBRAMANIAN and LAUGHLIN [10] proposed an assessed phase diagram of which Fig. 1 is a slightly modified version. The modifications were introduced in order to attempt to incorporate the very recent results of BRODDIN *et al.*[11]. The heavy solid lines in the figure indicate equilibria between the f.c.c. solid solution (α) and the f.c.c.-derived ordered phases. Much of this diagram is of speculative nature as no firm evidence for the peritectoid reactions exist. Nevertheless, certain basic features are well established. The phase region (α') of the simple ordered structure $L1_2$ (Cu_3Au type) peaks not at the expected stoichiometric composition of 25%, but at around 15% Pd. Near stoichiometry, one-dimensional long-period superstructures (LPS) (α_1 region) are found, being replaced at higher Pd content by two-dimensional LPS (α_2 region). At still higher Pd content, a (B2) b.c.c. superstructure (β') becomes stable. Stable B2-related equilibria are indicated by light lines in Fig. 1. The dashed lines are *ordering spinodals*, to be discussed later. The two open circles shown on the diagram indicate the presence of one-dimensional LPS according to BRODDIN *et al.* [11]. Clearly, in that region, the one-dimensional LPS may well be metastable with respect to the two-dimensional LPS, and both one- and two-dimensional LPS are certainly metastable with respect to the stable $\alpha_2 + \beta'$ equilibrium.

Recently, there has been a resumption of interest in alloys with LPS since it was suggested [12-16] that generalizations of the so-called axial next nearest neighbor Ising (ANNNI) model could serve as simple theoretical paradigms for these systems. The original explanation for the stability of LPS, based on Fermi surface considerations [17], is certainly still considered valid, especially now that GYORFFY and STOCKS [18] have given the idea a firmer quantitative basis. Nevertheless, viewing these alloys as Ising systems with competing interactions has proven to be a useful approach. Polytypes encountered in Ag_3Mg [16] and Al_3Ti [14], for example, agree well with ANNNI model predictions.

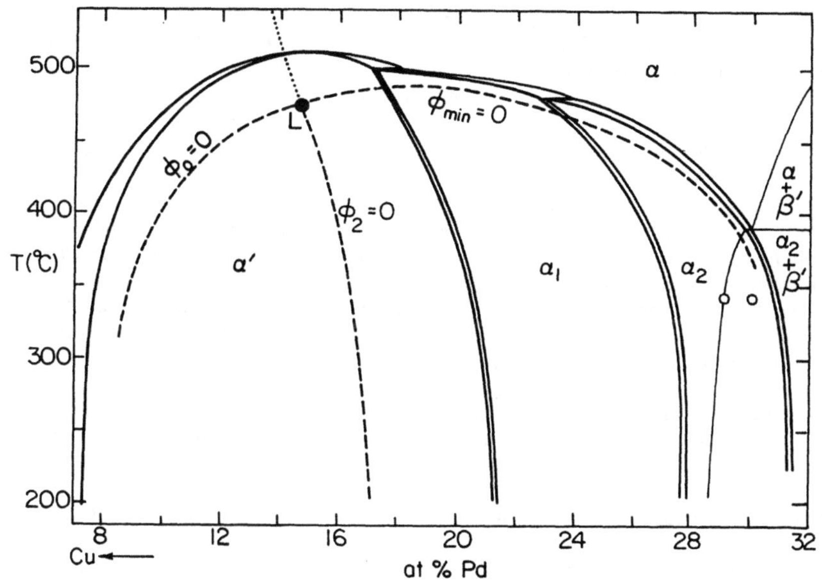

Fig. 1. Cu rich side of the Cu-Pd phase diagram

The experimental Cu-Pd phase diagram certainly merits further study. However, kinetics in this system are extremely sluggish and virtually nonexistent near room temperature. Consequently, statements concerning equilibrium configurations at these temperatures must be made with caution and are difficult to verify. In an attempt to increase the rate of kinetic processes we performed *in situ* irradiation of several Cu-Pd specimens in a 1.5 MeV transmission electron microscope (TEM).

In general, irradiation of metals by high-energy electrons results in a number of different processes including the replacement and displacement of atoms. There is a consequent increase in the concentration of vacancies and interstitials. In long-range ordered (LRO) or short-range ordered (SRO) alloys, destruction of order may result at sufficiently low temperatures. On the other hand, ordering during irradiation can occur if the temperature is sufficiently high to allow the thermal motion of the irradiation-induced point defects. (A summary of these effects can be found in URBAN, BANERJEE and MAYER [19] and references cited therein.) In the case of Cu-Pd, it was hoped that the excess vacancies induced by the high-energy electron beam would enhance the ordering kinetics at room temperature. However, instead of obtaining a long-range ordered state, modulated short-range order was obtained, even in regions where an *unmodulated* long-range ordered structure (i.e., $L1_2$) was the expected equilibrium state. By "modulated short range order" we mean the presence in diffraction data of broad, diffuse peaks of intensity located away from one of the special points of high symmetry in the Brillouin zone. The displacement of the peaks from the special point is the modulation wave vector. This is discussed further in Section 3 below. The unexpected occurrence of this SRO state has revealed something of the nature of the thermodynamic instabilities which underlie the first order transitions occurring in this system. We present in this paper a description of this result.

2. Experimental Results

Alloys of 16, 18, 20, 22, and 26 at.% Pd were prepared by arc-melting under an Ar atmosphere. The ingots were held just below the melting point for several hours. Pieces of the original ingots were then remelted and splat cooled through the courtesy of L. E. Tanner at Lawrence Livermore

Laboratory. The purpose of the splat cooling was to obtain a pronounced <100> texture. The splat cooled foils were then homogenized in sealed quartz tubes at 1270 K under one atmosphere of Ar for one day. Solutionized foils were quenched in ice water. Disks for TEM observation were punched and electrolytically polished in a methanol solution of 15% HNO_3 at about -60° C.

After being quenched, the specimens were observed in a conventional TEM (100 kV) and then transferred to the 1.5 MeV machine at the National Center for Electron Microscopy at Lawrence Berkeley Laboratory. Use of both low and high-temperature double tilting stages provided a temperature range of about 90 K to over 770 K. Diffraction patterns of all samples were taken of the as-quenched samples in the [100] orientation. For some specimens, electron irradiation was then performed at 90 K at which temperature all traces of SRO are destroyed. Subsequent evolution of the disordered regions under irradiation at room temperature could then be followed *in situ*. In addition, the heating stage was used to perform *in situ* irradiation at temperatures from about 470 K up to the disordering temperature.

In what follows, all temperatures cited are those of the thermocouple on the specimen stage of the microscope. The actual temperatures of the irradiated areas were certainly higher, perhaps by a few tens of degrees.

The most extensive observations were made on the 20% Pd specimen. We begin with a description of these results. As a convenient reference, Fig. 2 is a schematic of electron diffraction patterns in [001] incidence from modulated LRO and modulated SRO specimens. Patterns (a), (b) and (c) show the three possible variants of the one-dimensional LPS which are, respectively, modulation along [100], [010] and [001]. Pattern (d) is a combination of (a) and (b), while (e) is a combination of all three variants. Pattern (f) is indicative of SRO. The small open circles in (f) are meant to indicate diffuse peaks as opposed to the sharp superstructure peaks in the LRO patterns. Note the absence of intensity at [100] and equivalent positions in (f). In all cases, the modulation can be characterized by the separation $2q$ of the satellite pairs. The quantity q is then the distance of the peaks from the [100] (or equivalent) position; that is, q is the modulation wave number. In contrast, Fig. 11 shows an example where the SRO or LRO peaks occur *at* the special point (i.e., [100] and equivalent).

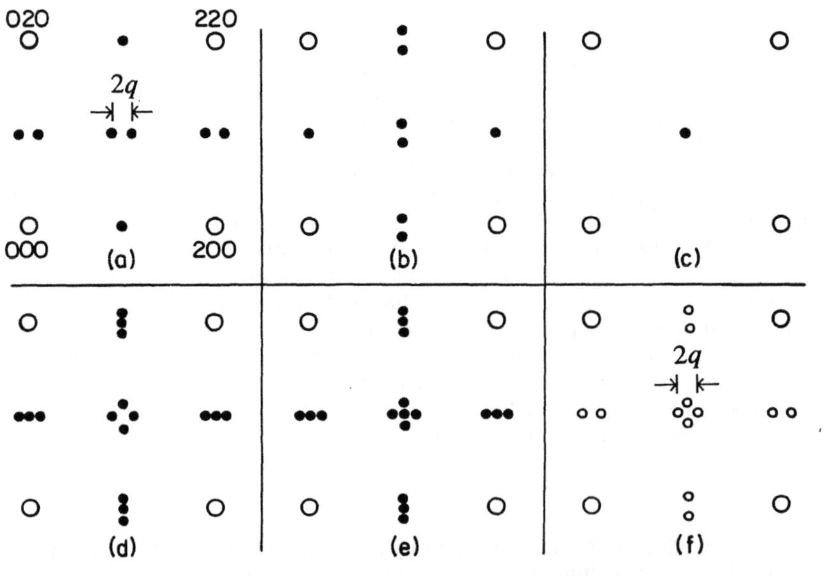

Fig. 2. Schematic of diffraction patterns from the one-dimensional LPS and from the SRO state

Figure 3(a) is an *in situ* high temperature pattern. It is clearly indicative of SRO fluctuations above the transition temperature. The long wavelength modulation characteristic of the SRO fluctuations in this alloy is in agreement with earlier observation [20, 21]. The diffraction pattern of the specimen as quenched from 1270 K is shown in Fig. 3(b). At 1270 K no SRO fluctuations are expected to be observable. The fact that they are present in the quenched specimen indicates that a certain amount of spinodal ordering (to be discussed later) has occurred during the quench. The intensity pattern of Fig. 3(b) is similar in character to that of Fig. 3(a) but with an increase in amplitude and sharpening of the satellites flanking the [100] (and equivalent) positions.

Irradiation of the as-quenched specimen at room temperature does *not* produce the long-range ordered structure which, according to the phase diagram of Fig. 1, is the $L1_2$. Rather, the state depicted in Fig. 3(b) essentially persists with a slight increase in satellite intensity and a small decrease in the value of q. This persistence of SRO is typical of all of our specimens at room temperature. The specimens of 18, 20, 22, and 26% Pd all displayed diffuse satellites around the [100] position. The 16% Pd specimen displayed diffuse scattering *at* the [100] position but, at room temperature, the intensity never became sharp enough to allow the structure to be characterized as long-range ordered. These results are in contrast to a previous HVEM study in which a 25% Pd specimen was irradiated at room temperature to produce the long-range ordered one-dimensional superstructure [22]. The difference between that experiment and the present one is in the energy of the electrons and the electron dose rate. Our experiment employed higher energy electrons (1.5 MeV as compared with 1 MeV) and a higher flux rate (an estimated 10^{21} electrons per cm^2 per second as compared with 2×10^{19} electrons per cm^2 per second).

In situ irradiation at 90 K completely destroys the short-range order as seen in Fig. 4(a). This effect is similar to that observed previously in Ni_4Mo by other investigators [22-25]. At this temperature, the high-energy electrons are inducing disorder through several different mechanisms including replacement collision sequences and random point-defect annihilation. The latter process is enhanced by the supersaturation of vacancies. Despite this supersaturation, the mobility of the vacancies is quite low at this temperature, and the *reordering* mechanism is effectively frozen.

When the temperature is increased to 290 K and irradiation resumed, the modulated SRO state reappears as seen in Figs. 4(b) through 4(c). Thus, despite an initial condition of *complete* disorder, the system still evolves toward a SRO state rather than the expected $L1_2$ equilibrium. As in the previous example, where irradiation was begun with the SRO state already present, continued irradiation does not promote LRO but causes some sharpening of and increase in intensity of the satellites accompanied by a small increase in the modulation wavelength, i.e., the satellites move toward the [100] position. The plot of Fig. 5 shows the modulation wave vector vs. time for the specimen irradiated at room temperature. There is also some indication that the value of

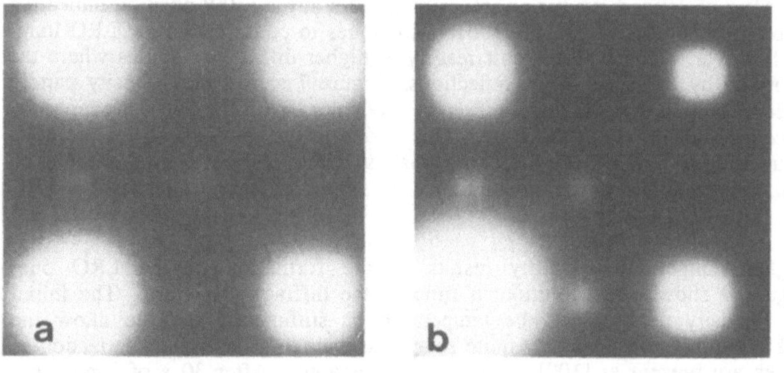

Fig. 3. Cu 20% Pd: (a) equilibrium fluctuations *in situ* at 810 K, (b) as quenched from 1270 K

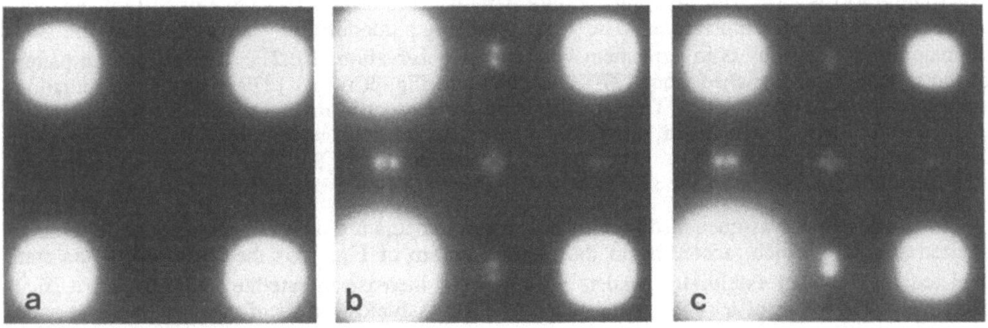

Fig. 4. Cu 20% Pd: (a) after irradiation at 90 K, (b) after 185 s of irradiation at room temperature, (c) after 720 s of irradiation

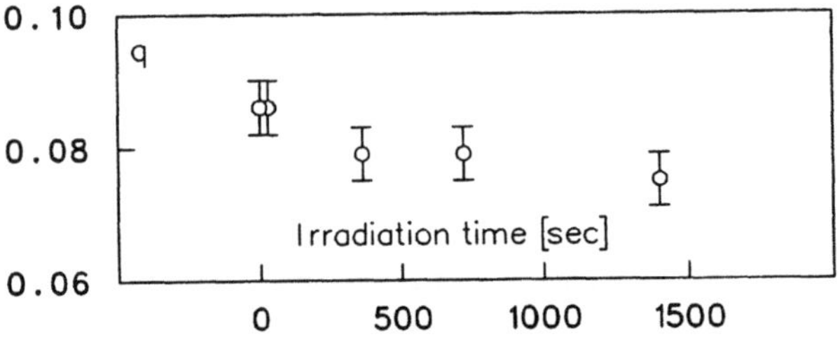

Fig. 5. Cu 20% Pd: wave number q as a function of irradiation time in the SRO state at room temperature

q at which intensity initially appears depends on temperature. A disordered specimen irradiated at 170 K exhibits an initial appearance of intensity at $q = 0.05 \pm 0.01$ as compared with an initial q of about 0.085 at 290 K as shown in Fig. 5.

Irradiation at 620 K produces what appears to be a mixed state. After irradiation at 90 K the specimen was heated to 620 K where weak SRO appears even before irradiation is begun (Fig. 6(a)). Irradiation causes a dramatic increase in the intensity and sharpness of the satellites after only 10 s (Fig. 6(b)). After prolonged irradiation intensity appears at the [100] position indicating that some LRO must be present (Fig. 6(c)). Weak SRO continues to coexist with the LRO long-period superstructure, however. This is seen most readily at higher diffracting angles where the film has not been saturated by the superlattice reflections. Figure 7 shows the intensity pattern around the [300] position. Notice that the diffuse intensity does not lie on the line determined by the sharp satellite reflections arising from the LRO. This is due to the tetragonality of the LRO phase. The broad streaking of the SRO scattering makes determination of its modulation wavelength impractical. Figure 8 shows the modulation wave vector vs. time. The onset of LRO is also indicated.

Irradiation at higher temperatures rapidly results in the formation of the LRO one-dimensional LPS. Figure 9 shows the evolution in time of the diffraction pattern. The initial state already exhibits intensity peaks since the temperature is sufficiently high to allow the occurrence of normal kinetics. These peaks are quite sharp (discouraging their characterization as SRO) although no peaks are present at [100] or equivalent positions. After 30 s of irradiation, intensity appears at [100], indicative of the normal one-dimensional LPS, accompanied by a decrease in the modulation wavelength (i.e., movement of the satellites away from [100]). This is

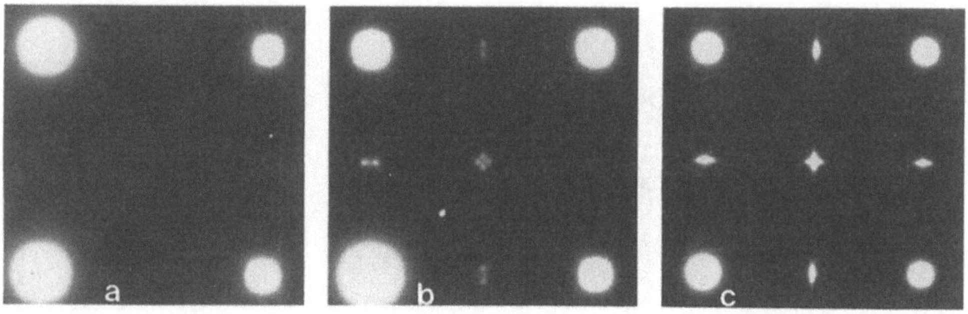

Fig. 6. Cu 20% Pd at 620 K: (a) weak SRO, (b) after 10 s of irradiation, (c) after 1440 s of irradiation

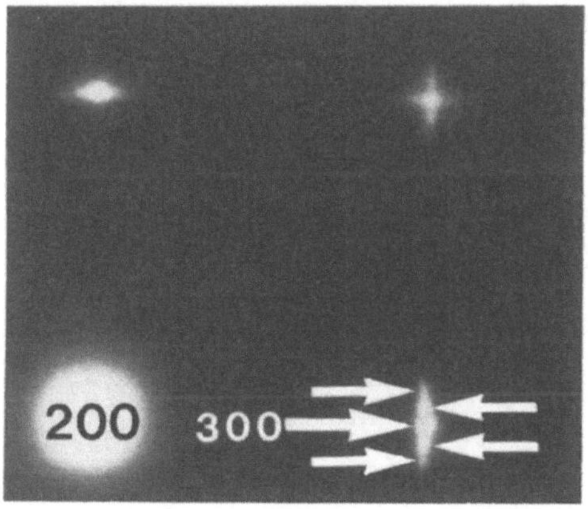

Fig. 7. Intensity surrounding the [300] position in Cu 20% Pd irradiated for 1440 s at 620 K

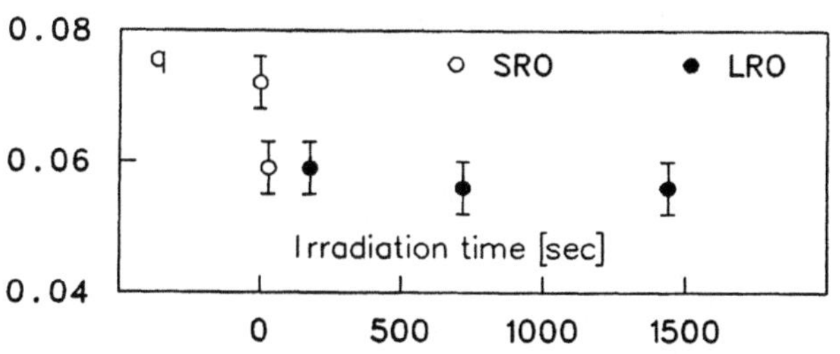

Fig. 8. Cu 20% Pd at 620 K: wave number q as a function of irradiation time

in contrast to the behavior of irradiated SRO at lower temperatures where the modulation wavelength *increases*. The change of wave vector vs. time is shown in Fig. 10.

Next, let us turn our attention to the 16% Pd specimen. In both the *in situ* high-temperature diffraction pattern and the as-quenched pattern no splitting or departure from spherical symmetry is detectable in the intensity at the [100] or [110] positions. If any satellites are present, their separation must be very small as may be seen by extrapolating the data of OHSHIMA and WATANABE [21] or the calculated values of GYORFFY and STOCKS [18]. The as-quenched specimen was disordered, as was the 20% specimen, at 90 K, and the temperature was then increased to room temperature whereupon, after prolonged irradiation, the spots are sufficiently

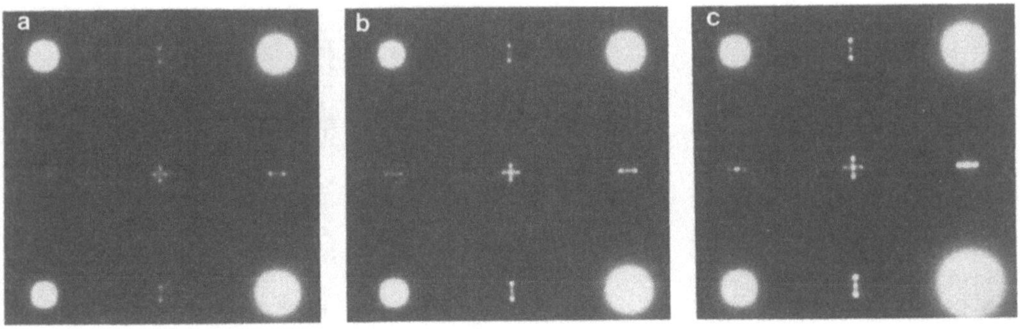

Fig. 9. Cu 20% Pd at 720 K: (a) before irradiation, (b) after 30 s of irradiation, (c) after 720 s

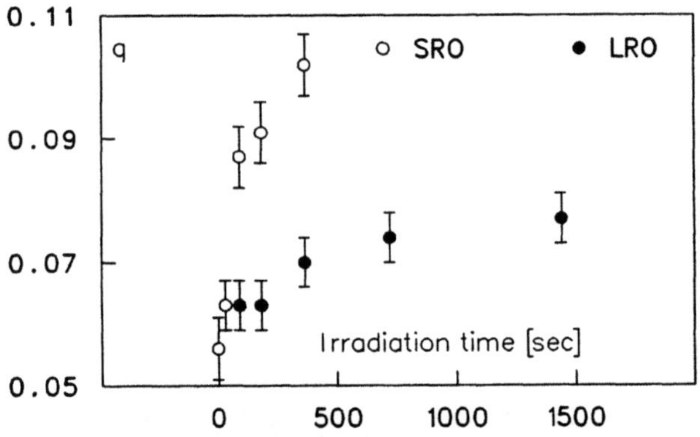

Fig. 10. Cu 20% Pd at 720 K: q vs. irradiation time

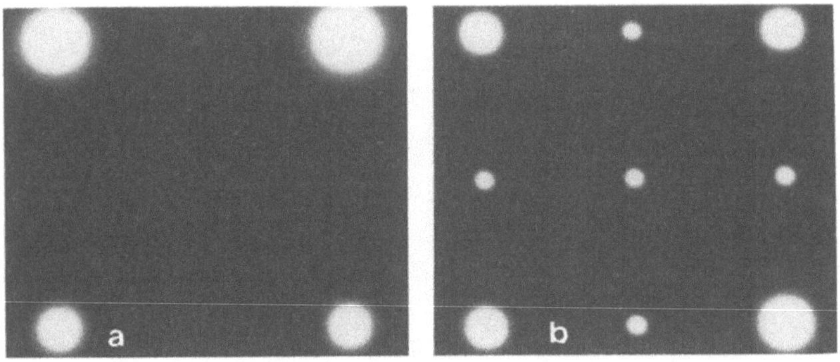

Fig. 11. Cu 16% Pd: steady states at (a) room temperature and (b) 520 K

intense to show clearly that no long wavelength modulation is present. Similarly, at 520 K and at 670 K no modulation is present. Figure 11 shows the steady states at room temperature and at 520 K.

For the 18% Pd specimen, destruction of SRO at 90 K followed by irradiation at room temperature results in the modulated SRO state with q smaller than that at 20%. Similarly, modulated SRO persists at 480 K (Fig. 12(a)). At 630 K, however, sharp superstructure peaks appear

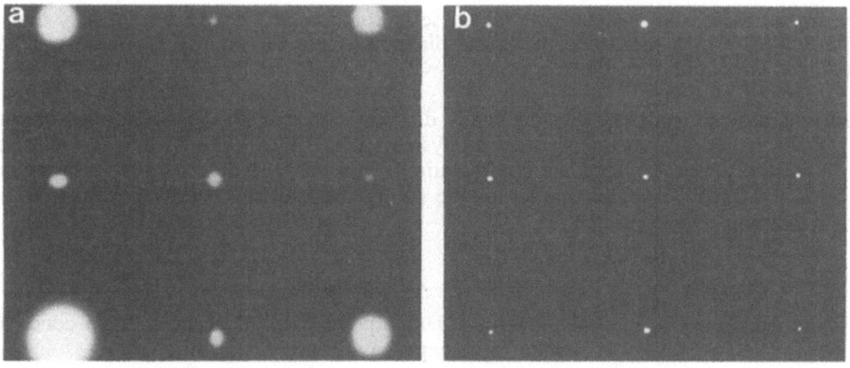

Fig. 12. Cu 18% Pd: steady states at (a) 480 K and (b) 630 K

only at the <100> positions, as seen in Fig. 12(b), indicating *unmodulated* LRO, the $L1_2$ structure, which is the expected equilibrium structure at this temperature. However, room temperature irradiation of the LRO $L1_2$ structure (which had been obtained by a furnace anneal at 623 K) did not result in modulated SRO. The superlattice peaks became broad and diffuse but maintained their spherical symmetry. This observation agrees with the interpretation of the modulated SRO as being spinodally developed since it appears that the system must first be in the disordered state before the modulated SRO appears. (See the discussion of spinodal ordering below.)

For specimens of 22 and 26% Pd, the general tendencies are the same as those of the 20% Pd specimen. Modulated SRO persists at room temperature with the satellite spacing becoming progressively larger as the Pd content increases, while at sufficiently high temperatures the one-dimensional LPS develops. The two-dimensional LPS, expected at 26% Pd, was not observed. This may be due to insufficient time to allow nucleation of this phase.

Before proceeding to the next section to discuss the theory of spinodal ordering, we remark on the possibility that local compositional changes may occur in a specimen as a result of irradiation by a strongly focused high-energy electron beam. Calculations have demonstrated the possibility of this phenomenon [26], and there is some experimental evidence for it as well [27]. We attempted to detect such a possible change of local composition using energy dispersive x-ray analysis. A specimen irradiated in the HVEM was transferred to a JEOL JEM 200CX analytical microscope equipped with an energy dispersive x-ray detector. No change of composition in the irradiated regions was detected, although it is possible that such a change would be below the detectability of such a system which is accurate to about 1%. In this regard, we note that the modulation wave vector of the 22% and 26% specimens does not change under irradiation at room temperature whereas those at 18% and 20% do change. A possible explanation for the change of wave vector in the 18 and 20% specimens is, of course, a local beam induced composition change. However, the lack of change with time in the 22 and 26% specimens argues against the occurrence of a composition change. It seems reasonable to assume that if a local composition change occurs it should do so for all of our specimens. Therefore, we suggest that the change in modulation wave vector in the 18% and 20% specimens is a real thermodynamic effect (possibly induced by the electron beam) and *not* the result of a composition change. Since the rate of change of q with respect to c is approximately constant from about 17 to 30% Pd, one would expect to see a change in wave vector in all the samples under irradiation provided that any possible composition change is of the same order of magnitude for all specimens. (The near constancy of $\partial q/\partial c$ can be deduced from previously published data such as that of OHSHIMA and WATANABE [21].)

3. Theory and Discussion

The present results are explainable, in part, in terms of the phenomenon of spinodal ordering. The theory of spinodal ordering as developed by DE FONTAINE [28, 29] is a theory of early stage kinetics in an unstable disordered solid solution. It should, in a phenomenological sense, explain why, for example, the unstable disordered solid solution of Cu 20% Pd initially evolves toward *modulated* SRO in a region of the phase diagram where the equilibrium ordering wave vector is an *unmodulated* one, i.e., the special point <100>.

The theory of spinodal ordering was originally developed to explain the appearance of peaks of diffuse intensity in rapidly quenched ordering systems [28]. This phenomenon is understood most easily in reciprocal space where the configuration of the alloy is described by a set of concentration waves [30-32]. To begin, one writes the configurational free energy in terms of local concentration deviations

$$\gamma(\mathbf{p}) = c(\mathbf{p}) - \bar{c},$$

where $c(\mathbf{p})$ is a suitably averaged concentration at lattice point $\mathbf{p}$ and $\bar{c}$ is the average concentration of the alloy. The free energy is written as an expansion in powers of the $\gamma(\mathbf{p})$:

$$F(\gamma(\mathbf{p}_1), \gamma(\mathbf{p}_2), \cdots, \gamma(\mathbf{p}_N)) = F_0 + F_1 + F_2 + F_3 + \ldots,$$

where N is the number of lattice sites. The terms F_n are given by

$$F_n = \frac{1}{n!}\left[\sum_{\mathbf{p}} \gamma(\mathbf{p}) \frac{\partial}{\partial \gamma(\mathbf{p})}\right]_0^n F, \tag{1}$$

where the subscript 0 indicates that the derivatives are to be evaluated in the disordered state. Since the disordered state corresponds either to a minimum of F (stable or metastable) or to a saddle point (unstable), then $F_1 = 0$. So we can say

$$\Delta F \equiv F - F_0 = F_2 + F_3 + F_4 + \ldots,$$

where ΔF is the difference in free energies between a given state (described by the set $\{\gamma(\mathbf{p})\}$ for all $\mathbf{p}$) and the completely disordered state. In Fourier space, the preceding expansion contains terms like the following:

$$F_2 = \frac{N}{2}\sum_{\mathbf{k}} f_2(\mathbf{k})\, |\Gamma(\mathbf{k})|^2,$$

$$F_3 = \frac{N}{3!}\sum_{\mathbf{k}_1,\mathbf{k}_2,\mathbf{k}_3} f_3(\mathbf{k}_1,\mathbf{k}_2,\mathbf{k}_3)\,\Gamma(\mathbf{k}_1)\,\Gamma(\mathbf{k}_2)\,\Gamma(\mathbf{k}_3)\,\delta(\mathbf{k}_1+\mathbf{k}_2+\mathbf{k}_3-\mathbf{g}), \tag{2}$$

where the $\{\Gamma(\mathbf{k})\}$ are the Fourier transforms of the $\{\gamma(\mathbf{p})\}$. In other words, $\Gamma(\mathbf{k})$ is the amplitude of the concentration wave of wave vector $\mathbf{k}$. The quantities $f_2(\mathbf{k})$ and $f_3(\mathbf{k}_1,\mathbf{k}_2,\mathbf{k}_3)$ are the Fourier transforms of the second and third derivatives, i.e., the derivatives that arise in the expansion as given in (1). For the third order term, if the three wave vectors $\mathbf{k}_1$, $\mathbf{k}_2$, and $\mathbf{k}_3$ sum to a reciprocal lattice vector $\mathbf{g}$, then the Kronecker delta in (2) equals unity.

It is the term F_2 which will interest us here. According to Landau theory, a second order transition occurs when $f_2(\mathbf{k})$ changes sign for the particular wave vector $\mathbf{k}^0$ at which $f_2(\mathbf{k})$ is a minimum. The temperature for the second order transition is then determined by the condition

$$f_2(\mathbf{k}^0, c, T) = 0, \tag{3}$$

where we have indicated that f_2, in addition to being a function of $\mathbf{k}$, depends on the concentration c and the temperature T. Equation (3) then defines a critical temperature $T_0(c)$ for each concentration c.

The system of interest here, Cu-Pd, undergoes not a second order transition but rather a first order one. In this case, one or more concentration waves combine to give negative contributions to some of the higher order terms $F_3, F_4, \ldots$ in the expansion. Thus, ΔF can change sign at some temperature $T_t > T_0$. In this case, T_0 is no longer a transition temperature but an instability, or *spinodal*, temperature below which the disordered state is unstable with respect to the spontaneous growth of concentration waves with wave vectors $\mathbf{k}^0$ and those obtained from $\mathbf{k}^0$ by the point group symmetry operations of the reciprocal lattice [28, 29]. This growth of concentration waves is what has occurred, for example, in the Cu-Pd specimens quenched from 1270 K. More significantly, it is this same phenomenon which was observed *in situ* in the specimens which had been disordered at 90 K and heated to room temperature where the kinetics (enhanced by the electron beam) allowed the development of the SRO state. This same explanation has also been successfully applied to the growth of the SRO state in irradiated Ni_4Mo [23-25]. That the SRO state below T_0 is qualitatively similar to the fluctuating state above T_t follows because in both cases it is the second order term in the free energy expansion which governs the configuration of the alloy. The SRO state below T_0 is energetically favorable provided the concentration wave amplitudes remain small. (Of course, higher order terms must prevent the unlimited growth of the concentration waves.) Likewise, above T_t the fluctuations in the alloy as measured by the diffracted intensity of x-rays, for example, are described by the fluctuation-dissipation theorem [29]:

$$I_{SRO}(\mathbf{k}) \propto \frac{k_B T}{f_2(\mathbf{k})}.$$

If one uses a Bragg-Williams (mean field) approximation to obtain a form for $f_2(\mathbf{k})$, one obtains the Krivoglaz-Clapp-Moss formula [31, 33].

The quantity $f_2(\mathbf{k})$ must exhibit the symmetry of the f.c.c. lattice and, consequently, must have extrema at certain special points, namely, <000>, <100>, $<1\frac{1}{2}0>$ and $<\frac{1}{2}\frac{1}{2}\frac{1}{2}>$ [28]. Often, the absolute minimum of $f_2(\mathbf{k})$ will be at one of these special points as it is, for example, in Ni_4Mo [23-25]. What distinguishes Cu-Pd in this regard is that for Pd content greater than about 16% $f_2(\mathbf{k})$ presents minima *away* from the special point [100] in the Brillouin zone; it is at the $[q\,10]$ and equivalent positions that the diffuse intensity appears as the disordered state evolves into the SRO state. In analogy to the definition of a Lifshitz point in modulated systems with second order transitions [32], we can define and attempt to locate a *metastable* Lifshitz point [16, 34]. Such a point is defined as follows: the metastable Lifshitz point L divides an ordering instability (spinodal) line into two segments on one of which the instability occurs at a special point (the [100] point in this example) and on the second of which the instability occurs at a wave vector $\mathbf{k}$ whose coordinates (in this example, $[q\,10]$) vary as some other system parameter (concentration) is varied. The point L is also the terminus of a second line which separates the phase diagram into two regions, one in which f_2 is minimized at the special point corresponding to the previously mentioned spinodal (i.e., [100] in this case) and the second in which f_2 is minimized at a point removed from the special point by some vector $\mathbf{q}$ (in this case $\mathbf{q} = [q\,00]$). In the Bragg-Williams approximation this second line must be vertical.

Any $\mathbf{k}$-space function such as $f_2(\mathbf{k}, c, T)$ having the required f.c.c. symmetry can be expanded in a sum of "shell functions" (one for each coordination shell). At the point $\mathbf{k} = [q\,10]$, for arbitrary q, if we make the definition

$$f_2(\mathbf{k}, c, T)\,|_{\mathbf{k}=[q\,10]} \equiv \Phi(q, c, T),$$

this expansion reduces to the following cosine series [35]:

$$\Phi(q, c, T) = \omega_0(c, T) + \sum_n \omega_n(c, T) \cos 2\pi n q. \tag{4}$$

In a Bragg-Williams approximation, the ω_n would be independent of T. Note that if the series is truncated at $n = 2$, one has the f.c.c. ordering analog of the ANNNI model. A reasonable model for Cu-Pd, however, would likely require terms beyond $n = 2$ [11, 15].

From (4), it is apparent that if we expand $\Phi(q, c, T)$ in a Taylor series about $q = 0$, only even powers of q appear:

$$\Phi(q,c,T) = \Phi_0 + \Phi_2 q^2 + \Phi_4 q^4 + \ldots \quad (5)$$

The Φ_n in (5) are linear combinations of the ω_n appearing in (4). It was shown elsewhere [13] that Φ_4 will most likely be positive so that when examining (5) one can neglect terms of order 6 and higher. (The expansion is uninteresting, of course, unless Φ_2 changes sign as c is varied.) Recall that the instability occurs when the minimum of f_2 changes sign. The minima of $\Phi(q,c,T)$ are at

$$q = 0 \qquad \text{for } \Phi_2 > 0,$$

$$q = \pm\sqrt{-\Phi_2/2\Phi_4} \quad \text{for } \Phi_2 < 0.$$

The values of Φ at these points are

$$\Phi_{\min}(c,T) = \begin{cases} \Phi_0 & \text{for } \Phi_2 > 0. \\ \Phi_0 - \frac{1}{4}\dfrac{\Phi_2^2}{\Phi_4} & \text{for } \Phi_2 < 0. \end{cases}$$

Thus the spinodal lines in the (c,T) plane on either side of the metastable Lifshitz point L are given by the conditions $\Phi_0 = 0$ for for the special point instability and $\Phi_2^2 - 4\Phi_0\Phi_4 = 0$ for the non-special point instability. The second of these two equations is valid only in the vicinity of the metastable Lifshitz point. In general, we can write simply $\Phi_{\min} = 0$ as the condition which determines the instability. Point L then has coordinates (c_L, T_L) determined by

$$\Phi_0(c_L,T_L) = \Phi_2(c_L,T_L) = 0,$$

which is the intersection of the special point spinodal and the non-special point spinodal.

Figure 1 shows estimates for the three metastable loci,

$$\Phi_0 = 0,$$

$$\Phi_{\min} = 0 \quad \text{for } \Phi_2 < 0,$$

$$\Phi_2 = 0,$$

indicated as dashed lines. The metastable Lifshitz point has been labeled as L. The loci have been drawn in a manner which best accounts for the available data and which is consistent with general theoretical knowledge of alloy phase diagrams and instabilities. The most significant result of the present data is the position of the $\Phi_2 = 0$ instability which is rather far to the left of the two-phase region which separates the α' and α_1 phases. The present data are lacking in detail since no composition between 16% and 18% Pd was examined, but at room temperature and up to about 520 K this instability must lie at or near 17% Pd since at these temperatures the SRO steady state under irradiation appears to be unmodulated for the 16% composition but modulated for the 18% composition. As mentioned above, for a GBW approximation the Φ_2 line would be vertical. Here, it was drawn curved for the following reasons. At low temperatures (room temperature to about 520 K) its position was estimated as just described. On the other hand, extrapolation of previous data [21] and first principals calculations [18] indicates that high temperature SRO fluctuations remain modulated for concentrations of Pd as low as 16% or even slightly lower, although this conclusion is not clear from the present data. Furthermore, as mentioned previously, we have seen some indication that the value of q at which the instability appears may decrease with decreasing T. Also, CVM calculations for two-dimensional models [36] clearly yield curved Φ_2 lines.

As for the other two instabilities, $\Phi_0 = 0$ and $\Phi_{\min} = 0$, they have been drawn close to the equilibrium first order transitions because experimental evidence [10] indicates very narrow two phase regions, so that the postulated peritectoids must be close to being critical endpoints through which instability lines must pass [37]. It must be emphasized, however, that the location of instability lines cannot be determined rigorously by experiment because the temperature at which the system becomes unstable to some set of concentration waves must depend upon the past history of the specimen, unlike an equilibrium transition which is independent of history. Nevertheless, approximate determination of stability limits is useful in understanding the behavior of quenched or irradiated solid solutions, as is clear from the present discussion.

4. Conclusion

We have reported the serendipitous discovery that room temperature irradiation of a completely disordered solid solution could produce steady state modulated SRO for Cu-Pd samples in regions of the phase diagram where only simple $L1_2$ ordering could be found at equilibrium. The authors believe this to be the first observation of its kind ever reported.

This effect could be explained qualitatively by applying spinodal ordering ideas to irradiated systems and by extending the model "beyond the Lifshitz point," i.e., to cases for which the harmonic coefficient f_2 of the free energy possesses minima away from the special points of high symmetry. These generalizations lead to the notion of special point (commensurate) and non-special point (incommensurate) stability limits and of a metastable Lifshitz point.

The experimental portion of this investigation was made possible by use of a high voltage TEM with which it is possible to study, *in situ*, various types of equilibria: (a) high-temperature equilibrium SRO fluctuations, (b) equilibrium LRO at various temperatures and (c) steady state SRO under irradiation at low temperatures. In addition, high-energy electron irradiation at very low temperatures may be, in some cases, the best (or perhaps only) way to obtain *completely* disordered systems.

Acknowledgements

The authors are grateful to Drs. A. Finel, P. Turchi and G.M. Stocks for helpful conversations. TEM work was performed on the 1.5 MeV microscope at the National Center for Electron Microscopy, Lawrence Berkeley Laboratory. The authors are indebted to Dr. K. Westmacott for his support and interest and to D. Ackland for invaluable technical help. One of us (S.T.) benefited from an MMRD collaborative fellowship. This work was supported by the Director, Office of Energy Research, Office of Basic Energy Sciences, Materials Sciences Division of the U.S. Department of Energy under Contract No. DE-AC03-76SF00098.

References

1. K. Schubert, B. Kiefer, M. Wilkens, R. Haufler: Z. Metallkd. **46**, 692 (1955)
2. M. Hirabayashi, S. Ogawa: J. Phys. Soc. Japan **12**, 259 (1957)
3. K. Okamura: J. Phys. Soc. Japan **28**, 1005 (1970)
4. O. Michikami, H. Iwasaki, S. Ogawa: J. Phys. Soc. Japan **31**, 956 (1971)
5. A. Souter, A. Colson, J. Hertz: Mem. Sci. Rev. Met. **68**, 575 (1971)
6. D. Watanabe, S. Ogawa: J. Phys. Soc. Japan **11**, 226 (1956)
7. M. Guymont, D. Gratias: Phys. stat. sol. (a) **36**, 329 (1976)
8. M. Guymont, R. Portier, D. Gratias: In *Electron Microscopy and Analysis 1981*, ed. by M.J. Goringe, Inst. Phys. Conf. Ser. No. 61 (Inst. Phys., Bristol and London 1981) p. 387
9. O. Terasaki, D. Watanabe: Jpn. J. Appl. Phys. **20**, 1381 (1981)
10. P.R. Subramanian, D.E. Laughlin: Bulletin of Alloy Phase Diagrams (to be published)
11. D. Broddin, G. van Tendeloo, J. van Landuyt, S. Amelinckx, R. Portier, M. Guymont, A. Loiseau: Phil. Mag. **A54**, 395 (1986)
12. J. Kulik, D. de Fontaine: In *Phase Transformations in Solids*, ed. by T. Tsakalakos, Materials Research Society Symposia Proceedings vol. 21 (North-Holland, NY 1984) p. 225
13. D. de Fontaine, J. Kulik: Acta metall. **33**, 145 (1985)
14. A. Loiseau, G. van Tendeloo, R. Portier, F. Ducastelle: J. de Physique **46**, 595 (1985)

15. D. de Fontaine, A. Finel, S. Takeda, J. Kulik: In *Noble Metal Alloys*, ed. by T. B. Massalski, W.B. Pearson, L.H. Bennett and Y.A. Chang (The Metallurgical Society, Inc., Warrendale, PA 1986) p. 49
16. J. Kulik, S. Takeda, D. de Fontaine: Acta metall. **35**, 1137 (1987)
17. H. Sato, R.S. Toth: In *Alloying Behavior and Effects in Concentrated Solid Solutions*, ed. by T.B. Massalski, Metallurgical Society Conferences vol. 29 (Gordon and Breach, NY 1965) p. 295
18. B.L. Gyorffy, G.M. Stocks: Phys. Rev. Letters **50**, 374 (1983)
19. K. Urban, S. Banerjee, J. Mayer: In *Phase Stability and Phase Transformations*, ed. by R. Krishnan, S. Banerjee and P. Mukhopadhyay, Materials Science Forum vol. 3 (Trans Tech Publications, Switzerland 1985) p. 335
20. K. Ohshima, D. Watanabe, J. Harada: Acta Crystallogr. A**32**, 883 (1976)
21. K. Ohshima, D. Watanabe: Acta Crystallogr. A**29**, 520 (1973)
22. G. van Tendeloo, S. Amelinckx: J. de microscopie et de spectroscopie electronique **6**, 371 (1981)
23. S. Banerjee, K. Urban, M. Wilkens: Acta metall. **32**, 299 (1984)
24. S. Banerjee, K. Urban: Phys. stat. sol. (a) **81**, 145 (1984)
25. J. Mayer, K. Urban: Acta metall. (to be published)
26. P.R. Okamoto, N.Q. Lam: In *Advanced Photon and Particle Techniques for the Characterization of Defects in Solids*, ed. by J.B. Roberto, R.W. Carpenter and M.C. Wittels, Mat. Res. Soc. Symposium Proceedings vol. 41 (Materials Research Society, 1985) p. 241
27. N.Q. Lam, G. K. Leaf, M. Minkoff: J. Nucl. Mater. **118**, 248 (1983)
28. D. de Fontaine: Acta metall. **23**, 553 (1975)
29. D. de Fontaine: Solid State Physics **34**, 73 (1979)
30. L.D. Landau, E.M. Lifshitz: *Statistical Physics* (Addison- Wesley, Reading, MA 1958)
31. M.A. Krivoglaz: *The Theory of X-ray and Thermal Neutron Scattering from Real Crystals* (Plenum, NY 1969)
32. A.G. Khachaturyan: Phys. stat. sol. (b) **60**, 9 (1973)
33. P.C. Clapp, S.C. Moss: Phys. Rev. **142**, 418 (1966)
34. R.M. Hornreich, M. Luban, S. Shtrikman: Phys. Rev. B**19**, 3799 (1979)
35. S. Takeda, J. Kulik, D. de Fontaine: Acta metall. (to be published)
36. A. Finel, D. de Fontaine: J. Stat. Physics **43**, 645 (1986)
37. S.M. Allen, J.W. Cahn: In *Alloy Phase Diagrams*, ed. by L.H. Bennett, T.B. Massalski and B.C. Giessen (North Holland, NY 1983) p. 195

Existence and Formation of LPAPB Structures in Pt_3V Studied Using High Resolution Electron Microscopy

D. Schryvers[1;3], *G. Van Tendeloo*[1], *and S. Amelinckx*[2]

[1]On leave from the University of Antwerp, RUCA, Groenenborgerlaan 171, B-2020, Antwerpen, Belgium
[2]University of Antwerp, also at SCK, B-2400, Boeretang, Mol, Belgium
[3]NCEM, Lawrence Berkeley Laboratory, Berkeley, CA 94720, USA under contract to Lawrence Livermore National Laboratory

1. INTRODUCTION

In recent years microstructural studies of ordered binary alloys have gained renewed attention due to the quest for adequate test systems for different calculation methods that seek to determine the phase diagram or at least the ordered ground state structures for a given system or for a set of different systems. A few examples are the Cu-Pd [1], Al-Ti [2] and Al-Mg [3] systems as described by the ANNNI model [4], the Cu-Au system as studied using the cluster variation method (CVM) [5], and different transition metal alloys (TMA) such as Pt-Ti [6,7] and Pt-V [7,8] investigated in terms of the generalized perturbation method (GPM) [9,10]. Since in many of these ordered alloys the separation between the stability regions of different microstructures with respect to composition, homogeniety and temperature is very crucial for the relation with the various calculations, a number of different experimental techniques, macroscopical as well as microscopical, have been applied. Here the high resolution electron microscopy (HREM) technique is a very powerful tool that can be used to solve different problems, such as the unravelling of complex diffraction patterns resulting from many different orientation variants and multiphase situations [11], the study of microstructural defects such as anti-phase boundaries (APB) and of several transformation mechanisms between different ordered structures down to the atomic level [8,12].

In the present paper the results of a HREM study of the different ordered phases that are stable at different temperatures in the Pt_3V alloy are presented. The existence of these phases will then be discussed briefly in terms of the calculation of the ordering energy using the Ising model and interaction parameters calculated by Bieber and Gautier [9]. The experimental results will be discussed in decreasing annealing temperature.

2. MICRODOMAINS

When the material is annealed at 1100°C, i.e., above the order-disorder (OD) transition temperature for one day and rapidly quenched in water at room temperature, images such as the one presented in Fig. 1 can be seen. Although some white dot constrast is revealed in this micrograph, clearly no long-range ordered (LRO) structures have developed yet. The present image is produced by including all diffracted intensity inside the objective aperture, as indicated by the white circle in the inset in Fig. 1. From the strong Bragg peaks one can conclude that the material has an FCC basic lattice (here viewed in a [100] orientation) while the diffuse intensity indicates that the present thermal treatment results in some short-range order (SRO). When the location of this

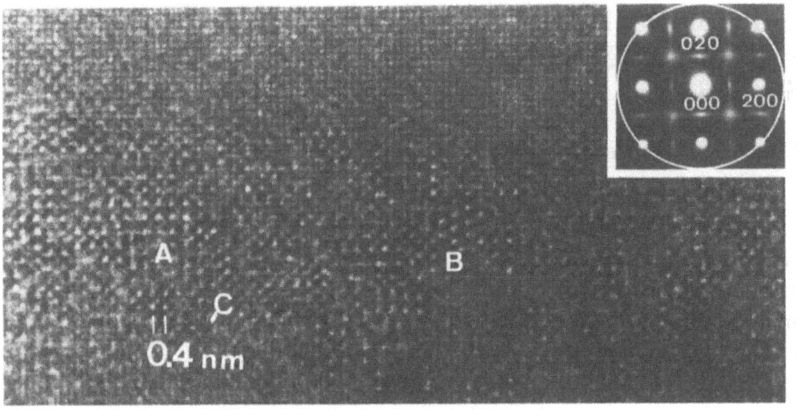

Fig. 1 HR micrograph of a Pt_3V sample quenched from 1100°C revealing microdomains of $L1_2$ (A) and DO_{22} (B). The inset shows the corresponding [100] electron diffraction pattern. The white circle indicates the used objective aperture

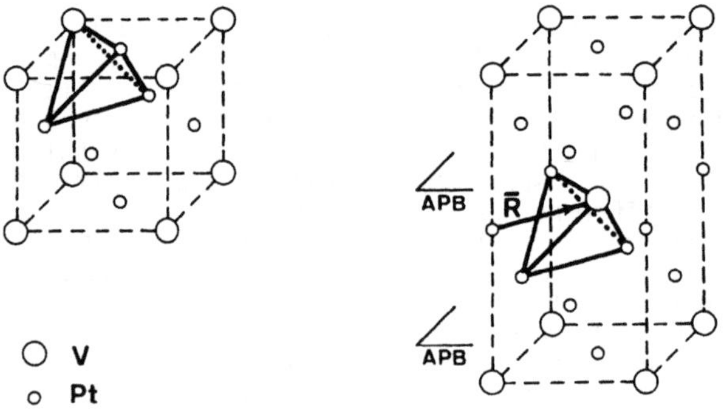

Fig. 2 Schematic representation of the samll tetrahedron with 3:1 composition with respect to the $L1_2$ (left) and DO_{22} (right) unit cell

diffuse intensity is interpreted using kinematical diffraction theory [8,13], it can be concluded that ordered tetrahedrons with the 3:1 composition will exist in the material. Such tetrahedrons are depicted in Fig. 2 with respect to the $L1_2$ and DO_{22} structures. These structures are column structures when viewed along the [100] direction; i.e., along each atomic column only one type of atoms is seen, Pt or V. When such structures are viewed in the HREM along this particular direction, they can show up as a pattern of white dots under suitable experimental conditions, i.e. defocus, thickness, etc.. The geometry of this pattern will represent the geometry of the minority atom columns, in this case V columns. For the $L1_2$ structure a square pattern will be recognized while for DO_{22} the white dots will be arranged in a triangular pattern. Both types of geometries are revealed in the HR image of Fig. 1: at A a microdomain having the $L1_2$ structure is seen, while at B the triangles representing half the DO_{22} unit cell can be recognized. It is believed that the rest of the material, not showing such clear ordered patches, is nevertheless also ordered in microdomains; overlapping of different microdomains and the positioning at different heights in the sample can result in lower contrast for several of these microdomains. This microdomain state is the only evidence that was found for the existence of $L1_2$ regions; no long-range ordered $L1_2$ domains could be produced by annealing the sample below the OD transition temperature.

3. LONG RANGE ORDERED STRUCTURES

When the sample is annealed for a few hours at 960°C or at 920°C the structures schematically depicted in Figs. 3a and 3b are formed respectively. The drawings show two-dimensional projections along the [100] orientation: both structures are one FCC unit cell thick. Again these structures are column structures and their respective HR images are shown in Figs. 4a and 4b. These structures can be described using $L1_2$ and DO_{22} unit cells as building blocks. Indeed, when one follows the images along an imaginary horizontal line from left to right, one encounters a long period sequence of one $L1_2$ and one DO_{22} unit cell (i.e., one square and two triangles) for the 960°C phase and one $L1_2$ and one

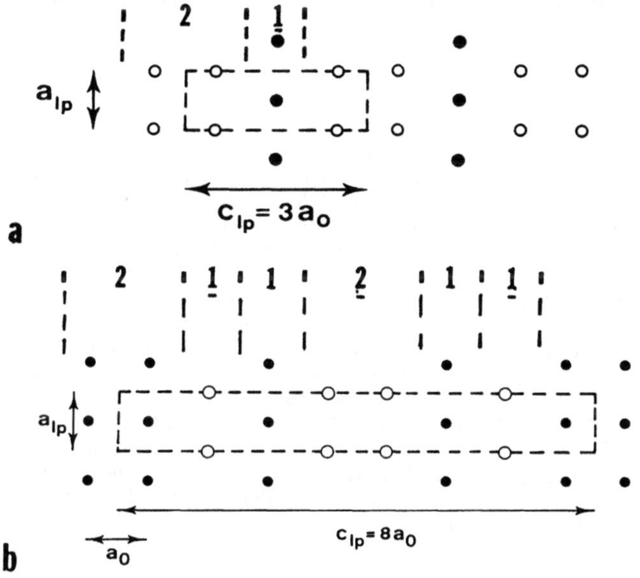

Fig. 3 [100] projections of the LPAPB phases as found in Pt_3V at a) 960°C $(2\bar{1})$ and b) 920°C $(2\bar{1}1\bar{2}1\bar{1})$. Only V atoms are shown

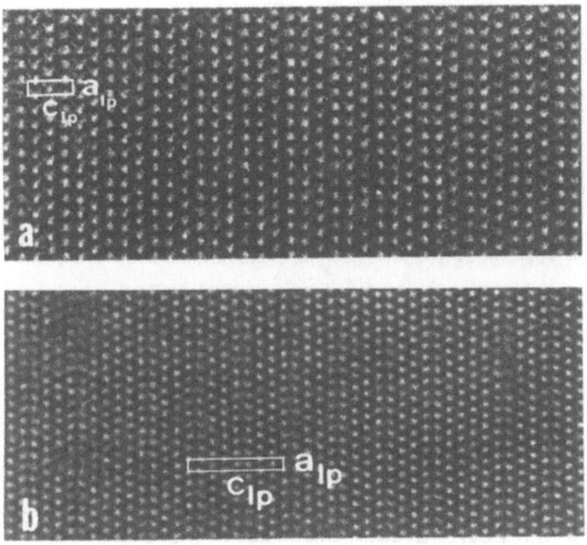

Fig. 4 IH images of the a) $(2\bar{1})$ and b) $(2\bar{1}1\bar{2}1\bar{1})$ LPAPB phases. The different sequences of $L1_2$ (squares) and DO_{22} (triangles) building blocks is clearly

and a half DO_{22} unit cell (i.e., one square and three triangles) for the 920°C phase. In the conventional description where the DO_{22} phase is regarded as a long period antiphase boundary (LPAPB) structure with respect to the $L1_2$ phase, these long period ordered structures can be denoted as $(\underline{21})$ and $(\overline{2}1\underline{1}2\overline{1}1)$ respectively, where 2 and 1 indicate the number of FCC unit cells in between two successive APB's. This is also indicated in Fig. 3. The M-value of the 960°C phase is 3/2, while that of the 920°C phase is 4/3.

It can thus be concluded that, by decreasing temperature, the long period ordered phases in the Pt_3V system will accommodate more and more DO_{22} unit cells and eventually, below 900°C, the material will become entirely ordered with the DO_{22} structure. This behavior suggests a competition between a second nearest neighbor pair interaction favoring the DO_{22} structure and an entropy effect tending to form long periodic structures.

4. TRANSFORMATION MECHANISMS

When samples that are ordered in one of the long period phases are further annealed below 900°C, the DO_{22} structure finally forms; the $L1_2$ layers have thus to disappear from the structure. From static HREM images two distinct mechanisms can be distinguished: they are recognized from their frozen-in situations as formed during the quenching. In the first mechanism, two adjacent $L1_2$ layers form a hairpin as indicated by arrows 1, 2 and 3 in Fig. 5. If the $L1_2$ layers are now interpreted as APB's with respect to the low-temperature DO_{22} structure, the hairpin formation results in a sum of two displacement vectors $R=(1/2)[110]$ (fcc units), equalling [110] which is a lattice vector of the ordered DO_{22} region; thus only the DO_{22} phase remains after the withdrawal of a hairpin. A few single APB's ($L1_2$ layers) will then remain in the DO_{22} domain. These APB's can move towards one another by means of ledges; at each ledge the habit plane of the APB is shifted over one fcc unit cell in the direction of the long period, i.e., perpendicular to the habit plane. This is seen at the small inclined arrows in Fig. 5. When two APB's come into close range, they can disappear by the hairpin mechanism.

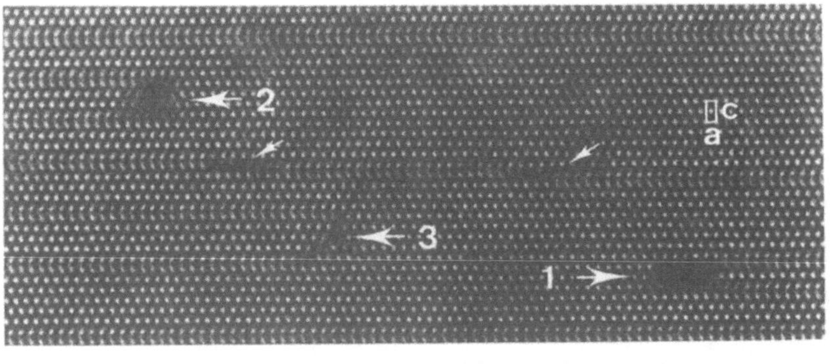

Fig. 5 HR image of a frozen-in situation of the transformation mechanism when a LP structure develops into the low-temperature DO_{22} structure (see also text)

5. CONCLUSIONS

From an extended investigation on the application of the Ising model for the calculation of the ordering energy and on the determination of the interaction parameters in the case of TMA such as Pt_3V, Bieber and Gautier [9] were able to conclude that the ordering energy can be written as a sum of effective pair interactions up to the fourth nearest neighbor, provided the interactions are a function of the composition. On the other hand, from the present experimental results it is clear that there exists an interesting competition between interaction forces favoring the $L1_2$ or the DO_{22} structure in this material. The energy difference between both structures as calculated in the Ising approach including the fourth nearest neighbor can be written as $(1/4)V_2 - V_3 + V_4$ where V_i is the i-th pair interaction parameter. This energy difference is also often called the APB energy and in the case of Pt_3V it becomes nearly zero (approximately $(1/2)V_4$, calculated using a tight-binding coherent potential approximation (TB-CPA) [14,15]). Small increases in annealing temperature can then result in the formation of different LPAPB structures of the type described above. Indeed, the two LPAPB phases are the same as the first two appearing at low temperature in the staircase in $Al_{72}Ti_{28}$ studied by A. Loiseau et al. [2]. Since the latter was used as experimental evidence for phase diagram calculations using the ANNNI model, we believe that the existence of similar structures in Pt_3V can also be of interest with regard to theoretical calculations involving anisotropic interactions. In the present case, however, the small temperature range (~100°) prohibits one from performing a large amount of different heat treatments, possibly resulting in more complicated LPAPB phases.

Work performed under the auspices of the association RUCA-SCK and with financial support from the IIKW.

References

1. D. Broddin, G. Van Tendeloo, J. Van Landuyt, S. Amelinckx, R. Portier, M. Guymont and A. Loiseau, Phil. Mag. A, 54, 395 (1986). S. Takeda, J. Kulik and D. de Fontaine, Acta Met. (in press).
2. A. Loiseau, G. Van Tendeloo, R. Portier and F. Ducastelle, J. de Physique 46, 595 (1985).
3. J. Kulik, S. Takeda and D. de Fontaine, Acta Met. 35, 1137 (1987).
4. D. de Fontaine and J. Kulik, Acta Met. 33, 145 (1985).
5. J. M. Sanchez and D. de Fontaine, Phys. Rev. B17, 2926 (1978); ibid B21, 216 (1980).
6. D. Schryvers, J. Van Landuyt, G. Van Tendeloo and S. Amelinckx, phys. stat. sol. (a) 76, 575 (1983).
7. D. Schryvers and S. Amelinckx, Res. Mech. (in press).
8. D. Schryvers, G. Van Tendeloo and S. Amelinckx, Acta Met. 34, 43 (1986).
9. A. Bieber and F. Gautier, Acta Met. 34, 2291 (1986).
10. D. Schryvers and D. Van Dyck, Phil. Mag. A (in press).
11. S. Amelinckx, G. Van Tendeloo and J. Van Landuyt, Ultramicroscopy 18, 395 (1985).
12. D. Schryvers, G. Van Tendeloo and S. Amelinckx, phys. stat. sol. (a) 87, 401 (1985).
13. R. De Ridder, G. Van Tendeloo and S. Amelinckx, Acta Cryst. A32, 216 (1976).
14. A. Bieber, F. Ducastelle, F. Gautier and G. Treglia, Sol. St. Comm. 45, 585 (1983).
15. P. Turchi, G. Treglia, A. Bieber, F. Gautier and F. Ducastelle, private communications.

Dense Packings of Hard Spheres

J. Villain, K.Y. Szeto, B. Minchau, and W. Renz

Institut für Festkörperforschung der Kernforschungsanlage Jülich, Postfach 1913, D-5170 Jülich, Fed. Rep. of Germany

In this contribution we investigate some properties of densely packed hard spheres. In the first section we define what we call hard spheres and what we call densely packed (the packing should be locally dense, in contrast to a ~~fcc~~ lattice for instance). In the second section the case of identical spheres is recalled. The next three sections are devoted to mixtures of two different species. The two-dimensional theory is given in Section 3, and applied to quasi-crystals in Section 4. The three-dimensional case is treated in Section 5.

1. Hard core interactions

One of the most fundamental problems in solid state physics is the determination of the ground state of a system of many atoms. This problem will be addressed here within the crudest possible approximations. Firstly quantum effects will be neglected, so that the ith atom has a well-defined position $\vec{r}_i$. Secondly, pair interactions between atoms will be assumed:

$$\mathcal{H} = \Sigma \, V_{ij} \, (|\vec{r}_i - \vec{r}_j|) \; . \tag{1}$$

Here we shall only consider the simplified case where each interaction is the sum of a hard core repulsion and a contact attraction

$$V_{ij}(r) = \infty \qquad (r < d_{ij}) \tag{2a}$$
$$V_{ij}(d_{ij}) = -E_{ij} < 0 \tag{2b}$$
$$V_{ij}(r) = 0 \qquad (r > d_{ij}) \; . \tag{2c}$$

This potential mimicks a Lennard-Jones interaction. It can in priciple not be used for ionic crystals. Models of this type (i.e. without elasticity) are often used, e.g. in the Eden model [1], diffusion limited aggregation [2], or in related problems (SOS model [3]). A solid described by (2) would dissociate at any finite temperature, but at T = 0 (and of course in the classical approximation) the model makes sense. Its advantage is that it allows one to replace the mechanical problem of minimizing (1) by a geometrical problem which is hopefully equivalent, namely the determination of dense packings of hard spheres. This geometrical problem is the one which will be addressed here in the case of two kinds of hard spheres, denoted B and s for big and small. There are therefore 3 distances d_{BB}, d_{ss}, d_{Bs}, but only 2 reduced geometrical parameters

$$x = d_{ss}/d_{Bs} \; , \qquad Y = d_{BB}/d_{Bs} \; .$$

Note that we do <u>not</u> impose $x + Y = 2$ as would be the case for true hard spheres. We would like, however, $x + Y$ to be close to 2.

Whether the densest packings do minimize $\mathcal{H}$ or not depends on the energies E_{BB}, E_{Bs} and E_{ss}, or rather on the reduced parameters

$$E_B = E_{BB}/E_{ss} , \qquad E_s = E_{ss}/E_{Bs} .$$

So, there <u>is</u> a mechanical problem to solve. This will not be done here, as we will be satisfied with the plausible assumption that densest packings do minimize $\mathcal{H}$ for some values of E_B and E_s.

The definition of the problem should be completed by a definition of what we call dense packings. A packing of hard spheres is called dense in the present work if the space is filled with tetrahedra formed by mutually touching nearest neighbours. The importance of this definition will become clearer the next section, where the well-known case of identical spheres will be recalled.

2. Systems of identical hard spheres

Let us first recall the two-dimensional case. The closest packing of hard circles is, of course, a triangular lattice and is a dense packing of equilateral triangles. This has important consequences:

i) There is only one close-packed lattice in 2 dimensions.
ii) Any finite, two-dimensional close-packed cluster is a subset of an infinite triangular lattice.
iii) Any growth model of hard spheres in two dimensions therefore generates a close-packed, triangular lattice. This lattice contains, in general, vacancies, but no dislocations.

The situation in 3 dimensions is different. Dense packings of regular tetrahedra (in the sense defined in section 1) do not exist because the dihedral angle of a regular tetrahedron:

$$\gamma_o = \cos^{-1}(1/3) \simeq 70{\cdot}5^o \tag{3}$$

is not a submultiple of 2π (Fig. 1).

The so-called close-packed lattices (fcc, hexagonal compact, etc) are the densest possible lattices if they are infinite (or with periodic boundary conditions). This follows from the fact that each atom has 12 neighbours and that 12 is the maximum. On the other hand, small clusters in their ground state are not subsets of one of these so-called close-packed lattices. This can be seen for a cluster of 6 atoms. A subset of 6 atoms in a fcc or hcp lattice cannot have more than 11 bonds, while a cluster a 6 hard spheres forming 3 regular tetrahedra has 12 bonds (Fig. 1) and can easily be seen to be the ground state for 6 spheres. Thus, the property (ii), which in the two-dimensional case could be deduced from the dense nature of the lattice, does not hold in three dimensions. Properties (i) and (iii) do not hold either: there is an infinity of three--dimensional close-packed lattices, obtained by stacking layers of triangular lattices. A growth model of identical hard spheres generally develops dislocations or stacking faults.

With 2 kinds of hard spheres it <u>is</u> possible to form a dense packing. A simple example is the NaCl structure (Fig. 2) which may be regarded as formed by Cl_4 and Cl_3Na tetrahedra. More complicated examples will be found in section 4, but the two-dimensional case will be treated first. It will be seen to have many solutions including quasi-crystalline ones.

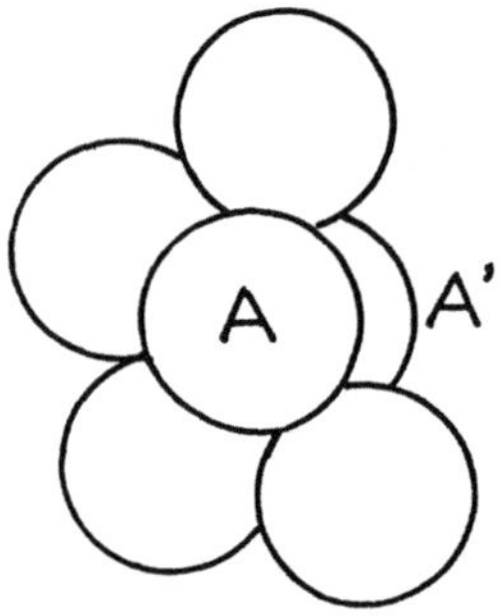

Fig. 1: The densest structure of 6 identical pingpong balls (i.e., that with the maximum number of bonds) is formed by 3 tetrahedra with a common edge AA'. Adding more balls will produce vacancies and cannot lead to a fcc of hcp lattice.

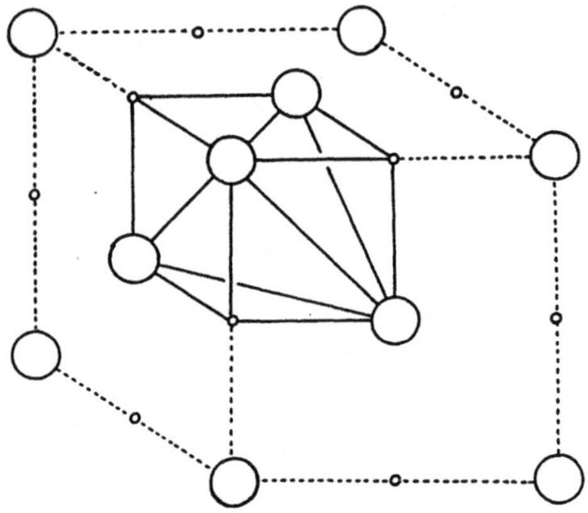

Fig. 2: The NaCl structure may be regarded as formed by Cl_4 and Cl_3Na tetrahedra (full lines). Big, open circles are Cl ions, dots are Na ions. Cl and Na ions correspond respectively to B and s spheres.

3. Two-dimensional arrays of hard circles of two different kinds.

The question we address now is whether it is possible to choose x and Y such that the two-dimensional plane can be filled by triangles formed by nearest neighbours. There are 4 sorts of triangles, BBB, sss, BBs and Bss. The last two triangles have angles $\hat{B}$, $\hat{S} = \pi - 2\hat{B}$, $\hat{s}$ and $\hat{b} = \pi - 2\hat{s}$ respectively, and these angles are functions of x and Y.

$$\sin\frac{\hat{S}}{2} = \cos\hat{B} = Y/2 \qquad (4a) \; ; \qquad \sin\frac{\hat{b}}{2} = \cos\hat{s} = x/2 \; . \qquad (4b)$$

Necessary conditions for dense packing to be possible are obtained if one writes that each s atom is common to p(sBB) triangles, 2p'(ssB) triangles and p"(sss) triangles, and that each B atom is common to q (Bss) triangles, 2q'(BBs) triangles and q"(BBB) triangles

$$p\,\hat{b} + 2p'\,\hat{B} + p''\,\frac{\pi}{3} = 2\pi \quad (5a); \quad q\,\hat{S} + 2q'\,\hat{s} + q''\,\frac{\pi}{3} = 2\pi \; . \qquad (5b)$$

Remembering that "s" atoms are small and that "B" atoms are big, it is reasonable to assume

$$\frac{\pi}{2} > \hat{s} > \pi/3 \qquad (6a) \qquad\qquad \pi > \hat{S} > \pi/3 \qquad (6b)$$

so that there is a finite number of sets {q,q',q"} = 0 allowed by (5b) and (6). They are listed in table 1.

Assuming (5b) to be satisfied by a single set Q for given x and Y (apart from Q = [006] which satisfies trivially) it is easy to

Table 1: Possible values of q q′ q" in 2 dimensions

q	1	**2**	**3**	5	0	1	2	**1**	2	0	0
q′	2	**1**	**1**	0	2	1	1	**2**	2	3	0
q"	0	**0**	**0**	0	1	1	1	**2**	2	3	6

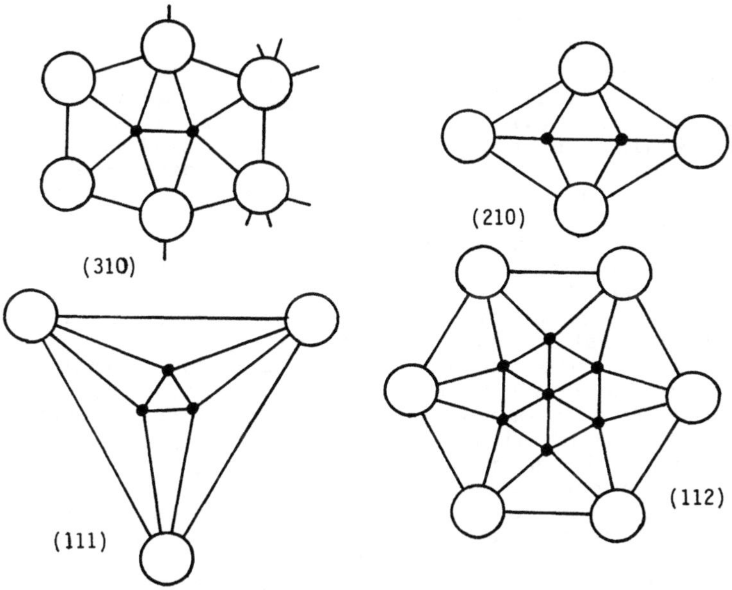

Fig. 3: Some two-dimensional solutions of the problem. The figures between brackets are (q q′ q"). (310) and (210) are quadratic, with a continuously variable distortion from the cubic phase, depending on x. (111) and (112) are decorated hexagonal lattices. B atoms are open circles and s atoms are dots.

construct the whole system from a given s atom and its environment. For the sets Q in thin letters in Table 1, a dense packing is obtained only for a finite set of values (x,Y). On the other hand, those in fat letters are compatible with a dense packing for an infinite, continuous set of (x,Y)'s (Fig. 3). Note that the number of corresponding sets P is finite.

It has been assumed above that a single set of Q satisfies (5b) for given x and Y. There are, however, values of x and Y such that (5) has more than one solution. A particularly interesting case discovered by LANÇON and BILLARD /4/ will be described now.

4. Quasi periodic solutions: The Lançon-Billard model /4,5,6/

The 3 boxed sets of values of (q q′ q") in table 1 have remarkable properties: i) all three satisfy (5b) for

$$x = 2\sin(\pi/10) \quad (7a)\ ; \qquad Y = 2\sin(\pi/5)\ , \quad (7b)$$

ii) for the same values of x and Y, (5a) is also satisfied for several sets of p's, namely

$$10\hat{b} = 2\hat{B} + 7\hat{b} = 4\hat{B} + 4\hat{b} = 6\hat{B} + b = 2\pi \tag{8a}$$

while the form of (5b) corresponding to boxed sets in table 1 is

$$5\hat{S} = 3\hat{S} + 2\hat{s} = \hat{S} + 4\hat{s} = 2\pi \;. \tag{8b}$$

The corresponding values of the angles are therefore

$$\pi - 2\hat{B} = \hat{S} = 2\pi/5 \quad (9a) \qquad \pi - 2\hat{s} = \hat{b} = \pi/5 \;. \quad (9b)$$

Since there are so many solutions of (5), the number of dense packings satisfying (7) is expected to be large. Actually, it <u>is</u> large. Some of the possible packings will now be discussed in the context of quasi-crystals. Indeed, hard circles satisfying conditions (7) have been used /4,5,6/ to realize microscopic models of quasi-crystals. Quasi-crystals are known to be quasi-periodic to a good approximation /7,8/. The problem is thus to fill the plane quasi-periodically with hard circles satisfying (7). This problem is easy to solve because quasi-periodic tilings of the plane can be realized, as recognized by PENROSE /9/, by rhombi of two types (with acute angles $\pi/5$ and $2\pi/5$ respectively) and Penrose rhombi can be decorated by hard circles satisfying (7) (Fig. 4). As seen from Fig. 4 a fat rhombus can be decorated in two ways, so that an infinite Penrose lattice can be decorated in an infinity of ways. However, one should avoid those decorations where 4 BB pairs form a lozenge (Fig. 4c) because this would violate our definition of a dense packing. It has not been demonstrated in a perfectly general way that all Penrose lattices can be decorated avoiding BBBB lozenges, but it can be shown in certain cases. For instance, it can be shown when the Penrose lattice has been generated using the matching rules of Fig. 5a. In this case it is indeed easily shown that each fat rhombus is adjacent to exactly two other fat rhombi. Thus, fat rhombi form rings or stars separated by thin rhombi. It is then sufficient (and easy) to prove that any ring or star of fat rhombi can be decorated avoiding BBBB lozenges (Fig. 5b).

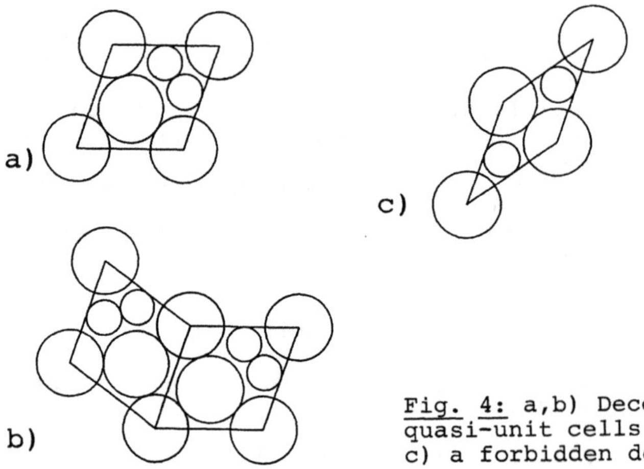

<u>Fig. 4:</u> a,b) Decoration of the two rhombic quasi-unit cells of a Penrose lattice; c) a forbidden decoration

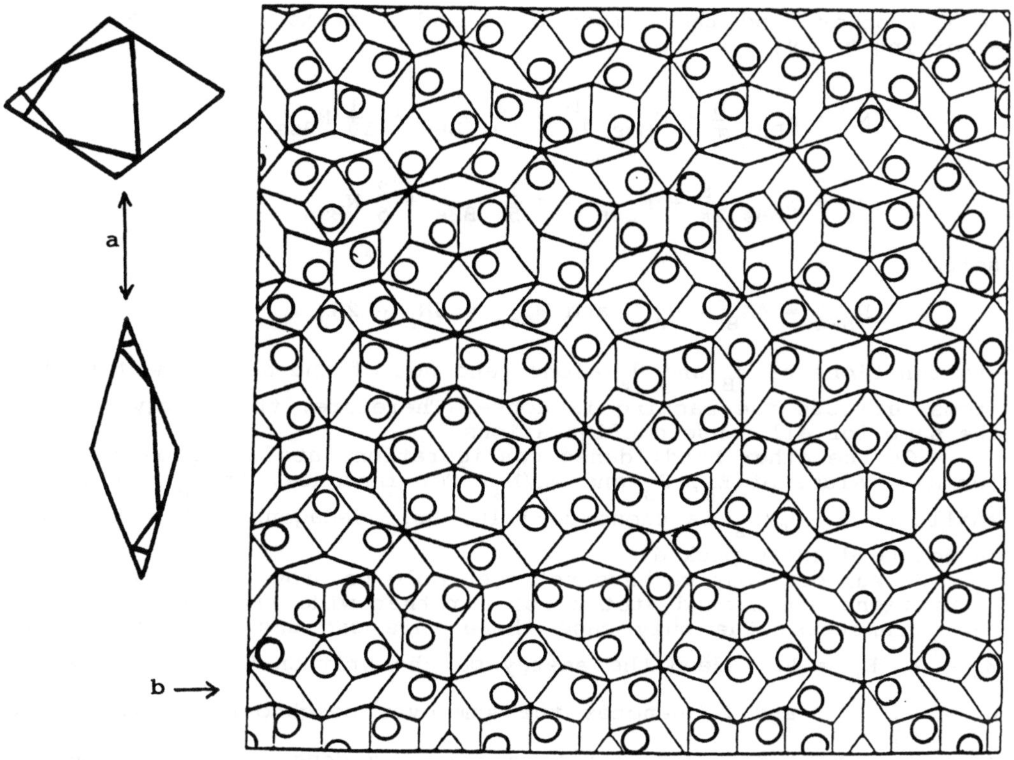

Fig. 5: a) Matching rules for building a quasi-periodic Penrose lattice: the lines should generate infinite straight lines. b) A Penrose lattice generated according to this rule (taken from ref. 7, Fig. 3) and decorated a la Lançon and Billard, but with configuration of Fig. 4c being avoided. Open circles denote B atoms. There are additional B atoms at all vertices, but those are not shown. s atoms have also been omitted but their location follows from Fig. 4.

The Lançon-Billard model of a quasi-crystal, described above, is probably not realistic. In real quasi-crystals like AlMn alloys, the distance AlMn fluctuates very strongly with the environment of the pair, so that the hard core potential model is probably not applicable. However, the Lançon-Billard model has the merit of simplicity and has been used for molecular dynamics studies /4/, Monte-Carlo simulations /5/ and in growth models /6/.

It turns out that, assuming interactions between nearest neighbours as in the first section, all dense configurations have the same energy for a given composition, as will now be seen. Their energy is $W = -(N_{BB}E_{BB}+N_{Bs}E_{Bs}+N_{ss}E_{ss})$ where N_{BB} is the number of BB bonds, etc. Now, since each BB bond is common to 2 BBs triangles, $N_{BB} = N_{BBs}/2$ where N_{BBs} is the number of BBs triangles. Similarly $N_{ss} = N_{Bss}/2$ and $N_{Bs} = B_{Bss}+N_{Bss}$.

Therefore

$$W = -N_{BBs}(E_{Bs}+E_{BB}/2) - N_{Bss}(E_{Bs}+E_{ss}/2) \ . \tag{10}$$

On the other hand the number of B and s atoms are respectively, according to (9),

$$N_B = N_{BBs} \frac{\hat{B}}{\pi} + N_{Bss} \frac{\hat{b}}{2\pi} = \frac{3}{10} N_{BBs} + \frac{1}{10} N_{Bss}$$

$$N_s = N_{BBs} \frac{\hat{S}}{2\pi} + N_{Bss} \frac{\hat{s}}{\pi} = \frac{1}{5} N_{BBs} + \frac{2}{5} N_{Bss}$$

so that

$$N_{BBs} = 4N_B - N_s \quad \text{and } N_{Bss} = 3N_s - 2N_B$$

are functions of N_B and N_s alone, and not of the arrangement of the atoms provided the packing is dense. Therefore, the energy (10) is the same for all dense configurations.

On the other hand, dense configurations are not ground states for all values of E_B, E_s and N_B/N_s. For instance, if E_B and E_s are small and if N_B/N_s is close to 1/10, Bs_{10} decagons will form which cannot be densely packed.

We suspect, but we have not been able to prove, that dense configurations are ground states for a rather wide range of values of the parameters. The only case where this can easily be proved is $E_B = 2$, $E_s = 1$. Indeed the energy can be written $W = \frac{1}{2} \sum_i w_i$ where $w_i = -\sum_j' E_{ij}$ and Σ' denotes the sum over neighbours. Now, in the case $E_B/2 = E_s = 1$, w_i is the same in any dense configuration for all B (or s) atoms, whatever their environment, and is smaller (more negative) than, or equal to any corresponding w_i in a non-dense configuration. Thus, dense configurations are ground states if $E_B/2 = E_s = 1$.

As in the case of Penrose tilings there are not only quasi--periodic dense fillings but also more random structures which can be obtained for instance by adding a finite density of "phasons" to a quasi-periodic structure.

5. Tiling the three-dimensional space /16/

We now wish to find x and Y, such that the space can be filled by tetrahedra of the five types BBBB, BBBs, BBss, Bsss, and ssss. Those tetrahedra have the following dihedral angles $\widehat{BB}$ and $\widehat{BS}$ or BBBs tetrahedra, $\widehat{Bb}$, $\widehat{Bs}$ and $\widehat{Ss}$ for BBss tetrahedra, $\widehat{bs}$ and $\widehat{ss}$ for Bsss tetrahedra and $\gamma_o = \cos^{-1}(1/3) \simeq 70 \cdot 5^o$ for regular tetrahedra.

Those dihedral angles are known functions of x and Y. With the notations $t_x = 1 - x^2/4$ and $t_Y = 1 - Y^2/4$ one easily finds

$$\cos \widehat{Ss} = 1 - Y^2/(2t_x) \quad (11a) \qquad \cos \widehat{ss} = \sqrt{\frac{x^2}{12\, t_x}} \quad (11b)$$

$$\cos \widehat{Bb} = 1 - x^2/(2t_Y) \quad (11c) \qquad \cos \widehat{BB} = \sqrt{\frac{Y^2}{12\, t_Y}} \quad (11d)$$

$$\cos \widehat{bs} = 1 - \frac{1}{2t_x} \qquad (11e) \qquad\qquad \cos \widehat{BS} = 1 - \frac{1}{2t_Y} \qquad (11f)$$

$$\cos \widehat{Bs} = \sqrt{\frac{x^2\, Y^2}{16 t_x t_Y}} \qquad (11g)$$

One can obtain a necessary condition for filling to be possible, if one writes that around any BB bond there are finite numbers p, 2 p′ and p″ of BBss, BBBs and BBBB tetrahedra. Similar conditions are obtained for BS and SS bonds. Thus there are 3 necessary conditions which replace the 2 conditions (5) in two dimensions, namely

$$p\, \widehat{Ss} + 2p'\, \widehat{ss} + p''\, \gamma_o = 2\pi \qquad (12a)$$

$$q\, \widehat{Bb} + 2q'\, \widehat{BB} + q''\, \gamma_o = 2\pi \qquad (12b)$$

$$r\, \widehat{bs} + 2r'\, \widehat{Bs} + r''\, \widehat{BS} = 2\pi \qquad (12c)$$

Here, p, p′, p″, q, q′, q″, r, r′, r″ are positive integers. Since, for each set of these integers there are 3 equations for only 2 unknowns x and Y, it might be expected that (12) has no solution. As will be seen, it has many, most of which however do not correspond to any dense filling of the whole space.

A special case is obtained when one type of bond is absent. Obviously it cannot be Bs. Let then ss bonds be assumed to be absent. Then p = p′ = p″ = q = r = r′ = 0. On upper limit of r″ is easily determined and for each of the other ones it is easy to determine Y and then to see whether (12b) has solutions in q′ and q″. The only solution is q′=q″=2, r″ = 4 and $Y = \sqrt{2}$. This corresponds to NaCl (Fig. 2). This value $\sqrt{2}$ is very large, so that this structure is mainly expected in ionic crystals, and only when a big anion coexists with a small cation. This is indeed the case in NaCl, where the ionic radii are $R(Cl^-) = 1.81$ Å and $R(Na^+) = 0.95$ Å, hence $Y = d(Cl\text{-}Cl)/d(Na\text{-}Cl) = 1.31$ which is smaller than $\sqrt{2}$. On the other hand the naive model (2) does not take Coulomb repulsion between Cl ions into account so that an effective value of $Y = \sqrt{2}$ is not too surprising. Although any chemist will certainly smile when reading the present exercise, it is interesting to see that it has something to do with reality. CsCl, for instance, crystallizes in a different structure because the ionic radius $R(Cs^+) = 1.69$ Å is close to $R(Cl^-)$.

There are, in fact, an infinite number of solutions corresponding to the same values of x, Y, p, p′ ... . This results from the fact that B atoms form a fcc lattice. They can be reorganized into a hcp-lattice or more complicated lattice by means of stacking faults along (111). Of course ions should be reorganized accordingly.

Inequalities, as in the two-dimensional case, make it unnecessary to play with an infinite set of p p′ p″ q q′ q″ r r′ r″. Apart from the obvious inequalities $p'' \leq 5$ $q'' \leq 5$ there are additional inequalities which depend on the values of x and Y. Remembering that B balls are big and s balls are small, it is reasonable to assume

$$x \leq 1 \leq Y\ . \qquad (13a)$$

In addition, s balls should be able to flow through the small hole between 3 neighbouring B balls. This implies

$$Y < \sqrt{3}\ . \qquad (13b)$$

From relations (11) and (13) one deduces

$$\pi/3 \leq \widehat{Ss} \leq \pi \qquad (14a) \qquad\qquad \gamma_o \leq \widehat{ss} \leq \pi/2 \qquad (14b)$$

$$\pi/3 \leq \widehat{bs} \leq \gamma_o \qquad (14c) \qquad\qquad \pi/3 \leq \widehat{BS} < \pi\ . \qquad (14d)$$

Relations (14) immediately provide upper bounds for $p(\leq 6)$, $p'(<\pi/\gamma_o)$, $r(\leq 6)$ and $r''(\leq 6)$. An upper bound for r' can be found by another method. It results from the fact that, if $\widehat{Bs}$ is too small (more precisely, if it is smaller than $\widehat{BS}$) two BBss tetrahedra are not allowed to share a common Bss triangle since, in this case, the other two B spheres would be at a distance shorter than d_{BB}. Thus, if $\widehat{Bs} < \pi/3$ (hence smaller that $\widehat{BS}$) each pair of $\widehat{Bs}$ angles around a given Bs pair should be separated by at least one $\widehat{bs}$ angle, so that $r' \leq r < 6$. On the other hand, if $\widehat{Bs} \geq \pi/3$, obviously $r' \leq 6$. Thus, in all cases, $r' \leq 6$. The result is that all 6 integers p p' p" r r' r" are smaller than or equal to 6 and therefore the set of their allowed values is finite. The method to find all solutions of (12) is then: i) enumerate all possible sets (p p' p" r r' r"). ii) For

Table 2: Solutions of Equations (12) for $0 < x \leq 1 \leq Y < \sqrt{3}$.

x	Y	p p'p"	q q'q"	r r'r"	Filling Three Dimensional Space
unspecified	$\sqrt{2}$	NO	022	004	Yes (Fig. 2)
$\sqrt{4/7}$	$\sqrt{12/7}$	400	600,410 220	012	Yes (Fig 6)
$\sqrt{8/11}$	$\sqrt{12/11}$	310 211 112	600	102	Yes (Fig. 6)
1	$\sqrt{2}$	210	022,211 400	040,121 202	No
$\sqrt{\frac{8}{11}}$	$\sqrt{2}$	013	022	040	No
$\sqrt{\frac{7}{4}-\sqrt{\frac{3}{2}}}$	1	112	311	121	No
$\sqrt{\frac{8}{11}}$	$\sqrt{\frac{32}{11}}$	013,111	113	302,211	No
$2\sqrt{\frac{\sqrt{2}-1}{3\sqrt{2}+1}}$	$2\sqrt{\frac{\sqrt{2}+1}{3\sqrt{2}+1}}$	400	800	211	No
1/2	$\sqrt{5/2}$	202	420	021	No
		111	q00	111	No

each of these sets solve (12a) and (12c) numerically and thus obtain x and Y. iii) Check whether these values of x and Y satisfy (12b) for some set (p p′ p″). Then, for any integer q, one can find values of x and y which satisfy (12).

The step (ii) requires a computer. Surprisingly, there is one set (p p′ p″ r r′ r″) which has an infinite number of solutions in x, Y, q, q′, q″. These solutions are

$$p = p' = p'' = r = r' = r'' = 1,\ q' = q'' = 0. \tag{15}$$

Then, for any integer q, one can find values of x and Y which satisfy (12).

Table 2 displays the solutions of (12) which satisfy (14). The numerical computer values have been translated into exact values of x and Y. Some solutions have been omitted because they are forbidden by trivial consistency relations.

The final step of the solution is to see whether the values of x and Y which satisfy (12) correspond to some dense filling of the space. The values which satisfy (15) require an analytic discussion to prove that q values larger than some upper limit are unacceptable. This proof /11/ is tedious but has no essential difficulty and will not be given here. For the remaining, finite set of (x,Y) we used the following pedestrian method: Starting with one tetrahedron, we add one atom after the other, each atom beeing chosen according to the following requirements: i) each new atom Z has to form at least one more tetrahedron with three already existing atoms L, M. N ii) the atoms L, M, N should be, if possible, such that the

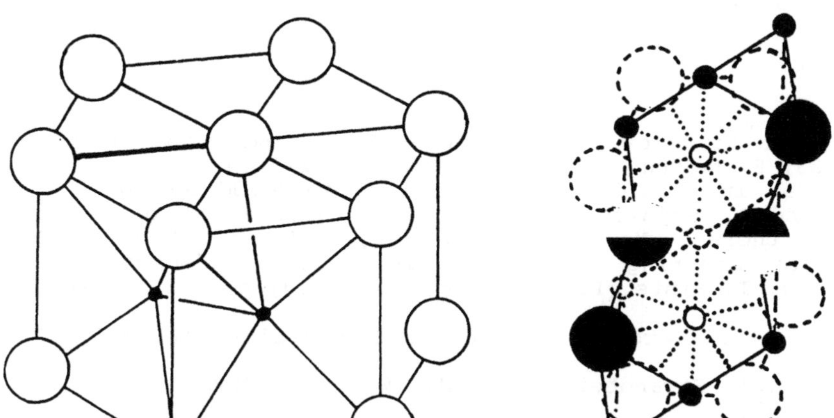

Fig. 6:

a)A part of a periodic, three-dimensional dense filling for $x = \sqrt{4/7}$ and $Y = \sqrt{12/7}$.

b) Projection of another three-dimensional solution, corresponding to $x = \sqrt{8/11}$ and $Y = \sqrt{12/11}$. Big circles are B atoms and small circles are s atoms. Black circles have integer height. Dashed circles have half-integer height. Open circles have height equal to 1/4 plus an integer. Horizontal bonds are full or dashed lines. Other bonds are dotted lines. The dash-dotted line shows the unit cell boundary.

(B or s) nature of the new atom Z is unambiguous. If it is impossible, both possibilities are considered. In most cases the computer announces after between 20 and 40 atoms that is cannot go further. If the computer does not capitulate after 40 atoms, one tries to build an infinite periodic set containing the cluster of 40 atoms already found. This method yields (in addition to NaCl) two other structures represented by Fig. 6.

The structure of Fig. 6a has formula Bs_2. It has the strange property to be both hexagonal and tetragonal since the faces of the hexagonal prism are squares. For this reason, one can produce a stacking fault by cutting the structure by a vertical plane and rotating both halves by 90° with respect to each other. Since one can produce an infinity of stacking faults, aperiodic structures (with randomly stacked layers) are also dense.

Finally, the structure of Fig. 6b also has formula Bs_2. It is monoclinic, but orthorhombic solutions are also possible. Here again, stacking faults and aperiodic structures are possible. It is remarkable that such a complicated structure can result from such simple interactions.

In conclusion, 3 sets of values of x and Y (those at the top of Table 2) permit a dense packing in 3 dimensions. In all three cases there are periodic solutions as well as solutions random in one direction, but no quasi-periodic solution is obtained.

As stated before, hard core potentials are poor models for most materials: ionic crystals, covalent solids with directed bonds, and often in metals, especially those which build quasi-crystals /10/. This exercise on hard spheres is but loosely related to real material science. Our excuse is that we are not the first to play this game /12-16,1-3/.

References

/ 1/ M. Eden, 4th Berkeley Symp. on Math. Stat. and Probability IV (1961) 233, F. Neyman, ed. (Berkeley: University)

/ 2/ L.M. Sander in "Scaling Phenomena in Disordered Systems" ed. R. Pynn and A. Skjeltorp (Plenum, New York, 1985) p. 31 and references therein. M. Kolb, R. Jullien, R. Botet, ibid, p. 71

/ 3/ J. Weeks in "Ordering in strongly fluctuating condensed matter systems" ed. T. Riste (Plenum, New York, 1980) p. 293 and refs. therein

/ 4/ F. Lançon, L. Billard, P. Chaudhari, Europhysics Letter 2, 625 (1986)

/ 5/ M. Widom, K.J. Strandberg and R.H. Swendson, Phys. Rev. Lett. 58, 706 (1987)

/ 6/ B. Minchau, K.Y. Szeto, J. Villain, Phys. Rev. Lett. 58, 1960 (1987)

/ 7/ D. Levin, P.J. Steinhardt, Phys. Rev. B 34, 596 (1986)

/ 8/ C.L. Henley, submitted to Comments in Cond. Matt. Phys. (1987)

/ 9/ R. Penrose, Math. Intell. 2, 32 (1979)

/10/ M. Audier and P. Guyot. Private communication and Phil. Mag. Lett. 53, L 43 (1986)

/11/ K.Y. Szeto, J. Villain, Phys. Rev. B 36, 4715 (1987)

/12/ J.D. Bernal and J. Mason, Nature 188, 910 (1960)

/13/ M.R. Hoare, J. Non-cryst. Sol. 31, 157 (1978)

/14/ M. Rubinstein, D.R. Nelson, Phys. Rev. B 26, 6254 (1982)

/15/ R. Mosseri, J.F. Sadoc, J. Physique Lett. 45, L 827 (1984)

/16/ F. Lançon, L. Billard, A. Chamberod, J. Phys. F. 14, 579 (1984)

Competing Interactions in Metallic Superlattices

C.M. Falco, J.L. Makous, J.A. Bell, W.R. Bennett, R. Zanoni, G.I. Stegeman, and C.T. Seaton

Optical Sciences Center and Department of Physics, University of Arizona, Tucson, AZ 85721, USA

1. Introduction

Metallic superlattices provide an excellent system to study in a controlled manner a variety of physical phenomena, including superconductivity, magnetism, and electrical transport properties. As will be discussed in this paper, changes in certain of these properties as a function of superlattice modulation wavelength Λ are found to be correlated with structural changes and elastic property anomalies. The properties of two particular metallic superlattices, Cu/Nb and Mo/Ta, will be discussed in this paper, as examples of how competing interactions manifest themselves in the physical properties of these superlattices.

An anomalous dependence of the Rayleigh wave velocity with modulation wavelength Λ was reported in Cu/Nb superlattices by Kueny et al. [1] Using a light scattering technique they found a minimum in the Rayleigh wave velocity for $\Lambda \approx 22$ Å, which was interpreted as a softening of ~35% in the C_{44} elastic constant. Similar measurements were subsequently made on metallic superlattices consisting of Mo/Ni [2], Fe/Pd [3] and Mo/Ta [4]. Mo/Ni was found to behave in a similar way to Cu/Nb, but in Fe/Pd the Rayleigh wave velocity was found to be insensitive to Λ. Very recently Mo/Ta was found to exhibit only a small decrease (~8%) in the C_{44} elastic constant near $\Lambda \approx 35$ Å. Figure 1 shows the elastic constant C_{44} vs. Λ for Cu/Nb and Mo/Ta, demonstrating the range of behavior which can be exhibited by metallic superlattices. The method used to obtain these data is described later in this paper.

In the remainder of this paper we describe the preparation and structural characterization of these metallic superlattices. From x-ray diffraction measurements we are able to extract how the strain in these materials depends on Λ. We describe how C_{44} is determined from light scattering measurements, and compare this behavior with that of the strain. We also show how another physical property, the temperature dependence of resistivity, is correlated with the structural and elastic properties. Finally, we summarize these results.

2. Sample Preparation and Characterization

The Cu/Nb and Mo/Ta superlattices were prepared using a magnetically enhanced, dc triode sputtering system previously described[5], in which multilayers were formed by alternately passing a rotating substrate platform over the targets. Microprocessor control of the substrate table and feedback control of the sputtering rates enabled us to keep the deposited layer

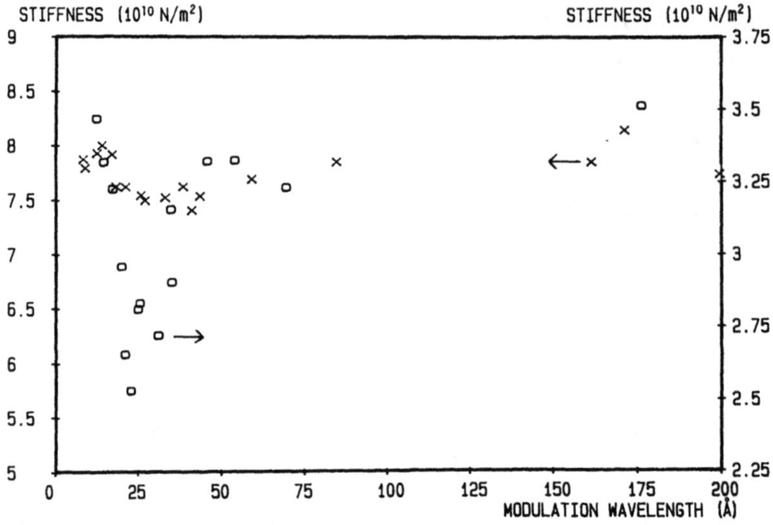

Figure 1. Modulation wavelength Λ dependence of the elastic constants C_{44} for Cu/Nb (ooo) and Mo/Ta (xxx) superlattices.

thicknesses constant to ±0.3%.[5,6] Base pressures before sputtering were typically in the low 10^{-7} torr range, and the argon pressure during sputtering was approximately 5 mtorr. The superlattices were deposited onto 90° single-crystal sapphire substrates at rates of ~10 Å/sec. Rutherford Backscattering Spectroscopy (RBS) and a number of x-ray diffraction techniques were used to characterize the chemical and physical structure of these samples. RBS verified the copper-niobium and molybdenum-tantalum ratios to an accuracy of better than 4%, and determined the atomic fraction of oxygen, the most significant impurity detected, to be less than 2% in the Cu/Nb, and 1% in the Mo/Ta. Also, Ar in the Mo/Ta films was less than 0.5%.

Cu is an fcc material and Nb is bcc, so no epitaxial interfacial registry can be expected. Nb/Cu samples of equal atomic fractions in each layer were grown. X-ray diffraction in the Bragg-Brentano geometry showed that the copper and niobium grew with the (111) and (110) planes approximately parallel to the surface. Broadening of the diffraction peaks indicated a loss of vertical coherence for samples with bilayer periodicities $\Lambda \leq 11$ Å. A series of samples with Λ varying from 8 Å to 320 Å with total thicknesses approximately 3000 Å were fabricated with 60% Cu by volume and 50% atomic fraction. This differs from the work of Kueny et al. in which the Cu and Nb films were equally thick, that is a 50% fraction by volume.

Unlike Cu/Nb, Mo and Ta are both bcc and have bulk lattice mismatch of only 5%. By control of the deposition process to ±0.3% we fabricated a series of Mo/Ta superlattices with wavelengths modulated by alternating integer, n, atomic planes of Ta and Mo.

That is, within ±0.3% the samples satisfied $\Lambda = n \times \{d_{110}(Mo) + d_{110}(Ta)\} = n \times (4.57$ Å), where n ranged from a "bulk" value of 153 to the monolayer limit of 1. The linewidth of the central Bragg diffraction peak remains very narrow over the entire range of Λ, down to the n = 1 monolayer limit. By deconvoluting the instrumental broadening and using the Scherrer equation,[7] it is possible to convert the measured linewidths into structural coherence lengths. In contrast to the case of Cu/Nb, for Mo/Ta the coherence length perpendicular to the growth direction is found to remain greater than 150 Å even for the n = 1 (Λ = 4.57 Å) superlattice.

The structural coherence length in the plane of the Mo/Ta films was determined using θ-2θ x-ray scattering from the (211) planes, and found to be in the range 200-300 Å. This is very much longer than the layer thicknesses of these samples.

The lattice constant perpendicular to the layers can be found from the x-ray diffraction measurements. In Figure 2 we show how the inverse lattice constant of Mo/Ta (which is proportional to strain) depends on Λ. Also shown in Figure 2 are the data for C_{44}.

The data in Figure 2 indicate that as strain builds up in the superlattice, probably due to lattice mismatch at the interface, the elastic constant C_{44} softens. However, for Λ < 40 Å the strain is relieved due to the introduction of misfit dislocations at the interface, and C_{44} resumes its "bulk" value. In Cu/Nb a similar behavior starts to occur. C_{44} softens (considerably more than in the case of Mo/Ta), but structural coherence is lost before the strain can be relieved.

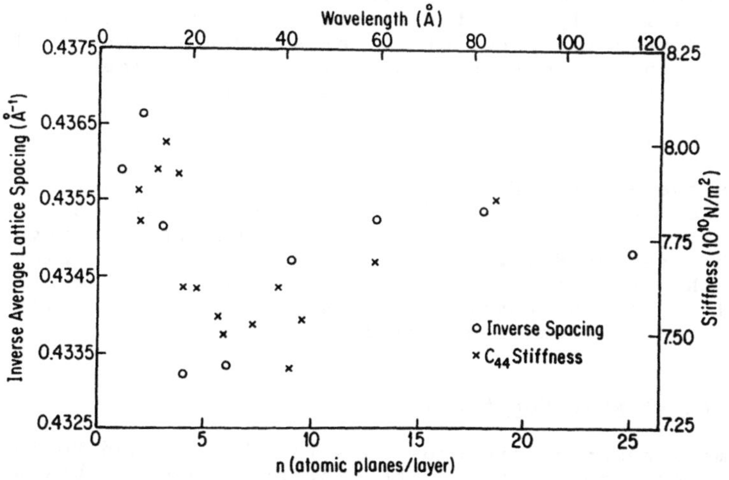

Figure 2. The dependence of inverse lattice constant (proportional to strain) and C_{44} on Λ for Mo/Ta.

3. Determination of Elastic Constants of Superlattice Films

Light scattering from metallic superlattices provides an excellent probe of the saggital-plane polarized acoustic waves. A surface corrugation is produced at the air-film interface by sound waves traveling both parallel to the surface and impinging onto it from the medium bulk [8,9]. Light is inelastically scattering from this surface corrugation, and can be analyzed to determine the energy and wavevector of the excitation responsible for the scattering.

The acoustic modes which will propagate in a superlattice film on a substrate such as sapphire are a Rayleigh wave, and a discrete number of Sezawa waves. The number of Sezawa waves depends on the superlattice film thickness, and the density and elastic constants of the film and substrate [10,11]. The elastic constants, including C_{44}, are obtained by least-squares fitting theoretical dispersion curves to the Rayleigh and Sezawa wave velocity measurements obtained using Brillouin scattering. The velocities measured to an uncertainty of approximately 0.3%.

4. Brillouin Scattering

The technique of Brillouin scattering from metal thin films is described elsewhere [12] and is only briefly summarized here. We use a tandem, triple-pass Fabry-Perot interferometer [13] of the basic design introduced first by Sandercock [12]. The cavities are scanned using a piezoelectric transducer which is linearized with feedback from a capacitive sensor. The tandem finesse and transmission is optimized using the window techniques described by May et al. [14]. A frequency stabilized argon ion laser was used for these measurements. The detector was a cooled ITT FW130 photomultiplier tube with dark count of 0.5 sec^{-1}. An acousto-optic modulator was used to attenuate the light by as much as 10^{-3} when scanning through the stray elastically scattered light peaks at the laser frequency.

Brillouin scattering from the surfaces of our metallic superlattices occurs primarily via the ripple mechanism described above [8,9]. The wavevector parallel to the surface is conserved so that for scattering geometries in which the incident and scattered wavevectors are coplanar with the surface normal,

$$k \sin\theta_s = k \sin\theta_i + Q_p$$

where θ_i and θ_s are the angles of incidence and scattering measured relative to the surface normal and $k = \omega/c$. We use a backscattering geometry where $Q_p = 2k\sin\theta_i$. The acoustic wave frequency shifts the scattered light by Ω, the measurement of which leads directly to the surface wave velocity $V_s = \Omega/Q_p$. By varying the scattering angles it is possible to map out the velocity dispersion curves as a function of normalized film thickness.

5. Evaluation of Film Elastic Constants

Light scattering measurements were made on sets of Cu/Nb and Mo/Ta superlattice films over a range of Λ. The procedure used to obtain the elastic constants from these measurements is explained elsewhere.[13] Briefly summarizing, we fixed C_{11} and C_{13} to their respective averages over the separate data sets and then allowed only C_{33} and C_{44} to vary freely over the least squares fitting. Varying C_{33} allowed us to search for a Λ-dependent strain normal to the surface. The departure of the measured velocities from those calculated by the fit are less than 0.5%.

The Λ dependence of C_{44} for Cu/Nb and Mo/Ta is shown in Fig. 1. For Cu/Nb the behavior is similar to that implied previously from the Kueny et al. experiments [1] with the following exceptions. The Rayleigh velocity minimum (relative to the velocity on the largest wavelength sample with Λ = 320 Å) is -13 ± 2%, and occurs at Λ = 38 Å. Kueny et. al found a -20 ± 4% minimum at Λ = 22 Å for their samples. This could be due to the small compositional difference between the two sets of samples used, or due to other factors such as deposition parameters, impurities, etc.

During the sample deposition process, each interface of the superlattice is exposed to trace impurities in the sputtering chamber. The relative amount of impurities which are incorporated depends on the relative time that a sample is exposed to background contaminants versus time spent directly over the sputtering targets. That is, the proportion of impurities is a function of the bilayer thickness Λ. The effect of atomic oxygen, the principal contaminant as determined by RBS, on the elastic constants was tested by growing a set of Cu/Nb samples with Λ = 20 Å in a much higher background pressure of oxygen. In this experiment the partial pressure of oxygen was 5×10^{-5} (normally it is $<4 \times 10^{-7}$ torr). When the oxygen content in the superlattice increased from 1.8 to 9.2% (as determined by RBS), the Rayleigh wave velocity increased by 8.8%. These results indicate that the dominant impurity oxygen is probably not responsible for the softening in C44.

6. Transport Properties

The temperature dependence of the resistivity of a series of Cu/Nb and Mo/Ni samples was previously reported. The absolute values of the resistivities of both of these materials increase to approximately 150 $\mu\Omega$-cm as Λ is decreased to around 20 Å. With a further decrease in wavelength the temperature dependence of resistivity changes from positive to negative at $\Lambda \approx 20$ Å for Nb/Cu[1] and at $\Lambda \approx 16$ Å for Mo/Ni.[2] This Λ coincides with the structural and elastic anomalies discussed above. Data for Cu/Nb is shown in Figure 3.

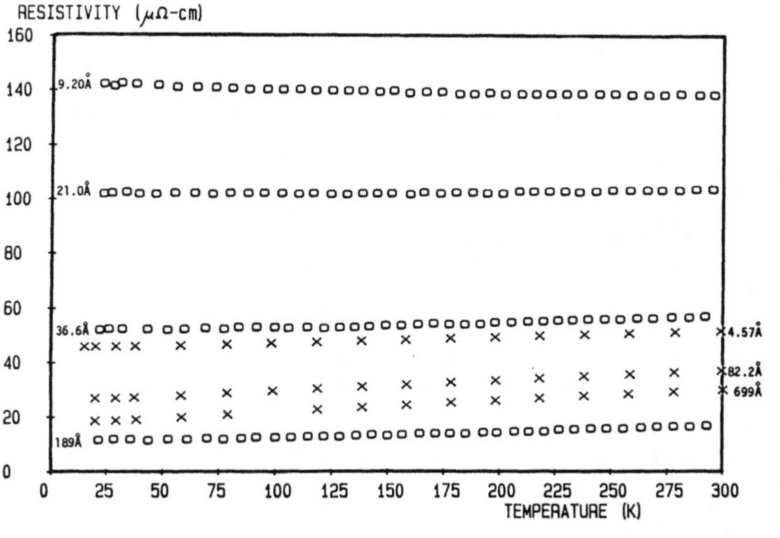

Figure 3. Temperature dependence of the resistivity of Cu/Nb (ooo) and Mo/Ta (xxx) samples.

We measured the resistivities of Mo/Ta superlattices from 300 K to 7.5 K using a four-probe technique. These results are also shown in Figure 3. Unlike the cases of Nb/Cu[1] and Mo/Ni,[2] these curves all have positive temperature coefficients of resistivity over the entire range of Λ. In addition, the highest resistivity value, occurring for the n = 1 monolayer superlattice, is only 52 μΩ-cm, which is well within the metallic regime.[15] This metallic behavior is consistent with the structural and elastic behavior discussed above.

7. Summary

Metallic superlattices provide excellent systems to study in a controlled manner a variety of physical phenomena, including superconductivity, magnetism, and electrical transport properties. We have discussed in this paper the specific example of how changes in the electrical resistivity as a function of superlattice modulation wavelength Λ are correlated with structural changes and elastic property anomalies. The data in Figure 2 indicate that as strain builds up in a metallic superlattice, the elastic constant C_{44} softens. However, for $\Lambda < 40$ Å in Mo/Ta the strain is relieved due to the introduction of misfit dislocations at the interface, and C_{44} resumes its "bulk" value. In Cu/Nb C_{44} also softens, but structural coherence is lost before the strain can be relieved.

Acknowledgements - This research was supported by the Joint Services Optics Program of the Army Research Office and the Air Force Office of Scientific Research (F4962-86-C-0123), and the U.S. DOE (DE-FG02-87ER45297).

8. References

1. A. Kueny, M. Grimsditch, K. Miyano, I. Banerjee, C. M. Falco and I. K. Schuller, Phys. Rev. Lett. 48, 166 (1982).
2. M. R. Khan, C. S. L. Chun, G. P. Felcher, M. Grimsditch, A. Kueny, C. M. Falco and I. K. Schuller, Phys. Rev. B 27, 7186 (1983).
3. P. Baumgart, B. Hillebrands, R. Mock, G. Guntherodt, A. Boufelfel and C. M. Falco, Phys. Rev. B Rapid Commun. 34, 9004 (1986).
4. J. A. Bell, W. R. Bennett, R. Zanoni, G. I. Stegeman and C. M. Falco, Phys. Rev. B Rapid Commun. 35, 4127 (1987).
5. C. M. Falco, J. Phys. Colloq. 45, C5-499 (1984).
6. W. R. Bennett, *The Growth, Structure and Electrical Transport Properties of Mo/Ta Superlattices*, PhD dissertation (University of Arizona, 1985).
7. H. P. Klug and L. E. Alexander, in *X-Ray Diffraction Procedures*, 2nd ed. (Wiley, New York, 1974), chap. 9.
8. R. Loudon, Phys. Rev. Lett. 40, 581 (1978).
9. N. L. Rowell and G. I. Stegeman, Solid State Comm. 26, 809 (1978).
10. G. W. Farnell and E. L. Adler, *Physical Acoustics*, Vol. IX, edited by W. P. Mason and R. N. Thurston (Academic Press, New York, 1972) chap. 2.
11. B. A. Auld, *Acoustic Fields and Waves in Solids*, Vol. 2 (Wiley, New York, 1973).
12. J. R. Sandercock, in *Light Scattering in Solids*, edited by M. Balkanski (Flamarion, Paris, 1971), p. 9.

13. J. A. Bell, *Brillouin Scattering From Metal Superlattices*, PhD dissertation (University of Arizona, 1987).
14. W. May, H. Kiefte, M. Clouter an G. I. Stegeman, Appl. Opt. 17, 1603 (1978).
15. A. F. Ioffe and A. R. Regel, Prog. Semicond. 4, 237 (1960).

Quasi-Two-Dimensional Phase Transitions in Graphite Intercalation Compounds

G. Dresselhaus[1], *M.S. Dresselhaus*[2], *and J.T. Nicholls*[2]

[1]Francis Bitter National Magnet Laboratory*,
[2]Department of Physics, Massachusetts Institute of Technology,
Cambridge, MA 02139, USA

Abstract

Graphite intercalation compounds (GICs) are chemically formed superlattices which can show extremely small interlayer interaction and correlation. These compounds provide good experimental evidence for quasi–two dimensional structural and magnetic phase transitions. An example of a commensurate–incommensurate structural phase transition is presented. A review is given of the evidence for 2D magnetism in graphite intercalation compounds (GICs) with particular reference to the possible 2D–XY phase transition in $CoCl_2$–GICs.

Introduction

The possibility of preparing single isolated layers either by intercalation or by atomic deposition processes has stimulated a great deal of experimental activity in the observation of highly anisotropic and 2–dimensional (2D) structural and magnetic phenomena. We discuss the two phenomena which have been given the most attention, namely 2D striped domains and magnetism.

In the case of graphite intercalation compounds (GICs), incommensurate structures occur when there is a large lattice mismatch between the lattice constants of the pristine intercalate material and the possible commensurate structures that are compatible with the graphite honeycomb structure (e.g., $(\sqrt{3} \times \sqrt{3})R30°$, $(2 \times 2)R0°$, $(\sqrt{7} \times \sqrt{7})R19.1°$, etc). For totally incommensurate GICs, the lattice constant of the intercalant usually remains essentially unchanged relative to the pristine parent material. In contrast, the nearly commensurate GICs frequently show structural commensurate–incommensurate phase transitions which are driven by the different coefficients of thermal expansion for the graphite and the intercalate layers. When the insertion of extra lattice planes at a macroscopically incommensurate interface is periodic, a striped domain phase is achieved.[1,2] Related discommensuration phenomena are frequently observed in alkali metal GICs.[3]

The quasi–two dimensional nature of these transitions is made possible by the phenomena of staging which appears to be unique to GICs. Staging consists of a periodic arrangement of the intercalate layers which thus occur as monolayers with separations of as much as 15 to 20Å. The stage index refers to the number of graphite layers between successive intercalate layers. X–ray measurements usually show 3D correlations for stage 1 materials and 2D correlations for stage 3 and higher compounds. The staging phenomena make the study of 2D magnetism possible.

* Supported by the NSF.

2D Structural Phase Transitions Studies with GICs

Though less thoroughly studied experimentally and theoretically than the adsorbed gas phases on graphite, the structural ordering of the intercalates in graphite show most of the same structures and phase transitions as the adsorbed systems. Examples of both commensurate–incommensurate and melting phase transitions are observed in GICs.

Graphite intercalation compounds provide an ideal medium for the study of phase transitions of current interest for two–dimensional systems.[3] A novel example of such a phase transition is found in the graphite–bromine system which exhibits a phase transition from a $(\sqrt{3} \times 7)R(30°, 0°)$ commensurate structure to an incommensurate striped domain phase.[1] At higher temperatures two–dimensional melting is observed.[2,4] The commensurate in–plane structure at room temperature for a stage 4 graphite–Br_2 compound is shown in Fig. 1, where the $\sqrt{3}a_0$ basis vector in the figure is at 30° with respect to one of the graphite basis vectors and the $7a_0$ superlattice basis vector is parallel to the other graphite basis vector. As the temperature is raised above room temperature, the much higher thermal expansion coefficient of solid bromine relative to graphite in the basal plane suggests that at some elevated temperature, a transition to an incommensurate intercalant phase should occur. A transition to an incommensurate phase has in fact been observed in the stage 4 graphite–Br_2 compound $C_{28}Br_2$ at $T = 69.05°C$, but to a highly unusual structure, called a striped domain incommensurate phase, first predicted by Frenkel and Kontorova in 1938 [5] and by Frank and Van der Merwe in 1949.[6] In this striped domain phase, the lattice constant remains commensurate in the $\sqrt{3}\ a_0$ direction. However, in the $7a_0$ direction, the lattice remains locally commensurate, but after N unit cells, a phase slip of $2b/7$ occurs, as indicated in Fig. 2, where $b = 7a_0$ is the lattice constant of the commensurate superlattice in the long direction and τ is the width of the domain wall. As T is increased above the transition temperature, the intercalant becomes increasingly incommensurate by decreasing N, which controls the size $L = Nb + \tau$ of the striped domain (see Fig. 2). The quantitative structure of the incommensurate striped domain phase is established by detailed high resolution x–ray scattering experiments. The results are shown schematically in Fig. 3 where

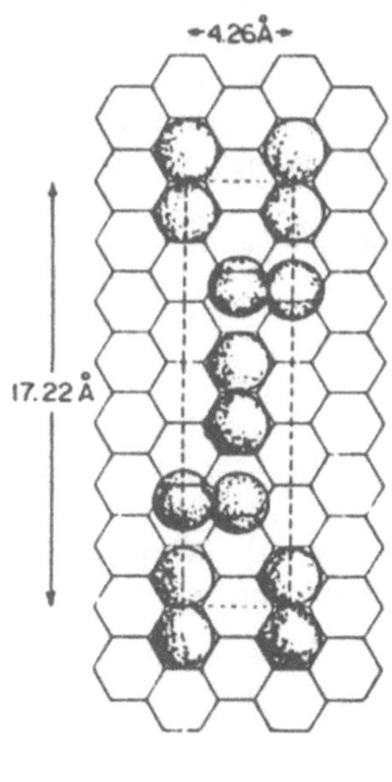

Figure 1: The 2D intercalate Br_2 lattice projected onto the hexagonal graphite network. The dashed lines indicate the centered $(\sqrt{3} \times 7)a_0^2$ rectangular unit cell, where $\sqrt{3}a_0 = 4.26Å$ and $7a_0 = 17.22Å$.[1]

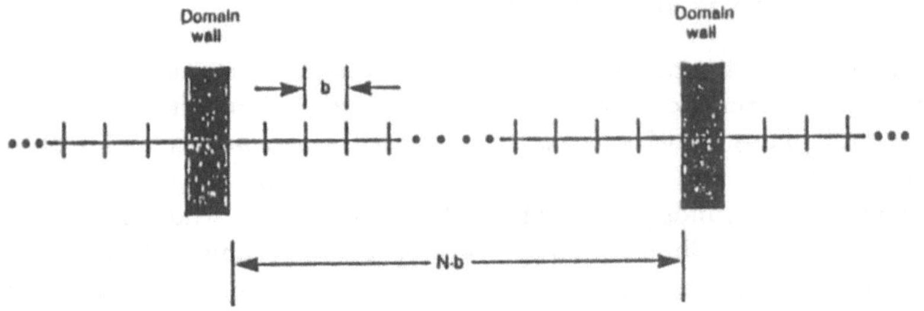

Figure 2: Schematic diagram for the striped domain phase in stage 4 graphite–Br_2.[1]

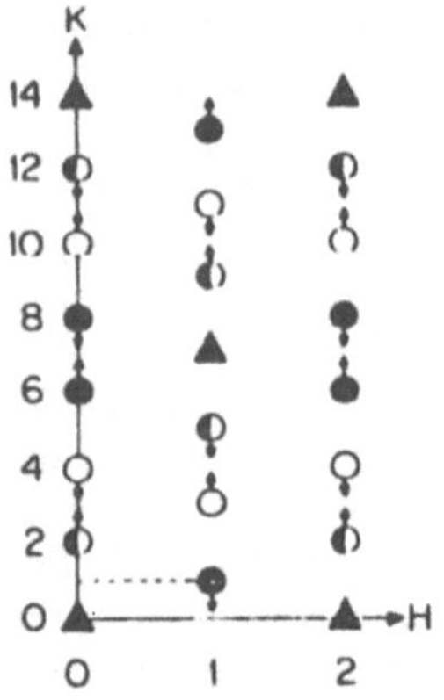

Figure 3: A quadrant of the $\ell = 0$ reciprocal plane of a Br_2 sublattice. The circles correspond to Bragg reflections originating from the 2D Br_2 superlattice. The triangles represent 3D graphite and bromine peaks. The arrows at each Br_2 peak show the directions of the principal peak position shifts during the transition from the commensurate to striped domain phase. The Bragg reflections represented by the open circles, solid circles, and half circles exhibit shifts of $\epsilon/2$, ϵ, $3\epsilon/2$, respectively, from their commensurate peak positions during the phase transition.[1]

the $(hk0)$ superlattice bromine spots (circles) are shown in relation to the graphite spots (triangles) for the commensurate structure. The transition to the incommensurate striped domain phase is observed as the onset of a modulated diffraction pattern in which the dominant bromine superlattice spots move in the 7–fold direction as indicated by the arrows in Fig. 3; no motion of the bromine superlattice spots in the $\sqrt{3}$–fold direction is observed, as is also the case for the diffraction spots for the graphite host. The relative motion is described by the incommensurability coefficient $\epsilon = 4\pi/7N$ where N is the number of unit cells between domain walls, as given below by Eq. 6.

The diffraction pattern can be understood in terms of the structure factor

$$F = \sum e^{ikx_n} \tag{1}$$

where the lattice positions are described by

$$x_n = x_m^0 + n_2 b + (Nb + \tau)n_3 \tag{2}$$

in which x_m^0 gives the location of the centers of each of the 7 bromine molecules within the unit cell, n_2 denotes the unit cell within the domain ($n_2 = 0, 1, 2, \ldots N-1$), and n_3 specifies the domain index ($n_3 = 0, 1, 2, \ldots p-1$). The x–ray intensity profile is then given by

$$I \sim |F|^2 = \left| \sum_{m=0}^{7} e^{ikx_m^0} \right|^2 \left(\frac{\sin^2 Nkb/2}{\sin^2 kb/2} \right) \left(\frac{\sin^2 p(Nb+\tau)k/2}{\sin^2 (Nb+\tau)k/2} \right) \tag{3}$$

where the first term comes from the summation of phase factors within the unit cell, the second from phase factors within the commensurate domains and the third from the domains themselves. The domain wall modulation of the third term gives rise to peaks whenever

$$k = n_3 \frac{2\pi}{Nb + \tau}. \tag{4}$$

The striped domain phase thus has domains of length Nb that are commensurate with the graphite structure separated by small phase slips τ, giving rise to a macroscopic incommensurate structure for the intercalate so that the average lattice constant for the bromine can increase by making the size of the domains smaller. The incommensurability ϵ which measures the deviation of the wave vector from the commensurate phase is then obtained from Eq. 4 by writing

$$k = n_3 \left[\left(\frac{2\pi}{Nb}\right) + \epsilon \right] \tag{5}$$

and noting that $\tau = 2b/7$ to obtain

$$\epsilon = \frac{2\pi(2b)}{7Nb} = \frac{4\pi}{7N}. \tag{6}$$

Thus as the size of the domain decreases (i.e., N decreases), the incommensurability ϵ will increase.

A detailed theory for a 1D commensurate–incommensurate transition in a 2D lattice has been given by Pokrovsky and Talapov.[7] They predict that the domain wall density and hence the incommensurability ϵ should initially rise as $(T - T_c)^{1/2}$ but should saturate as the domain wall density increases. The experimental results for the stage 4 Br_2–GICs show just this behavior near the transition temperature $T_c = 69.05$°C with an exponent of $\beta = 0.50 \pm 0.02$, as shown in Fig. 4 for several Br(h, k, ℓ) reflections. The fit here was made to the functional form for the incommensurability

$$\epsilon = \epsilon_0 \left[\frac{(T - T_c)}{T_c} \right]^{\beta} \tag{7}$$

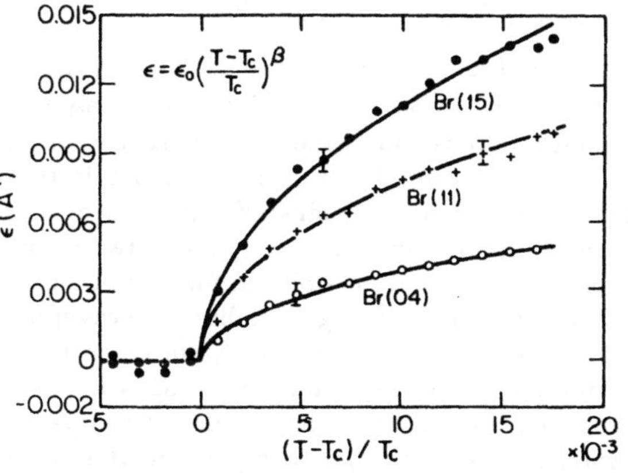

Figure 4: The shifts of the principal peak position as a function of reduced temperature for the Br(0,4,0), Br(1,1,0) and Br(1,5,0) peaks. The solid curves show the power law fits to the model.[1]

for the Br(1,1,0) Bragg peak, and excellent agreement was obtained for the temperature dependence of the Br(1,5,0) and Br(0,4,0) peaks with no adjustable parameters. It is found that the incommensurate peak positions and intensities are well described by a sharp domain wall model with domain wall widths less than one unit cell and with a displacement at the walls of $2b/7$, where b is the lattice constant in the 7-fold direction. As the temperature of the stage 4 sample is further increased, the intercalate layer undergoes a continuous anisotropic two-dimensional melting transition at 102.25 °C, also identified by analysis of the high-resolution x-ray scattering lineshapes.[8]

Although some intercalants exhibit large in-plane defect-free regions (such as the KHg and Br_2 intercalants), other intercalants show island structures. Such island structures are commonly found for intercalants forming incommensurate in-plane structures, and those exhibiting disproportionation phenomena. Disproportionation phenomena relate to the intercalation of one chemical species which results in the formation of new chemical species inside the graphite host material. For example, it is believed that the introduction of $SbCl_5$ into graphite results in the formation of $SbCl_3$, $SbCl_6$ in addition to $SbCl_5$ between the graphite layers. The presence of intercalate island structures is expected to lead to line broadening in spectroscopic studies and to additional scattering channels for transport measurements.

2D Magnetism Studies with GICs

By varying the separation between sequential magnetic layers (i.e., the stage of the compound) through the intercalation mechanism, the interplanar magnetic coupling in graphite intercalation compounds can be reduced in a controlled manner to very small values, thereby providing a convenient system for studying 2D magnetism and the cross-over between 3D and 2D magnetic systems. In addition, by proper choice of intercalant, the spin dimensionality can be varied, as can be seen by considering the magnetic Hamiltonian

$$\begin{aligned} H = & -J\sum_{\langle i,j\rangle}\vec{S}_i\cdot\vec{S}_j - J_A\sum_{\langle i,j\rangle}S_{iz}\,S_{jz} \\ & -J'\sum_{\langle i,k\rangle}\vec{S}_i\cdot\vec{S}_k - J_A{}'\sum_{\langle i,k\rangle}S_{iz}\,S_{kz} \\ & +g_{aa}\mu\sum_{\langle i\rangle}(\vec{H}_6+\vec{H})\cdot(\vec{S}_i-\hat{k}S_{iz})+g_{cc}\mu\sum_{\langle i\rangle}H_zS_{iz} \end{aligned} \quad (8)$$

in which J and J_A are the in-plane exchange and anisotropy energies, J' and J'_A are the inter-plane exchange and anisotropy energies, H_6 is the six-fold in-plane anisotropy field, $\hat{k}$ is a unit vector along the c-axis and g_{aa} and g_{cc} are the in-plane and c-axis components of the g-tensor.[9] By increasing the stage index, the interplanar exchange and anisotropy terms J' and J'_A can be made sufficiently small so that the spins effectively form a 2D spatial system of isolated planes of magnetic spins. The exchange coupling between three dimensional spins $\vec{S}$ on a single magnetic plane has several possibilities for this 2D spatial system. When the anisotropy term J_A is small, then the exchange couplings are essentially isotropic and a Heisenberg spin system results; an example of such a system is the $FeCl_3$-GIC system. The parameters of the spin Hamiltonian of Eq. 8 for $CoCl_2$-GICs are listed in Table 1. If on the other hand, J_A and J have opposite signs and are of comparable magnitude, then only the x and y spin components contribute to Eq. 8 and the system is described by an XY model whereby the spins lie in the magnetic plane; examples of a 2D-XY system are the $CoCl_2$-GIC and the $NiCl_2$-GIC systems. Finally, if J_A is large and positive, then the z components of the spin dominate in Eq. 8 and the system is described

Table 1: Values of the exchange parameters for $CoCl_2$–GIC. The values represent the best estimates to date. The effective spin is $|\vec{S}| = 1/2$.

Stage	J (K)	J_A (K)	J' (K)	$J_A'^{(a)}$ (K)	$H_6^{(b)}$ Oe	g_{aa}	g_{cc}
$0^{(c)}$	$28.5^{(d)}$	$-16.0^{(d)}$	$-2.2^{(d)}$	$1.2^{(d)}$	10	$4.97^{(e)}$	$2.63^{(e)}$
1	28.5	$-14.3^{(f)}$	$-0.02^{(g)}$	0.01	10	$4.6^{(f)}$	$2.80^{(f)}$
2	28.5	$-10.0^{(f)}$	$-0.002^{(h)}$	7.0×10^{-4}	10	$4.02^{(f)}$	$2.86^{(f)}$
$3^{(i)}$	28.5	-10.0	-1.0×10^{-4}	3.5×10^{-5}	10	4.0	2.8

$^{(a)}$ For the intercalation compound, one assumes Lines's[10] condition ($J_A/J = J_A'/J'$).
$^{(b)}$ Estimated from gap in spin–wave spectrum.[11]
$^{(c)}$ Pristine $CoCl_2$.
$^{(d)}$ Inelastic neutron scattering.[12]
$^{(e)}$ Crystal field theory calculation.[13]
$^{(f)}$ High field magnetization experiment.[9] Somewhat higher values for g_{cc} were inferred from the high temperature susceptibility measurements.[14]
$^{(g)}$ Estimated from the reduction in critical field.
$^{(h)}$ Neutron scattering data gives $|J/J'| \sim 10^{-4}$.[15]
$^{(i)}$ All numbers for stage 3 are estimated by extrapolation from the lower stage compounds.

by an Ising model whereby the spins lie perpendicular to the magnetic plane; the $FeCl_2$–GIC system is believed to be an example of a 2D–Ising system. As the temperature and magnetic field are varied, unusual magnetic phase transitions are observed for the various types of spin alignments that can arise in these low dimensional systems. Of particular interest are the high stage compounds which exhibit 2D behavior and the stage 1 magnetic GICs, where cross–over behavior between 3D and 2D spatial dimensionality may dominate.

The number of magnetic GICs that have been synthesized to date is small, and most of these have relatively low magnetic transition temperatures. Both donor (Eu) and acceptor ($FeCl_3$, $FeCl_2$, $CoCl_2$, $NiCl_2$, $MnCl_2$, $MoCl_5$) magnetic GICs have been synthesized and investigated with regard to their magnetic properties. The magnetic transition temperatures in the intercalation compounds tend to be significantly lower than in the pristine parent materials because the number of magnetic nearest neighbors is reduced by the planar geometry; in the intercalation compound, the c–axis nearest neighbors are at much larger distances.

Experimental Evidence for 2D–XY Behavior in GICs

Of the various possible types of 2D magnetic systems, the 2D–XY magnetic systems have been most widely studied, because of the special role these systems play in theoretical studies of critical phenomena, particularly with regard to their unusual magnetic phase transitions. The absence of long-range order in 2D–XY systems[16] which show a magnetic phase transition was reconciled by Kosterlitz and Thouless[17,18,19] who introduced the idea of spin vortices which form bound vortex pairs at the phase transition.

If the interaction between these spin vortices within the two–dimensional plane is governed by a $(1/r)$ force, then the energy of a single vortex has a logarithmic decay

$$E = 2\pi J \ell n(\frac{L}{a}) \tag{9}$$

where J is the exchange energy, a is the nearest neighbor distance and L is the sample size. Kosterlitz and Thouless showed that if a pair of vortices of opposite vorticity are brought together, they will bind at a temperature $T_{KT} \sim \pi J/k_B$ to produce a lower energy state with only short range correlations. For an ideal 2D–XY system, the magnetic susceptibility is expected to diverge at the Kosterlitz–Thouless transition temperature T_{KT}, following the functional form

$$\chi \sim \exp(b/t^{\frac{1}{2}}) \qquad t > 0 \tag{10}$$

where b is a constant and $t = (T - T_{KT})/T_{KT}$ is the reduced temperature for $T > T_{KT}$ in the free vortex phase.

Experimental verification of the Kosterlitz–Thouless predictions requires the synthesis of a truly isolated magnetic layer of infinite extent. The difficulty with the experimental realization of such a system has led theorists to extend the Kosterlitz–Thouless concept to finite size systems[20] and systems with in–plane symmetry–breaking fields,[21] associated with crystalline ordering in the magnetic planes.

From the experimental side, magnetic graphite intercalation compounds have been proposed as experimental systems which approximate 2D–XY behavior. To optimize the XY approximation, the intercalant is chosen to maximize the magnitude of (J_A/J) for a magnetic system where J_A and J have opposite signs, and to achieve 2D spatial behavior it is necessary that high stage compounds can be synthesized. Approximations to such magnetic phases can be achieved in magnetic GICs over a limited range of temperatures and magnetic fields, as described below.

Of the various magnetic GICs that have been synthesized, the $CoCl_2$–GICs exhibit the largest c–axis anisotropy ($J_A/J = 0.56$), and therefore have been selected by several research groups for the study of 2D–XY behavior. (Using a Monte Carlo calculation, Kawabata and Bishop[22] have shown 2D–XY behavior for $J_A/J < 1$.) Furthermore, it is possible to synthesize $CoCl_2$–GICs for a variety of stages, and thus to investigate stage dependent effect and 2D–3D spatial cross–over phenomena. Unfortunately, the magnetic cobalt planes are not quasi–infinite (i.e., the magnetic planes do not extend throughout the sample), but rather break up into separated magnetic islands. However, the structural coherence lengths in the intercalate $CoCl_2$ islands are large enough to contain a statistically meaningful number of spins ($\sim 4 \times 10^3$ spins/island). The island microstructure is very suggestive of the model used in the Monte Carlo simulation of the spin configuration in a magnetic field by Kawabata et al.[23,24]

The experimental work on magnetic GICs exhibiting 2D–XY behavior dates back to the early 1970s when Karimov and others first discovered anomalies in the magnetic susceptibility of a variety of transition metal chloride–GICs.[11,25,26,27] In this pioneering work, attention was also given to the identification of temperature ranges where spontaneous magnetization could be found and where anomalies in the specific heat could be correlated with anomalies in the susceptibility. For most acceptor 2D–XY systems that have been studied, two critical temperatures have been identified (T_{cl} and T_{cu}), thereby delineating three temperature regimes, each exhibiting distinct magnetic behaviors: (1) $T < T_{cl}$, (2) $T_{cl} < T < T_{cu}$ and (3) $T > T_{cu}$. At high temperature ($T \gg T_{cu}$), Curie–Weiss behavior is found.[14,28,29]

Assuming that a Kosterlitz–Thouless transition occurs in high stage magnetic acceptor GICs, experimental studies indicate that the upper critical temperature T_{cu} corresponds to T_{KT} so that the region $T > T_{cu}$ is expected to exhibit free vortex behavior with the binding of the vortices occurring at T_{cu}. Because of the finite size of the 2D magnetic domains and other weak perturbations such as the in–plane symmetry–breaking 6–fold crystalline field anisotropy and the finite interplanar exchange coupling, the observed susceptibility for magnetic GICs does not become infinite at T_{KT}, and therefore experimental studies have been directed to other evidence for 2D–XY behavior. Three measurements in particular support this point.

Firstly, strong evidence that the high stage $CoCl_2$–GICs exhibit 2D–XY behavior comes from comparison of the experimental magnetic susceptibility for a stage 3 compound to the high-temperature series expansion calculation for different spin dimensionalities.[30] Using previously determined values of C and J for pristine $CoCl_2$,[12,31] dimensionless plots of $\chi T/C$ vs. k_BT/J were calculated for a triangular lattice for the 2D–Heisenberg, 2D–XY and 2D–Ising models, using no adjustable parameters, as shown in Fig. 5. The expansions were calculated up to powers of $(J/k_BT)^{10}$ and the coefficients for each power of J/k_BT were found to increase as the exponent increases. The difficulty in obtaining higher order terms in the series is a result of the large number of diagrams that must be counted to calculate their coefficients. A good fit to the experimental data is obtained for the 2D–XY model, while the fits to the 2D–Heisenberg and 2D–Ising models are poor.

Further evidence in support of the 2D–XY model comes from the extension of the Kosterlitz–Thouless calculations to include (1) finite size effects imposed by the observation of magnetic intercalate islands, (2) 6–fold in–plane crystalline symmetry–breaking fields and (3) weak interplanar coupling between spins on different magnetic planes.[30] The number of spins contained in a magnetic island was determined by comparison of the calculated susceptibility and experimental measurements over a wide temperature range.

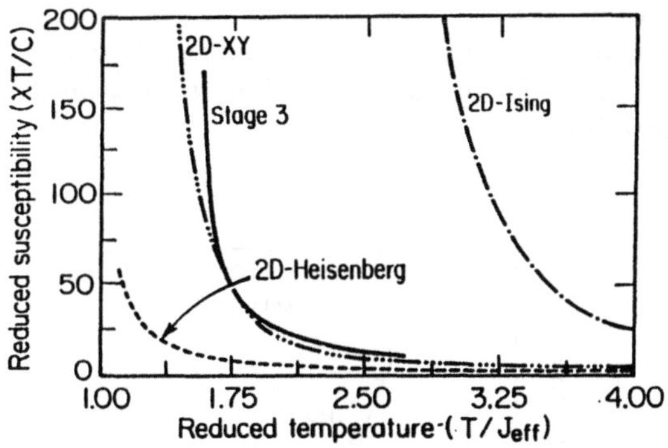

Figure 5: High temperature series expansion. The solid curve represents the experimental data which has not been scaled. The series expansions use the value $J_{eff} = JS^2$ where J is listed in Table 1.[32]

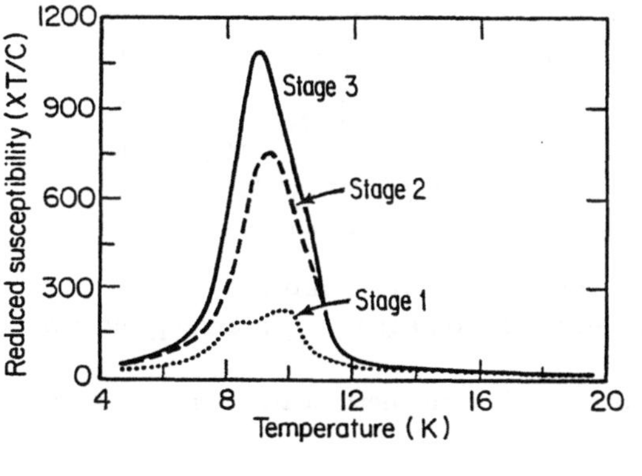

Figure 6: Stage dependence of the low temperature magnetic susceptibility for stage 1, 2, and 3 $CoCl_2$–GICs. (From the work of Chen et al.[33])

Further refinements in the calculation are needed to yield quantitative agreement between the calculated and measured temperature dependence of the susceptibility.

Inelastic neutron scattering measurements on a stage 2 $CoCl_2$–GIC also provide strong experimental evidence in support of the 2D–XY model.[15] In this experiment the spin configuration is probed directly. In addition to the magnetic Bragg peaks at $(0,0,\frac{1}{2})$ associated with the spin ordering below T_{cl} into an antiferromagnetic stacking of ferromagnetically ordered planes of spins, broad inelastic neutron scattering is observed. As shown in Fig. 7, the integrated intensity of the Bragg scattering peak vanishes at T_{cl} while that for the broad diffuse background scattering falls rapidly as T_{cu} is approached (identified with the short range correlation of the vortex pairs) and a more gradual approach to zero scattering intensity is found above T_{cu} (identified with the much weaker scattering of the free vortices above T_{cu}). The availability of a theoretical model for the temperature dependence of the diffuse scattering for free and bound vortices would be especially helpful for a definitive identification of the binding of vortices at T_{cu}.

The neutron scattering studies show no dependence of the diffuse scattering intensity on wave vector along the c–direction in the temperature range $T_{cl} < T < T_{cu}$, consistent with the spins lying in the basal plane in a bound vortex configuration with strong in–plane spin–spin correlation but negligible interplanar correlation. Above T_{cu}, the absence of a remanent magnetization and a linear dependence of magnetization on magnetic field[34] identifies the phase above T_{cu} with paramagnetism, consistent with short-range order of the spins in a free vortex configuration.

Weak perturbations, such as the 6–fold in–plane symmetry–breaking crystal field H_6 and the interplanar magnetic interaction J' become important at lower temperatures and give rise to a 2D–3D cross–over transition at T_{cl}. The spontaneous magnetization due to the coupling of the spins to H_6 produces ferromagnetically ordered in–plane domains below T_{cl}. At zero external magnetic field, the antiferromagnetic interplanar exchange interaction couples these ferromagnetic planar domains in an antiferromagnetic stacking arrangement, as implied by the Bragg peaks in neutron scattering experiments.[15]

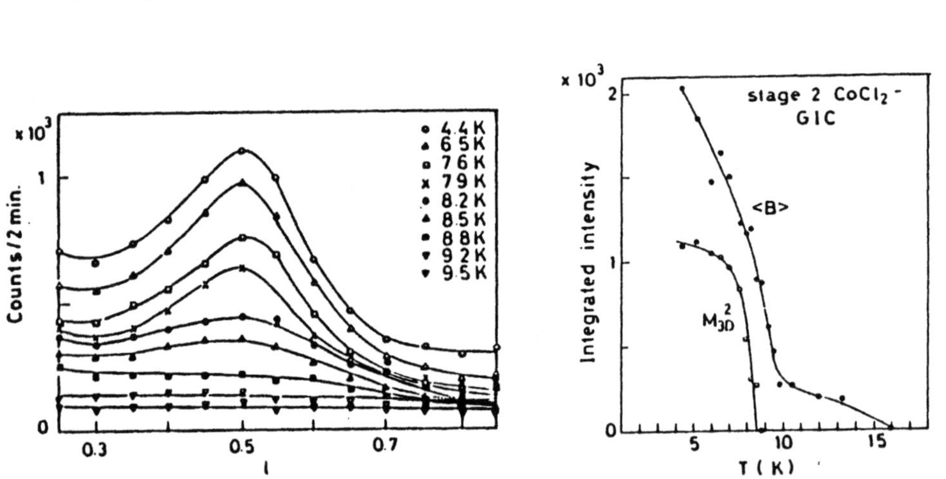

Figure 7: Magnetic neutron scattering from a stage 2 $CoCl_2$–GIC sample.
(a) The peak occurs at $(0,0,\frac{1}{2})$, consistent with an antiferromagnetic spin arrangement on adjacent planes.
(b) Plot of the peak intensity and background from (a). The magnetic scattering background is identified with scattering from bound spin vortices.[15]

Since the magnitudes of these magnetic perturbations are very small, weak magnetic fields have a major effect on the magnetic properties of the $CoCl_2$–GICs. For example, magnetic fields of a few Oe are sufficient to suppress the K–T transition and to establish long-range spin order. At lower temperatures, small externally applied magnetic fields also compete with the small 6–fold in–plane anisotropy field in establishing the spin direction of the magnetic domains. Different magnetic field dependences are found at T_{cl} and T_{cu} for the various magnetic properties such as the magnetic susceptibility, magnetization and heat capacity, indicative of the different origins of these phase transitions. The magnetic phases are also observed to be stage dependent, with the behavior of the stage 1 $CoCl_2$–GICs distinctively different from that of the higher stage compounds.

We now discuss the interesting magnetic phases observed at low temperature in the stage 1 $CoCl_2$–GIC as a function of magnetic field. Assuming that the spins in a given magnetic plane are ferromagnetically aligned along an easy in–plane axis, with the antiferromagnetic interplanar exchange interaction strong enough to produce an antiferromagnetic arrangement of the ferromagnetically aligned planes, theoretical calculations predicted two phase transitions as a function of magnetic field.[30] At the lower critical field, a transition from an antiferromagnetic to a spin–flop arrangement of the ferromagnetic planes was predicted, followed by a transition to a ferromagnetic interplanar alignment of the ferromagnetic planes. Experiments subsequently carried out on the magnetic field dependence of the magnetic susceptibility at constant temperature for a stage 1 $CoCl_2$–GIC exhibited two peaks for temperatures below T_{cl}. A magnetic phase diagram was constructed based on the spin flop transitions as determined from susceptibility, as shown in Fig. 8. In the absence of sufficient interplanar coupling J', the magnetization in the ferromagnetic planes would be aligned along any one of the 6 easy axis orientations with equal probability, and only a single broad peak would be expected, associated with the flipping of the magnetization from any one of the easy axis directions to alignment along the applied field direction. Such behavior is observed experimentally for $CoCl_2$–GICs with stages ≥ 2.[34] Magnetic field–dependent behavior qualitatively similar to that for the stage 1 $CoCl_2$–GIC has also been reported for stage 2 $NiCl_2$–GICs,[35,36] suggesting larger J' values for the $NiCl_2$–GICs relative to $CoCl_2$–GICs of similar stage.

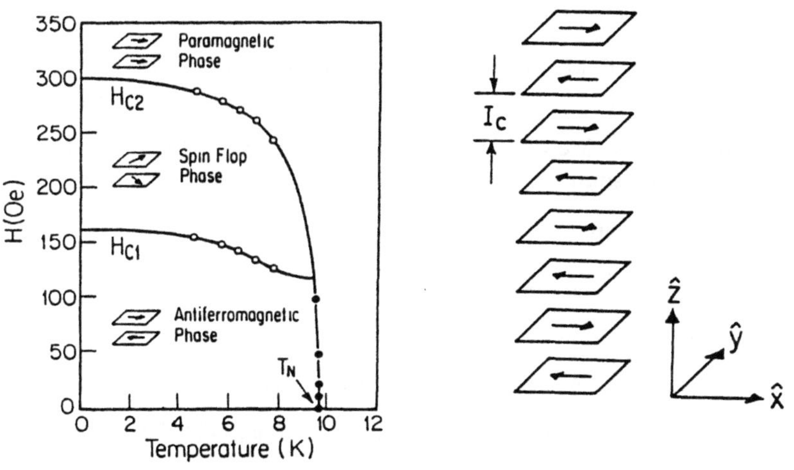

Figure 8: Magnetic phase diagram for a stage 1 $CoCl_2$–GIC . The magnetic phase diagram is derived from the peaks in the magnetic susceptibility.[30]

For $CoCl_2$–GICs, the intraplanar ferromagnetic coupling constant J (28.5 K) is much larger than the interplanar antiferromagnetic coupling constant J' (2.1 K for stage 1). Because the nearest neighbor in–plane coupling is ferromagnetic, we can assume that spins in the same magnetic plane all point in the same direction at low temperature (e.g., $T < 0.1\ J$) and behave like a single spin. Because of the large XY anisotropy ($J_A/J \sim 0.56$), we can also assume that all spins lie in the magnetic plane. The magnetic problem for the low-temperature phase ($T < T_{cl}$) can thus be approximated by a planar rotator model with space dimensionality equal to one (along the c–axis) and spin dimensionality equal to two (spins lie in the c–plane).

Acknowledgments

The authors wish to thank Prof. R.J. Birgeneau, Drs. S.T. Chen, and K.Y. Szeto for many important discussions. This work was supported by AFOSR contract #F49620–83–C–0011.

References

1. A. Erbil, A. R. Kortan, R. J. Birgeneau, and M. S. Dresselhaus. *Phys. Rev.*, **B28**, 6329, (1983).
2. S. G. J. Mochrie, A. R. Kortan, R. J. Birgeneau, and P. M. Horn. *Phys. Rev. Lett.*, **53**, 985, (1984).
3. R. Moret. In M. S. Dresselhaus, editor, *Intercalation in Layered Materials*, page 185, Plenum Press, New York, NY, (1987).
4. A. Erbil. *Phase Transitions in the intercalated Graphite–Bromine system.* PhD thesis, Massachusetts Institute of Technology, June 1983. Department of Physics.
5. J. Frenkel and T. Kontorova. *Phys. Zeits. d. Sowjetunion*, **13**, 1, (1938).
6. F. C. Frank and J. H. Van der Merwe. *Proc. Roy. Soc. London, Ser. A*, **198**, 216, (1949).
7. V. L. Pokrovsky and A. L. Talapov. *Phys. Rev. Lett.*, **42**, 65, (1979).
8. S. G. J. Mochrie, A. R. Kortan, P. M. Horn, and R. J. Birgeneau. *Phys. Rev. Lett.*, **58**, 690, (1987).
9. J. T. Nicholls, Y. Shapira, E. J. McNiff, Jr., and G. Dresselhaus. *Synthetic Metals*, (1987). 4th International Symposium on Graphite Intercalation Compounds, Jerusalem.
10. M. E. Lines. *Phys. Rev.*, **131**, 546, (1963).
11. Y. S. Karimov. *Soviet Phys. JETP*, **39**, 547, (1974).
12. M. T. Hutchings. *J. Phys. C*, **6**, 3143, (1973).
13. G. Mischler, D. J. Lockwood, and A. Zwick. *J. Phys. C*, **20**, 299, (1987).
14. D. G. Wiesler, M. Suzuki, P. C. Chow, and H. Zabel. *Phys. Rev.*, **B34**, 7951, (1986).
15. M. Suzuki, D. G. Wiesler, P. C. Chow, and H. Zabel. *J. Mag. Mag. Materials*, **54–57**, 1275, (1986).

16. N. D. Mermin and H. Wagner. *Phys. Rev. Lett.*, **17**, 1133, (1966).

17. J. M. Kosterlitz and D. J. Thouless. *J. Phys. C*, **6**, 1181, (1973).

18. J. M. Kosterlitz and D. J. Thouless. In D. F. Brewer, editor, *Progress in Low Temperature Physics*, page 395, North–Holland, Amsterdam, (1978).

19. J. M. Kosterlitz and M. A. Santos. *J. Phys. C: Solid State Phys.*, **11**, 2835, (1978).

20. K. Y. Szeto and G. Dresselhaus. *Phys. Rev.*, **B32**, 3186, (1985).

21. J. V. José, L. P. Kadanoff, S. Kirkpatrick, and D. R. Nelson. *Phys. Rev.*, **B16**, 1217, (1977).

22. C. Kawabata and A. R. Bishop. *Solid State Commun.*, **60**, 167, (1986).

23. C. Kawabata, M. Takeuchi, and A. R. Bishop. *J. Mag. Mag. Materials*, **54–57**, 871, (1986).

24. C. Kawabata and A. R. Bishop. *Z. Phys. B, Condensed Matter*, **65**, 225, (1986).

25. Y. S. Karimov, M. E. Vol'pin, and Y. N. Novikov. *Zh. Eksp. Teor. Fiz. Pis. Red.*, **14**, 217, (1971).

26. Y. S. Karimov, A. V. Zvarykina, and Y. N. Novikov. *Fiz. Tverd. Tela.*, **13**, 2836, (1971).

27. A. V. Zvarykina, Y. S. Karimov, M. E. Vol'pin, and Y. N. Novikov. *Fiz. Tverd. Tela.*, **13**, 28, (1971).

28. Y. S. Karimov. *JETP Lett.*, **41**, 772, (1976).

29. A. V. Zvarykina, Y. S. Karimov, M. E. Vol'pin, and Y. N. Novikov. *Sov. Phys. -Solid State*, **13**, 21, (1971).

30. K. Y. Szeto, S. T. Chen, and G. Dresselhaus. *Phys. Rev.*, **B32**, 4628, (1985).

31. M. K. Wilkinson, J. W. Cable, E. O. Wollan, and W. C. Koehler. *Phys. Rev.*, **113**, 497, (1959).

32. K. Y. Szeto. *Two–Dimensional XY Models and Its Application to Graphite Intercalation Compounds.* PhD thesis, Massachusetts Institute of Technology, January 1985. Department of Physics.

33. S. T. Chen, K. Y. Szeto, M. Elahy, and G. Dresselhaus. *J. de Chimie Physique*, **81**, 863, (1984).

34. S. T. Chen. *Magnetic Properties of Graphite Intercalation Compounds.* PhD thesis, Massachusetts Institute of Technology, November 1985. Department of Physics.

35. M. Matsuura, Y. Murakami, K. Takeda, H. Ikeda, and M. Suzuki. *Synthetic Metals*, **12**, 427, (1985).

36. I. Oguro, M. Suzuki, and H. Yasuoka. *Synthetic Metals*, **12**, 449, (1985).

Structure and Modeling of 2D Alkali Liquids in Graphite

S.C. Moss[1], X.B. Kan[1], J.D. Fan[1], J.L. Robertson[1], G. Reiter[1], and O.A. Karim[2*]

[1]Department of Physics, University of Houston, Houston, TX 77004, USA

[2]Department of Chemistry, University of California, Berkeley, CA 94720, USA

1. Introduction

When graphite samples are exposed at elevated temperature to the vapor phase of either alkali metals or a variety of molecular species, they form compounds in which the normal sequence of stacked hexagonal graphite layers is regularly interrupted by layers of the (so-called) intercalated species. These graphite intercalation compounds, or GIC's /1,2/, thereby provide us with very interesting examples of two-dimensional (2D) systems of interacting particles in a periodic host lattice. The competing interactions in the GIC's that determine the structure and dynamics of the intercalated layer are a) the interatomic or intermolecular potential and b) the host-intercalant potential. There are also, of course, the interlayer interactions responsible for the regular sequencing of the intercalated species and these are repulsive due both to coulombic and elastic effects /3/.

We shall here concern ourselves with the 2D layer and its immediate graphite bounding planes. In particular we will consider the liquid state of the alkali metals, Rb and K, intercalated to stage 2 (two graphite planes separating every alkali layer). This liquid state prevails at room temperature, is highly correlated within the layer and is incommensurate with its graphite host; it is uncorrelated from layer to layer /4-6/. On cooling, the stage 2 alkali liquids eventually freeze into incommensurate solids, generally believed to be composed of a domain-discommensuration state /7-10/ whose domain size is directly related to the alkali planar density or alkali/graphite spacing mismatch /7-10/. The melting is thought to proceed via the disordering or thickening of domain walls to give a loss of "solid-like" behavior with increasing temperature. Evidence for this model comes both from static structural studies of the phase transition /5,11,12/ and from extensive neutron studies of the dynamical response of the 2D intercalant layer through the melting transition and well into the liquid state /12/.

The structural studies taken together provide an extraordinarily rich collection of data on which to base a detailed model of the liquid state and its phase transitions. The issues are not so different from those encountered in the melting of rare gases on the surface of graphite, to which so much attention has been directed over the past several years /13,14/. Certainly the interactions here are different, particularly for the alkali metals which are donor species contributing essentially one electron per alkali atom to the neighboring graphite layers /1/. This results in a screened (repulsive) alkali-alkali planar interaction and an attractive alkali-graphite interaction which assumes the periodicity of the graphite latttice. It is the balance between these two that determines the structure and dynamics of the layer. However, where the weakly interacting rare gases on graphite offer a relative paucity of direct structural data on the 2D liquid state, the situation with the GIC's is more favorable. This is due to the availability of large samples of highly-oriented-pyrolytic-graphite (HOPG) and excellent (small) graphite single crystals both of which may be intercalated to provide an extended stack of 2D liquids. We may thus study lattice dynamics /15/ and diffusion /12,16/ in intercalated HOPG along with the X-ray scattering in HOPG and single cyrstal samples. In particular, the observation of 2D alkali liquid scattering well above the melting transition in a graphite single crystal /17,18/ would currently be difficult to observe with gases adsorbed on graphite.

While the experimental situation reviewed above for the alkali GIC's is excellent, the theoretical understanding is less advanced. We have either incomplete theoretical fits to structure factor data on the ordered phases /10/ or phenomenological fits, of varying degrees of success /19,20/, which suffer from being somewhat ad-hoc. Calculating the dynamical response function for the modulated liquid phase has, so far, not been done. An alternate approach is to use molecular dynamics (MD) to simulate the 2D liquid state with appropriate values of the competing interaction potentials as inputs. This approach relies on earlier analytical work by PLISCHKE /21/ on an unmodulated liquid which was extended by PLISCHKE and LECKIE /22/ to a modulated system using a Monte Carlo procedure in which the screened alkali-alkali potential was derived from the work of VISSCHER and FALICOV /23/. The attractive substrate-alkali interaction, which we refer to here as the graphite modulation potential, was chosen as a simple sine wave of assumed amplitude.

The aim of our own program has been to extract the alkali-graphite interaction, or graphite modulation potential, directly from experimental data using a method developed by REITER and MOSS /24/ and subsequently applied to a stage 2 Rb GIC by MOSS et al. /25/. This attractive potential may then be introduced into the MD calculation, along with a parameterized Visscher-Falicov potential /23/, to calculate initially the structure factor S(q) for the 2D alkali liquid for which synchrotron X-ray data is available /25,26/. Once the measured static structure factor has been well fit, our task is to generate $S(q,\omega)$, the dynamical structure factor from which both the vibrational density of states and the diffusion may be evaluated. Finally, with these experimental comparisons in hand we will freeze the liquid and/or melt the solid on the computer to generate the temperature-dependent domain-discommensuration state, and investigate the suggested /12/ domain-wall melting, should such be appropriate to our system. (When the domains are small the distinction between boundary and interior is difficult to make).

The present paper is a progress report on this work. Section 2 reviews the theory of the modulated liquid state /24/ in which the Bragg contributions of alkali to graphite are outlined. Section 3 presents prior results on Rb /25/ together with results of a recent analysis of single crystal X-ray data on a stage 2 K GIC (~ $C_{24}K$) to be reported in detail elsewhere /27/. A comparison of the modulation potentials for the two alkalis is given. In Section 4, we present new results comparing the experimental S(q) for the stage 2 Rb liquid with the MD simulation using directly the experimentally determined modulation potential. The agreement there is encouraging.

2. Theory of the Modulated Liquid

REITER AND MOSS /24/ have developed the theory for the X-ray scattering from a 2D liquid modulated by its periodic host. This theory is applicable to all 2D systems within or on crystalline substrates and it would be quite interesting to apply it to the disordered state of the rare gases and molecular species on graphite. Be that as it may, we confine our attention here to alkali metals intercalated in graphite. The major effects encountered in the diffraction patterns of these 2D liquids, for which the theory /24/ was developed, are:

a) The circular average of the alkali liquid structure factor from intercalated HOPG shows a familiar liquid-like diffuse scattering whose Fourier

(Bessel) transform yields a coordination number of six and an alkali metal nearest-neighbor spacing of ~ 5.90 Å - 6.10 Å /11,28/. This liquid is incommensurate with its graphite host whose lattice parameter is 2.46 Å.

b) The liquid scattering pattern from a single crystal /17,18/ shows that circularly averaged diffuse scattering, observed in HOPG, is quite anisotropic; it is modulated in a six-fold angular fashion by the subtrate field and is also replicated, as halos, about the $\{10.0\}$ graphite reciprocal latttice points.

c) The alkali metal makes an appreciable δ-function-like contribution to the graphite Bragg peaks /6/ through the presence in the liquid of static density waves with the graphite periodicity.

While the angular modulation of the liquid scattering and the halos about graphite $\{10.0\}$ positions supply the most visible evidence for a substrate or host modulation of the liquid, these effects appear to higher order in the relevant Fourier coefficients of the modulation potential /24/. They are, therefore, more difficult to calculate. The Bragg contributions arise directly from a scattering off induced static density waves and may be calculated in a more straightforward fashion. We therefore have used the theory to extract an accurate modulation potential $V(\vec{r})$ from experimental graphite Bragg intensities. This potential is then used in the MD simulation to generate the diffuse liquid scattering, $S(\vec{q})$, to compare with X-ray data.

The planar modulation potential, $V(\vec{r})$, may be written

$$V(\vec{r}) = \sum_{H,K} V_{HK} \exp(i\vec{q}_{HK}\cdot\vec{r}), \tag{1}$$

where $\vec{q}_{HK}$ is a basal plane vector of the graphite reciprocal lattice. We wish to evaluate $\langle\rho_q\rangle$, the Fourier transform of the alkali metal number density which will be non-zero at the graphite reciprocal lattice points in the presence of the periodic potential in (1). If this potential is quite small (i.e. all $\beta V_{HK} = V_{HK}/kT \le 0.1$), linear response theory /24/ may be employed to evaluate $\langle\rho_q\rangle$. If, in addition, the planar liquid structure factor $S(\vec{q})$ for the metal alkali is unity at all the graphite HK positions, a particularly simple form for the HK.L structure factor for a single layer or sandwich of C-alkali-C is obtained /24/, where we shall use Rb below to stand for the alkali species:

$$F(HK.L) = 4f_C e^{-M_C} \{\cos \frac{2\pi}{3} (H+2K)\cos(\pi L C_1)\} + \frac{X'}{6} f_{Rb} e^{-M_{Rb}^{\perp}} (\delta_{q_\parallel,0} - \beta\delta_{q_\parallel,q_{HK}} V_{HK}) , \tag{2}$$

where C_1 is the ratio of the sandwich thickness to the total c axis, X' is the ratio of the ideal concentration (1/24) to the actual value and $M_{Rb}^{\perp}$ is the Rb Debye-Waller factor along the c axis. In fact, the potential in the alkali GIC's is not so small and the liquid structure factor, $S(\vec{q})$, is appreciably different from 1.0 at the 10.0 graphite reciprocal lattice point. In this case, more care must be taken, but an expression similar to Eq. (1) results /24/:

$$F(HK.L) = 4f_C\, e^{-M_C} \{\cos \frac{2\pi}{3} (H+2K) \cos (\pi L C_1)\} + \frac{X'}{6} f_{Rb}\, e^{-M_{Rb}^{\perp}} \langle\rho'_{q_{HK}}\rangle . \tag{3}$$

$\langle\rho'_{q_{HK}}\rangle$ is simply the normalized value of $\langle\rho_q\rangle$ evaluated at the Bragg peaks:

$$\langle \rho'_{q_{HK}} \rangle = \frac{1}{N} \langle \sum_i e^{i\vec{q}_{HK}\cdot\vec{r}_i} \rangle \tag{4}$$

where the $\vec{r}_i$ are the in-plane positions of Rb atoms. $\langle \rho'_{q_{HK}} \rangle$ is given by

$$\langle \rho'_{q_{HK}} \rangle = [F_1(\vec{q}_{HK})/F_1(0)]\, e^{F_2(\vec{q}_{HK}) - F_2(0)} \quad , \tag{5}$$

where

$$F_1(\vec{q}_{HK}) = \int_0^a \int_0^a e^{i\vec{q}_{HK}\cdot\vec{r}} \exp[-\beta \sum_{H'K'} V_{H'K'} \exp(i\vec{q}_{H'K'}\cdot\vec{r})] d^2r \tag{6}$$

and

$$F_2(\vec{q}_{HK}) = \sum_{H'K'} F_1(\vec{q}_{H'K'}) F_1(-\vec{q}_{HK} + \vec{q}_{H'K'})$$
$$\times [S(\vec{q}_{H'K'}) - 1]/F_1(0)F_1(\vec{q}_{HK}) \quad . \tag{7}$$

If the value of $S(\vec{q}) = S(\vec{q}_{HK})$ is unity at all values of $\{HK\}$, $F_2(\vec{q}_{HK}) = F_2(0) = 0$, and we need only the integrals for $F_1(q)$. The effect of correlation in the liquid on the estimate of the δ-function-like Bragg part arises because the response of the liquid is not that of a random dense gas. The $S(q_{HK})$ that enters Eq. (7) is actually that of the unperturbed (isotropic) liquid, which is not experimentally accessible. The actual $S(\vec{q})$ will have features induced by the modulation /17,18,24/. In particular the circular average, $S(q)$, measured in HOPG, shows bumps or shoulders attributable to the halos /6/. In MOSS et al. /25/ these features were smoothed out to give an $S(q)$ in Fig. 1(a) of Ref. /25/ more appropriate to the unperturbed liquid; but it is not crucial for the evaluation of $\langle \rho'_{q_{HK}} \rangle$ because at the values of q_{HK} of interest, numerical work /29/ shows that the modulated and unmodulated liquids have essentially the same $S(q)$.

3. Experimental Evaluation of V(r)

The procedure for evaluating βV_{HK}, using Eqs (5) - (7), requires experimental values of $\langle \rho'_{q_{HK}} \rangle$, the alkali metal contribution to the total structure factor at graphite Bragg positions. In the case of Rb, these have been obtained in two ways. In the first instance /6/, we measured the integrated intensities of 85 Bragg reflections from $C_{24}Rb$ in HOPG in which an alkali contribution was formally included. The standard crystallographic least squares procedure then yielded values for the sandwich thickness, the alkali concentration, the Debye-Waller factors for both carbon and Rb and the $\langle \rho'_{q_{HK}} \rangle$. [Actually the original paper of OHSHIMA et al. /6/ treated the alkali atoms as fractionally registered after NIXON et al. /30/. Because the analysis gave a physically absurd planar Debye-Waller factor for the (registered) Rb atoms, the data were re-analysed to provide values for $\langle \rho'_{q_{HK}} \rangle$ reported in Ref /25/]. This procedure, while fine in principle, had to be checked because an appreciable stacking fault density introduced on intercalation made the isolation of the separate Bragg peaks problematical. We have since overcome this problem by doing continuous profile fittings of those reflection sequences affected by faulting /27,31/; it was nonetheless felt that an alternate method for separately identifying the Rb contribution would be useful. We have thus employed the anomalous scattering method at the Stanford Synchrotron Radiation Laboratory (SSRL) together with profile fitting /31/ to estimate directly $\langle\rho'_{q_{HK}}\rangle$ for $C_{24}Rb$. Both of the above methods /6,31/, done on separate samples, gave remarkably good agreement /25/.

For K-intercalated graphite we have recently collected Bragg data using a small single crystal intercalated by P. Chow and H. Zabel of the University of Illinois. These data were taken on the Oak Ridge National Laboratory beam line at the National Synchrotron Light Source (NSLS) and a report of the experiment and data analysis is currently being prepared /27/ . Approximately 60 reflections were collected but, in this case, the stacking of the graphite-alkali-graphite sandwiches was essentially random at every layer (either hexagonal or cubic sequences). This extreme of faulting made it quite easy to fit the L-dependent profiles taken at a particular value of HK for the fault-broadened reflection sequences (i.e. $H-K \neq 3m$, m = integer). With L as a continuous variable in Eq. (3) it is clear, for example, that there will be zeros in F(HK.L) that depend very sensitively on $\langle\rho'_{q_{HK}}\rangle$. The unbroadened profiles were fit in a more conventional fashion and the 00.L reflections were used, as before /6,31/, to assign structural parameters other than $\langle\rho'_{q_{HK}}\rangle$.

Table I presents the results of our evaluation of $\langle\rho'_{q_{HK}}\rangle$ for stage 2 samples of both Rb-and K-intercalated graphite. The evaluation of βV_{HK} introduces these experimental values into Eqs. (5)-(7) along with a trial set of $\beta V_{H'K'}$ and the experimental $S(q_{H'K'})$ on the right-hand side. First, the sum of the squared differences between the experimental and calculated $\langle\rho'_{q_{HK}}\rangle$ is minimized with the use of Newton's method. Then nonlinear regression analysis is used to refine the true position of the equilibrium minimum. The speed of convergence and agreement between input and output $\langle\rho'_{q_{HK}}\rangle$ indicated a successful set of βV_{HK}'s and these are included in Table I.

The comparison between Rb and K results in Table I is instructive. For Rb, which is the slightly larger atom, the potential can be well approximated by one set of sine waves along the six {10.0} directions. Indeed, in a MD simulation using only V_{10} from Table I, we were able to reproduce accurately the set of measured $\langle\rho'_{q_{HK}}\rangle$ /25/. The modulation of the K liquid appears to require values for the higher Fourier coefficients and this will introduce additional structure

TABLE I

Experimental values of the alkali metal contribution to the graphite structure factor together with derived values of βV_{HK}

	Rb (ref. /25/)		K (ref. /27/)	
HK	$\langle\rho'_{q_{HK}}\rangle$	βV_{HK}	$\langle\rho'_{q_{HK}}\rangle$	βV_{HK}
10	0.48±.03	-0.45	0.43±.04	-0.34
11	0.21±.03	-0.06	0.26±.03	-0.16
20	0.14±.02	-0.01	0.18±.04	-0.05
21	0.04±.02	0.03	0.05±.03	0.06
30	0.04±.03	-0.01	0.07±.04	-0.03
22	0.01±.03	-0.02	--------	-----

into $V(\vec{r})$. This is clear in the comparison of the contour plots for Rb and K in Fig. 1. The Rb modulation potential in Fig. 1(a) shows a barrier height of ΔV = 0.092 eV between the hexagon center and the saddle point between 2 carbon atoms. For K this barrier height is 0.085 eV which is in qualitative agreement with the fact that the K liquid freezes at $T_M \sim$ 123K while the Rb liquid freezes at $T_M \sim$160K. [Equating the ratio of T_M's to the ratios of ΔV's would not be correct because the repulsive part of the potential is not included; but it would be surprising if the Rb ordered at a lower temperature than the K]. The K potential also shows a local minimum over the carbon atom sites indicating a relative attraction. Formally, of course, this appears because a value of $V_{10} < 0$ produces a minimum over the hexagon center and a maximum over the carbon while $V_{11} < 0$ produces minima over both sites. If the Fourier coefficients fall off in a physically reasonable way (i.e. monotonically and rapidly), there will, thereby, always be a local minimum over the carbon site. It is also interesting that βV_{21} in both of the $C_{24}Rb$ samples /25/ and in the $C_{24}K$ sample shows a small sign reversal from the other coefficients.

The usefulness of these experimentally derived potentials comes both from direct comparison with other experiments and from their further use in the simulation studies. In the former, as we have noted earlier /25/, the values of ΔV fall within the values for the diffusional barrier height obtained from quasielastic neutron scattering /16,32/.

(a) (b)

Figure 1. Contour maps of the substrate modulation potential, $V(\vec{r})$, for stage 2 alkali liquids in graphite. The change from attractive [$V(r) < 0$] to repulsive is noted by a dashed contour: (a) the Rb potential in which the contours are in steps of 0.0037 eV and the barrier height, ΔV, between hexagon center and bond midpoint, is 0.092 eV; (b) the K potential with contour steps of 0.005 eV and ΔV = 0.085 eV.

4. Molecular Dynamics (MD)

Our MD work on these 2D liquids is currently in progress /29/, and a separate report on the simulation of the static structure is in preparation /33/. In MOSS et al. /25/ we reported on the use of the MD calculation to retrieve $\langle\rho_q\rangle$ for Rb in graphite from the average over several configuration flash-shots using only βV_{10} (T = 300 K) and a repulsive alkali-alkali potential as inputs. The agreement between the experimental set of $\langle\rho'_{qHK}\rangle$ and the MD output was quite good. We describe here the results for S(q), the liquid structure factor, which is obtained via a Bessel function transform of G(r), the circular average of the normalized pair distribution function for the modulated 2D liquid. The circular average is required to compare with data on HOPG samples in which the graphite grains are randomly oriented about a common c axis.

We have worked with a number of system sizes up to ~ 800 atoms, but most of our studies have been on a 200 atom array in a two-dimensional box composed of 40 x 30 graphite hexagonal unit cells (2.46 Å lattice parameter). By varying the Rb number density we could also check the dependence of the nearest-neighbor distance in G(r) - and the position of the first maximum in S(q) - on the in-plane alkali atom density. As inputs to the MD calculation we now use the full set of Fourier coefficients of the Rb modulation potential listed in Table I, or in Ref. /25/, setting β equal to its room temperature value. The intraplanar repulsive interaction is again of a screened coulomb form but is more complicated than a simple Yukawa potential because the experimental situation really consists of a layer of Rb ions between two sheets of donated charge. VISSCHER and FALICOV /23/ have treated this situation and we have used their formalism, as noted earlier. The only adjustable parameter in their potential is the planar dielectric constant which may be viewed as an adjustment for the effective charge transfer. In our case a value of ε = 2.0 was found to work ($K_0 = 4\pi\varepsilon$). This value is equivalent to a diminshed charge transfer of 0.7 rather than 1 full electron per alkali atom. The other variable in the repulsive potential is the sandwich layer thickness which is experimentally fixed at 5.71 Å /26/ for the stage 2 Rb GIC.

The MD output was evaluated for a total of 600 flash shots each separated by 10 time intervals of $\Delta t = 2 \times 10^{-3}$ ps. A pair distribution function G(r) was calculated for each shot and an average over the total was taken. This time-averaged G(r), for X-ray scattering, approaches the spatial average for a large system.

At this stage of the analysis there are two aspects of the G(r) which must be dealt with in taking a proper transform. The first is the familiar problem of taking an infinite transform of data in a finite box. It is not so severe here because our box is ~ 16 atoms x 12 atoms (it is not rectangular; for a square box we would have ~ 14 atoms per side). It is therefore about three times larger than the correlation range and the normal truncation problems are easy to deal with. The more demanding feature is the presence of a periodic modulation in G(r) which would give rise to a δ-function Bragg contribution at the graphite position if the system were infinite. This modulation does not decay with distance and the transform of the finite-sized liquid pattern will thus contain broad Bragg features and significant ripple or ringing. Suffice it to note here that there are a couple of ways of dealing with this and we have chosen one which cleanly removes the periodic component in G(r), leaving intact, of course, all of the other influences of the substrate on G(r) or S(q). Details of this procedure will be given elsewhere /33/.

In Fig. 2 we compare the MD simulated S(q) for the 2D Rb stage 2 liquid with reduced and normalized data from a HOPG sample /26/. The shape and position of the first peak in the experimental S(q) /6,26/ are well reproduced. Other experimental features are also apparent in the simulation. These include the general period of the familiar damped oscillatory structure factor as well as the subsidiary shoulders at $q \simeq 1.90$ Å^{-1} and $q \cong 4.05$ Å^{-1}. As we have discussed before /6/, these two features represent the circular average of the contribution of the inner and outer portions of the halos about {10.0} in the single crystal patttern. They are, therefore, direct evidence of the influence of the substrate on the liquid structure factor. That they appear in the simulation at the correct place in q and with about the correct amplitude is, encouraging. We feel that the agreement in Fig. 2 is good and we are now calculating the liquid pattern for a single crystal to compare with available X-ray data /17,18/. Once we are satisfied that our potentials can reproduce the

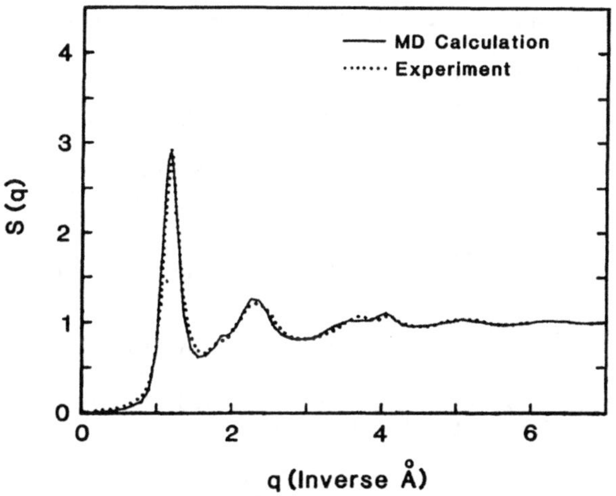

Figure 2. S(q), the circular average of the 2D liquid structure factor, for Rb intercalated in HOPG. The experimentally determined plot /26/ has been reduced and normalized to compare with the MD simulation; note the subsidiary maxima at $q \cong 1.90\ Å^{-1}$ and $q \simeq 4.05\ Å^{-1}$ which arise from the circular average of the modulation halos /6,17,18,24/. The Bragg contribution has been removed from both experimental and simulated plots (see text).

details of the anisotropic static structure factor we will proceed to an evaluation of $S(q,\omega)$ [in MD one gets them both at the same time].

5. Summary

Using experimental X-ray data on the Bragg and liquid-like components of the modulated 2D alkali liquid scattering pattern, we have extracted meaningful graphite modulation potentials for both Rb and K liquids. Molecular dynamics has then been employed, using the measured graphite-Rb potential, to generate a liquid S(q) in good agreement with experiment.

Acknowledgements:

We thank C. J. Sparks and K. Ohshima for their collaboration on the $C_{24}K$ single crystal study. O.A.K. wishes to express his appreciation to J. A. McCammon of the Chemistry Department, University of Houston, for advice and for support of the program of molecular dynamics used here. Work on the intercalates is supported by the NSF under grant no. DMR-8603662.

* Research performed in the Chemistry Department, University of Houston

References

1. S. A. Solin: Adv. Chem. Phys. 49, 455 (1982)
2. R. Moret: in Intercalation in Layered Materials, ed. by M. S. Dresselhaus, NATO ASI Series, Series B Physics 148, 185 (1986)
3. G. Kirczenow: Phys. Rev. Lett. 52, 437 (1984); Phys. Rev. B31, 5376 (1985)
4. G. S. Parry and D. E. Nixon: Nature 216, 909 (1967)
5. H. Zabel, S. C. Moss, N. Caswell and S. A. Solin, Phys. Rev. Lett. 42, 1552 (1979)
6. K. Ohshima, S. C. Moss and R. Clarke: Synth. Met. 12, 125 (1985)
7. R. Clarke, J. N. Gray, H. Homma and M. J. Winokur: Phys. Rev. Lett. 47, 1407 (1981)
8. Y. Yamada and I. Naiki: J. Phys. Soc. Japan 51, 2174 (1982)
9. M. Suzuki and H. Suematsu: J. Phys. Soc. Japan 52, 2761 (1983)
10. M. Suzuki: Phys. Rev. B33, 1386 (1986).
11. R. Clarke, N. Caswell, S. A. Solin and P. M. Horn: Phys. Rev. Lett 43, 2018 (1979)

12. H. Zabel, S. E. Hardcastle, D. A. Neumann, M. Suzuki and A. Magerl: Phys. Rev. Lett. 57, 2041 (1986)
13. E. D. Specht, M. Sutton, R. J. Birgeneau, D. E. Moncton and P. M. Horn: Phys. Rev. B30, 1589 (1984); K. L. D'Amico, D. E. Moncton, E. D. Specht, R. J. Birgeneau, S. E. Nagler and P. M. Horn: Phys. Rev. Lett. 53, 2230 (1984)
14. S. K. Sinha: in Neutron Scattering, ed. by D. L. Price and K. Skold in the Series: "Methods of Experimental PHysics", Springer-Verlag, New York (1987), in press.
15. W. A. Kamitakahara and H. Zabel: Phys. Rev. B32, 7817 (1985)
16. H. Zabel, A. Magerl, A. J. Dianoux and J. J. Rush: Phys. Rev. Lett. 50, 2094 (1983)
17. G. S. Parry: Mater. Sci. Eng. 31, 99 (1977)
18. F. Rousseaux, R. Moret, D. Guerard, P. Lagrange and M. Lelaurain: Synth. Met. 12, 45 (1985)
19. M. Mori, S. C. Moss, Y. M. Jan and H. Zabel: Phys. Rev. B25, 1287 (1982)
20. M. J. Winokur and R. Clarke: Phys. Rev. Lett. 54, 811 (1985)
21. M. Plischke: Can. J. Phys. 59, 802 (1981)
22. M. Plischke and W. D. Leckie: Can. J. Phys. 60, 1139 (1982)
23. P. B. Visscher and L. M. Falicov: Phys. Rev. B3, 2541 (1971)
24. G. Reiter and S. C. Moss: Phys. Rev. B33, 7209 (1986)
25. S. C. Moss, G. Reiter, J. L. Robertson, C. Thompson, J. D. Fan and K. Ohshima: Phys. Rev. Lett. 57, 3191 (1986)
26. C. Thompson: Thesis for the Ph.D. degree, Department of Physics, University of Houston, May (1987); C. Thompson and S. C. Moss, to be published
27. X. B. Kan, J. L. Robertson, S. C. Moss, K. Ohshima and C. J. Sparks; to be published
28. H. Zabel, Y. M. Jan and S. C. Moss: Physica 99B, 453 (1980)
29. J. D. Fan: Thesis for the Ph.D. degree, Department of Physics, University of Houston (in progress)
30. D. E. Nixon, K. M. Lester and B. C. Levene: J. Phys. C2, 2156 (1969)
31. C. Thompson, M. E. Misenheimer and S. C. Moss: Acta Cryst (in press)
32. D. P. DiVincenzo and E. J. Mele: Phys. Rev. B32, 2538 (1985)
33. J. D. Fan, Omar A. Karim, G. Reiter and S. C. Moss: to be published

Microstructures of Rare-Gas Films Adsorbed on Graphite: Classical and Quantum Simulations

F.F. Abraham

IBM Almaden Research Center,
650 Harry Road, San Jose, CA 95120, USA

1. Introduction

Using the computer simulation techniques of classical and quantum statistical mechanics, we have studied the microstructures of rare-gas krypton and helium monolayer films physisorbed on graphite. While the adsorbate-adsorbate and adsorbate-substrate interactions are short-range and simple, the incommensurability of the unrelaxed rare-gas film and the graphite substrate gives rise to a relaxed modulated film structure exhibiting the symmetry of the substrate face with characteristic length scales spanning tens to thousands of angstroms. The distinct advantage of computer simulation is that these microstructures may be directly "seen" at the atomistic level.

Yet severe constraints of size, time and physical complexity exist when thinking of studying a physical phenomenon by computer simulation. Let us digress by expanding on this statement. By assigning *complexity attributes* (i.e., n_1, n_2, n_3, n_4) associated with certain features of a physical problem, Table 1 allows us to qualitatively estimate the *computational complexity* of a physical problem using the following ad hoc relation:

$$\text{computational complexity} = 100^{(n_1+n_2+n_3+n_4)} . \quad (1.1)$$

Table 1. Complexity attributes n_1, n_2, n_3, n_4 associated with various features defining a physical problem

	Mechanics		Potential		Behavior		Correlation
n_1		n_2		n_3		n_4	
"0"	Classical (CL)	"0"	Spin, (HS) Hard Spheres	"0"	Equilibrium Thermo (EQ)	"0"	Solid, (SL) Liquid
			Empirical N-body (N-B)	"1"	Relaxation (RX)		Composite (CM) Length Scale
"1"	Semi- Classical (SCL)	"1" "2"	Short (SR) Long (LR)			"0" "1" "2"	×10 ×100 ×1000
"2"	Quantum Mechanical (QM)	"3"	Ab Initio (AI)	"2"	Chaotic (CH)		

For sake of reference, we note that a problem dealing with classical mechanics/hard sphere potential/equilibrium/liquid (CL/HS/EQ/SL) has a computational complexity measure equal to unity. For a "real liquid" simulation (CL/SR/EQ/SL), the complexity is one hundred times greater than the simple hard-sphere example. Historically, much of our knowledge of the liquid state evolved from such simulations, and until recently this represented the common practice for the simulation of classical systems. In Sec. 2, we discuss the simulation of classical krypton on graphite where the spatial scale changes a thousandfold. Hence, the system is defined by (CL/SR/EQ/CM) where $n_4 = 2$ and the complexity is one million, or ten thousand times greater than simulating a simple krypton liquid. Similarly, in Sec. 3, we describe the simulation of quantum helium on graphite. But in this case the system is small ($n_4 = 0$) and defined by (QM/SR/EQ/CM), giving a computational complexity of one million, or the same as classical krypton on graphite.

2. The Incommensurate Honeycomb Phase of Krypton on Graphite

A high-density surface layer of krypton physisorbed on graphite is a good model system to study the properties of the incommensurate solid phase and the commensurate-incommensurate (C-IC) transition. The krypton-carbon interaction favors regularly spaced adsorption sites at the graphite surface, and it is well known from experiment [1,2] that monolayer krypton on graphite forms a solid in registry with the underlying substrate. As a consequence of the size of the krypton atom, only one-third of the adsorption sites are occupied, and there exist three energetically degenerate commensurate sublattices. With increasing coverage, it is no longer possible for all of the krypton atoms to occupy adsorption sites, and the system becomes more and more incommensurate, approaching a lattice constant representative of bulk krypton. However, in the transition region, the krypton solid is significantly modulated due to the krypton-graphite interaction.

The nature of the incommensurate phase of krypton on graphite and its transition to the $\sqrt{3} \times \sqrt{3}$ R30° commensurate phase are being extensively investigated by laboratory experiment [1-9] by theory [10-20] and by computer simulation [21-27].

Theories have been developed for the incommensurate phase at zero temperature which are based on the model of incommensurate domain walls separating large commensurate regions such that the distance between domain walls is large compared to the width of the walls. For krypton on graphite, the domain walls may be in three different directions because of the hexagonal substrate structure. If wall intersections are energetically unfavorable, a *striped* phase might be expected where walls are only in one direction. However, Villain has noted that a *honeycomb* array of walls has a degeneracy in which the hexagons of the array can expand or contract without changing the total wall length or the number of nodes, i.e., wall crossings [12,13]. Hence, Villain argues that this additional contribution to the entropy stabilizes the honeycomb phase relative to the striped phase.

We now describe the observed structure of the incommensurate krypton film adsorbed on graphite as a function of temperature and coverage using molecular dynamics. This work has been done in collaboration with S. W. Koch (University of Arizona), W. E. Rudge and D. J. Auerbach (IBM ARC), and M. Schoebinger (University of Frankfurt). We have used systems as large as 103,041 and 161,604 krypton atoms [23]. With this many atoms, graphite substrate dimensions up to 1700Å are realized and are comparable to present-day laboratory capabilities. We observe that the incommensurate phase consists of commensurate islands separated by a interconnecting network of incommensurate domain walls, the structure of this network being a sensitive function of temperature and coverage. At low temperatures, the honeycomb network of domain walls is observed for all coverages. An incommensurate striped phase is not seen. With increasing temperature, distortions from the perfect honeycomb structure become more prevalent. At high temperatures, the individual domain walls fluctuate significantly from the symmetry directions while possessing boundary roughness and a greater wall thickness.

The molecular dynamics simulation technique yields the motion of a given number of atoms governed by their mutual interatomic interactions and requires the numerical integration of Hamilton's equations of motion. The dynamical equations are

$$\frac{d\vec{r}_i}{dt} = \frac{\partial \mathcal{H}}{\partial \vec{p}_i} = \vec{p}_i/m\,, \tag{2.1}$$

$$\frac{d\vec{p}_i}{dt} = -\frac{\partial \mathcal{H}}{\partial \vec{r}_i} = -\frac{\partial U(\vec{r})}{\partial \vec{r}_i}\,, \tag{2.2}$$

where $\mathcal{H}$ is the Hamiltonian of the system of interest. In the traditional molecular dynamics experiment, the total energy E for a fixed number of atoms N in a fixed volume V is conserved as the dynamics of the system evolves in time, and the time average of any property is an approximate measure of the microcanonical ensemble average of that property for a thermodynamic state of N,

V, E. The simplest and earliest procedure for simulating constant temperature takes the *instantaneous* temperature T_I defined by

$$\frac{3}{2} NkT_I = \sum_i p_i^2/2m \tag{2.3}$$

to equal the desired temperature T; the atomic velocities are renormalized at every time interval τ_T so that the *instantaneous* mean kinetic energy corresponds to the chosen temperature T. While naive, this ad hoc scaling procedure is favored by many because of its simplicity, low computational overhead, numerical stability and overall proven success.

In our studies, we have adopted the Lennard-Jones 12:6 pair potential to represent the interaction between the various atoms of the krypton-graphite system;

$$u(r) = 4\varepsilon\left\{\left(\frac{\sigma}{r}\right)^{12} - \left(\frac{\sigma}{r}\right)^{6}\right\}, \tag{2.4}$$

where r is the interatomic separation. The range of interaction is normally cutoff at some distance to lessen the computational burden of summing over all atoms, the cutoff typically being at 2.5σ. When using the Lennard-Jones potential, we normally express quantities in terms of *reduced units.* Lengths are scaled by the parameter σ, the value of the interatomic separation for which the LJ potential is zero, and energies are scaled by the parameter ε, the depth of the minimum of the LJ potential. Reduced temperature is therefore kT/ε Simple pairwise additivity of the atomic interactions is assumed, and the carbon atoms defining the semi-infinite solid are fixed at their lattice sites. In this case, the total potential energy U for a given configuration of N krypton atoms $\vec{r}(i)$, i = 1, ... N, above a *fixed* configuration of carbon atoms defining the graphite semi-infinite solid $\vec{R}(j)$, j = 1, ... ∞, has the form

$$U = \sum_{i>j}^{N} u_{Kr-Kr}(|\vec{r}(i) - \vec{r}(j)|) + \sum_{\substack{i=1 \\ \{Kr\}}}^{N} \sum_{\substack{j=1 \\ \{C\}}}^{\infty} u_{Kr-C}(|\vec{r}(i) - \vec{R}(j)|). \tag{2.5}$$

The krypton-graphite interaction may be evaluated analytically by expanding this potential as a Fourier sum in the surface reciprocal lattice vectors in order to lessen the computational burden.

In order to reduce the computer memory requirement, we constrained the movement of the krypton atoms to a plane parallel to the graphite surface. The external field of the graphite is reproduced to a very good approximation by the expression

$$\varphi_{Kr-G} = -V_g[\cos 2\pi s_1 + \cos 2\pi s_2 + \cos 2\pi(s_1 + s_2)], \tag{2.6}$$

where $V_g$equal$0.08\varepsilon_{Kr-Kr}$ and s_1, s_2 are the basis vectors of the graphite unit cell.

Figure 1 shows the principal results of our simulations for the 103,041 atom system. The incommensurate and commensurate regions are shown as solid black and solid white areas, respectively; in actual fact, we plotted the incommensurate atoms as points but the lack of graphical resolution merged the points to make a solid region. Lets first consider the case where the temperature is fixed at the low value of 0.05 and the coverage is varied. We note that for all coverages a honeycomb network of domain walls is established, the network consisting of straight walls with smooth boundaries which are aligned to the three symmetry directions of the graphite substrate. The commensurate regions form an array of honeycomb domains, the individual hexagons not being identical in size and shape. This honeycomb domain structure with *breathing* freedom is direct conformation of Villain's picture of the incommensurate phase, and this is the first direct observation of this structure. At fixed temperature, the percentage of krypton atoms that are commensurate (%C) decreases linearly with increasing coverage (90%, 80%, 75%, 60% and 37%, respectively), while the domain-wall thickness remains essentially constant at 18Å. This %C decrease is associated with an increase of the total length of domain walls, and this gives rise to smaller and more numerous commensurate domains. We simulated as large a system as we felt was practical within the constraint of our computer resources — a 161,604 krypton atom system. The temperature and coverage are 0.05 and 1.005, respectively. In Fig. 2, we again see the incommensurate honeycomb structure at this low temperature. We do not observe the *striped* phase.

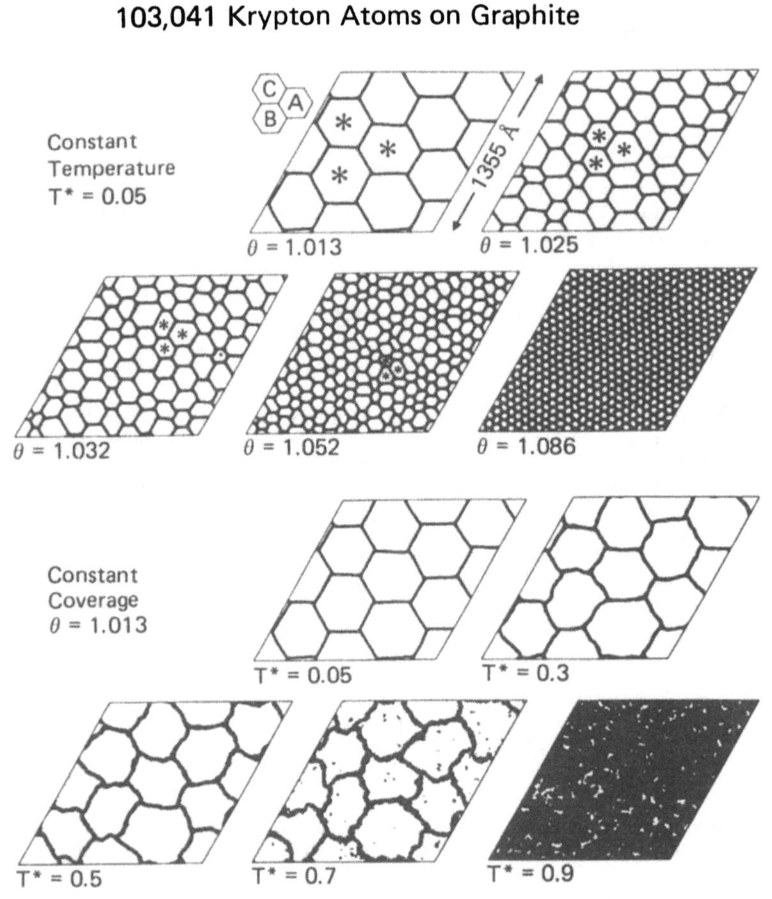

Fig. 1. Pictures of the domain-wall network for an equilibrium configuration of the incommensurate phase as a function of coverage θ at fixed temperature $T = 0.05$ and as a function of temperature at a fixed coverage $\theta = 1.013$.

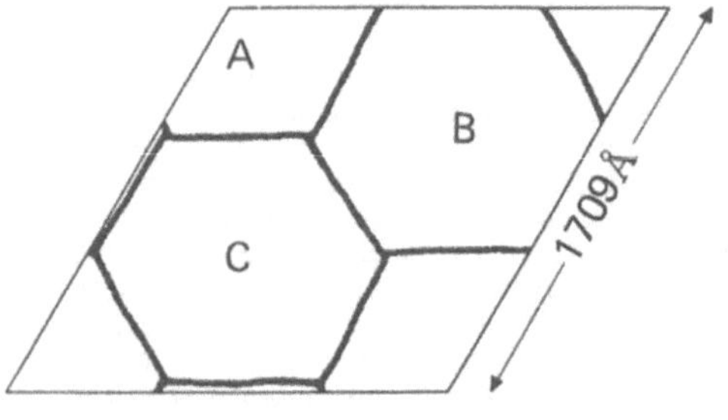

Fig. 2. A picture of the domain-wall network for an equilibrium configuration of the incommensurate phase simulated using 161,604 krypton atoms on graphite at a coverage $\theta = 1.005$ and temperature $T^* = 0.05$. The percentage of commensurate atoms %C is 96.

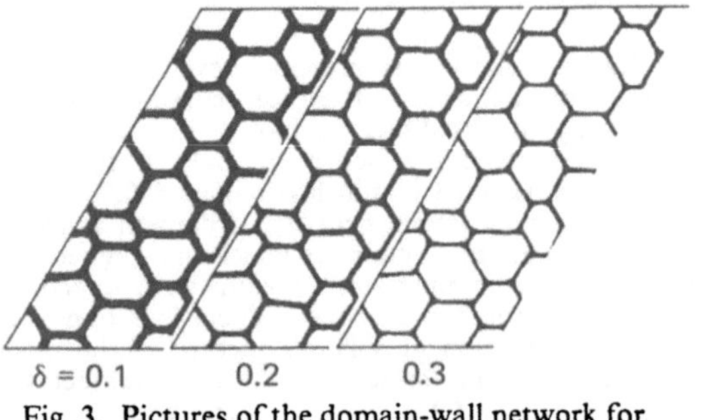

Fig. 3. Pictures of the domain-wall network for an equilibrium configuration of the incommensurate phase at a coverage $\theta = 1.025$ and temperature $T^* = 0.05$, for various values of the commensurate-radius cutoff criterion δ.

One can visualize two extremes for the atomic microstructure of the incommensurate state: (1) the krypton atoms are near registry except for thin domain walls [28], or (2) the krypton monolayer is a lattice that is weakly modulated by the substrate field [29]. In Fig. 3, we present the honeycomb picture for $T^* = 0.05$, $\theta = 1.025$ and for the three commensurate-radius cutoff criteria $\delta = 0.1, 0.2$ and 0.3. We conclude that at this low temperature, the krypton lattice is modulated according to the McTague-Novaco picture. With increasing coverage, this modulation will decrease, i.e., when the wall separation approaches the thickness of an individual wall.

Returning to Fig. 1, we now consider the series of simulations where the coverage was held fixed at 1.013. With increasing temperature, the overall appearance of the incommensurate phase remains that of the domain-wall network which becomes increasingly distorted, the walls becoming broader and the wall boundaries roughening considerably. Also, we note the marked increase in the wall thickness with increasing temperature, the wall thickness at $T^* = 0.7$ being approximately twice the wall thickness at $T^* = 0.05$. This is consistent with the gradual decrease of percentage of commensurate atoms (90%, 88%, 85%, 81%, respectively). At the highest temperature, the system is mainly incommensurate since the krypton film has melted into a liquid.

To classify the domain walls for krypton on graphite, one can distinguish between two configurationally distinct types of walls, which we will call *heavy (h)* and *light (l)* walls [17]. In some other references, e.g., KARDAR and BERKER [18], these walls are labelled "super heavy" and "heavy" walls, respectively. The classification is determined by the orientation of the particular wall and by the two different sublattices separated by this wall. Analyzing the pictures in Fig. 1, we find only heavy walls at the lower temperatures. At temperatures above $kT/\varepsilon = 0.5$, the wall orientation can be characterized as a distribution with peaks around the three symmetry directions. Those walls which can be classified according to the above discussed scheme are almost exclusively ($\gtrsim 95\%$) heavy walls. Only sometimes some short wall segments fulfill the condition required for light walls. In Fig. 4, we note the beautiful temporal *breathing* of the honeycomb domain structure, and this is the first direct observation of this temporal behavior.

To study the energetics of the weakly incommensurate phase, we have simulated a *quasi two-dimensional* system of 20,736 krypton atoms on a graphite substrate, and the results are discussed in reference [30].

Computer simulations of classical [molecular] physisorbed films are now being undertaken and are leading to unique insights into their associated structural phase transitions (e.g., MIGONE et al. [31]; TALBOT et al. [32]; SOKOLOWKSI and STEELE [33]; Piper et al. [34]; PETERS and KLEIN [35,36]). For example, the effect of compressing monolayers of N_2 physisorbed on graphite has been studied by PETERS and KLEIN [36]. They find that axial compression of a commensurate ordered herringbone monolayer initially generates striped domain walls about 35Å wide rather than a uniform incommensurate structure. Also, orientational order in the domain walls persists to higher temperatures than in the commensurate regions. The molecular structure provides a greater richness of ordering phenomena.

3. The Incommensurate Striped Phase of Helium on Graphite [37]

This study is in collaboration with J. Q. Broughton (S.U.N.Y., Stony Brook) and was largely inspired by the beautiful phase diagram of helium on graphite which was presented by M. SCHICK [38] as his interpretation of experimental data, and by our desire to see if Quantum Monte Carlo [39,40] could verify the diagram and give important information on the microstructure of the adsorbed phases. Schick's phase diagram is presented in Fig. 5, but we note that the β-phase was not part of his original picture. In fact, our expectation was to find a typical high-density fluid in the β-phase region. Up to the present, there have been a few significant successes of studying the quantum many-body problem at finite temperature using computer simulation [39,40]. However, we would highlight the very impressive He^4 simulation of Bose-Einstein condensation by CEPERLEY and POLLOCK [41]. We selected He^3 for our study so as to attack the Fermi statistics problem which can plague attempts to obtain reasonable averages basic to the Monte Carlo sampling procedure. It was only after we had implemented a scheme for accounting for Fermi statistics did we learn from computer experimentation that exchange was not important at the temperatures and densities of our study. This can be appreciated by the observation that the phase diagrams of He^3 and He^4 on graphite (Fig. 5) are very similar.

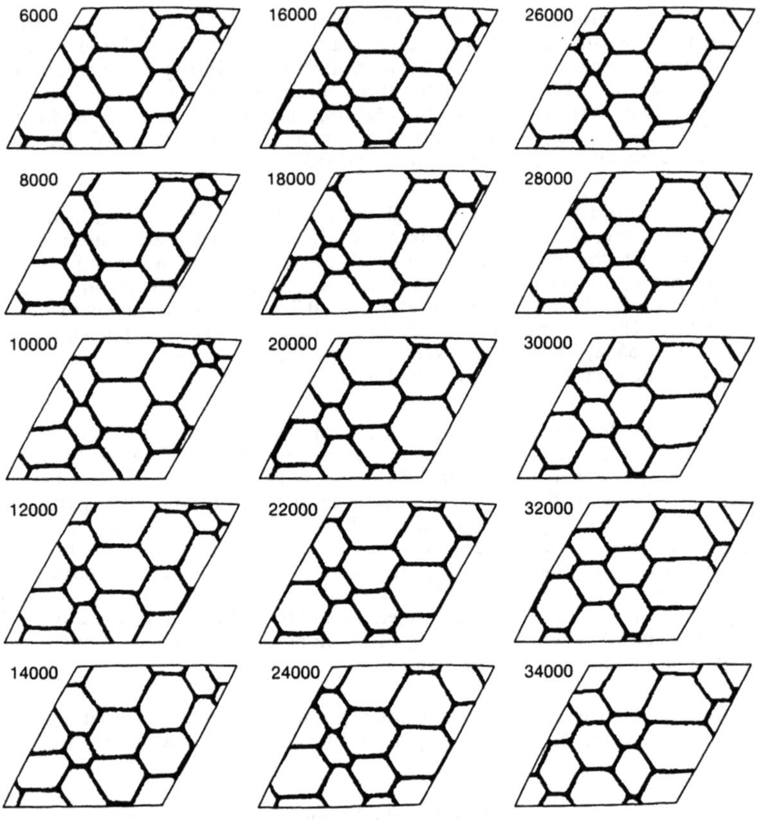

Fig. 4. Temporal "breathing" of the equilibrium honeycomb domain-wall structure. The numbers denote time in units of 100 time-steps.

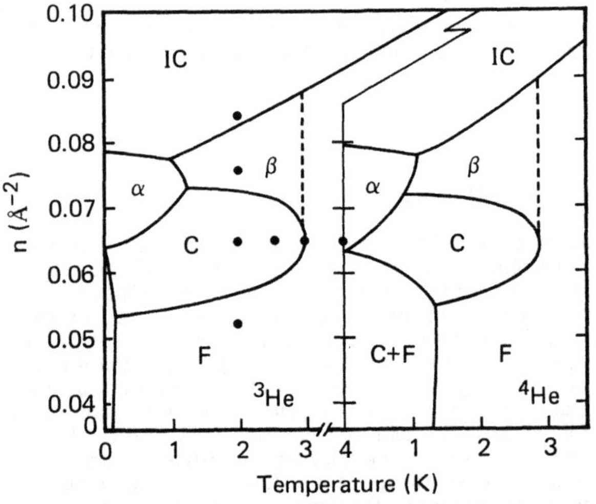

Fig. 5. Phase diagram for He^3 and He^4 on graphite [38]. The beta-phases The β-phases were not a part of the original published diagrams and were taken from [42]. The solid circles denote temperature-density locations where the Monte Carlo simulations were performed.

We will demonstrate that the phases of He^3 adsorbed on graphite can be accurately simulated by the Feynman path-integral Monte Carlo method, realistic potential functions for the substrate-adsorbate and adsorbate-adsorbate interactions, and a three-dimensional geometry. The fluid, commensurate solid, incommensurate solid, and reentrant fluid (β) phases are found and are in agreement with the experimental phase diagram. The microscopic structure of the re-entrant fluid is observed to be a striped domain-wall liquid, in agreement with the experimental interpretation of MOTTELER [42] and the striped helical Potts model calculation of HALPIN-HEALY and KARDAR [43]. It was only after our simulation of the *fluid* in the β-phase region were we made aware of the fact that it had been suggested earlier [44].

In the Feynman path-integral representation, a single quantum particle is isomorphic to a classical cyclic polymer chain of M beads in which each bead j interacts with its neighboring beads $j-1$ and $j+1$ through a harmonic force constant $mM/\hbar^2\beta^2$ and experiences a reduced external potential $V(\vec{r}(j))/M$ (the particle mass is m, and β is inverse temperature.) The harmonic coupling arises from the free particle propagator f describing the quantum mechanical contribution of kinetic energy to the density matrix:

$$f\left(\vec{r}(j+1), \vec{r}(j); \frac{\beta}{M}, L\right) \equiv \left(\frac{Mm}{2\pi\beta\hbar^2}\right)^{3/2} \times \sum_{n} \exp\left(-\frac{Mm}{2\beta\hbar^2}(\vec{r}(j+1)-\vec{r}(j)-n\cdot L)^2\right). \tag{3.1}$$

This representation is exact only in the limit of M going to infinity. For a given temperature and density, one has to determine empirically that M beyond which the thermodynamic properties do not effectively change. The lower the temperature, the larger M must be. We have adopted a higher order correction to this "high temperature approximation." It takes the form of a simple modification to the potential energy V [45]:

$$V'(\vec{r}(j)) = V(\vec{r}(j)) + \frac{1}{24}\frac{\hbar^2}{m}\left(\frac{\beta}{M}\right)^2\left(\frac{\partial V}{\partial \vec{r}(j)}\right)^2. \tag{3.2}$$

We refer the reader to the papers of TAKAHASHI and IMADA [45-47] for a detailed description of the path-integral Monte Carlo method that we implemented. Unless the He^3 atoms were treated as distinguishable, direct calculation of the determinant of free particle propagators was performed in evaluating the Monte Carlo weight function $|W|$ for the spinless fermion system of N atoms [48]:

$$W = (N!)^{-M} \prod_{j=1}^{M} \det A(j+1, j) \exp\left(-\frac{\beta}{M}\sum_{j=1}^{M} V'(j)\right), \tag{3.3}$$

and

$$\{A(j+1)\}_{k,l} = f(\vec{r}_k(j+1), \vec{r}_l(j)),\ 1 \le k, l \le N, \tag{3.4}$$

where $V'(j) = V'(\vec{r}_1(j), \ldots, \vec{r}_N(j))$ is defined by (3.2), and $A(j+1,j)$ is an $N \times N$ matrix of the one particle propagator matrix elements $f(\vec{r}_k(j+1), \vec{r}_l(j))$. The absolute value of W is taken since the weight function in importance sampling should be positive. Two kinds of displacements of coordinates are adopted for importance sampling; a "microscopic" displacement of an individual bead and a "macroscopic" displacement of all of the beads in a cyclic polymer chain according to the recipe of TAKAHASHI and IMADA [45]. One Monte Carlo move is defined as N attempted macroscopic displacements, each one made after M trials cf the microscopic displacements. Primitive displacement parameters were adjusted so that the acceptance ratios for microscopic and macroscopic displacements are approximately one-half.

In our Monte Carlo simulations, the number of atoms varied from 36 to 42, depending on the coverage of interest, chosen dimensions of the graphite substrate, and compatibility with periodic conditions for the initialized triangular lattice of a helium solid and the graphite lattice. Periodic boundary conditions were imposed at the four faces of the computational cell which pass through the sides of the basal plane at normal incidence to the surface. A reflecting wall was placed at the top of the computational box beyond the second layer height, but no atom was promoted to the second layer in any of the simulations. We adopted the Lennard-Jones 12-6 pair potential to represent the interaction between helium atoms and helium-carbon atoms, and the potential parameters are taken from COLE and KLEIN [49]. Similar parameters have been shown to describe the phase diagram

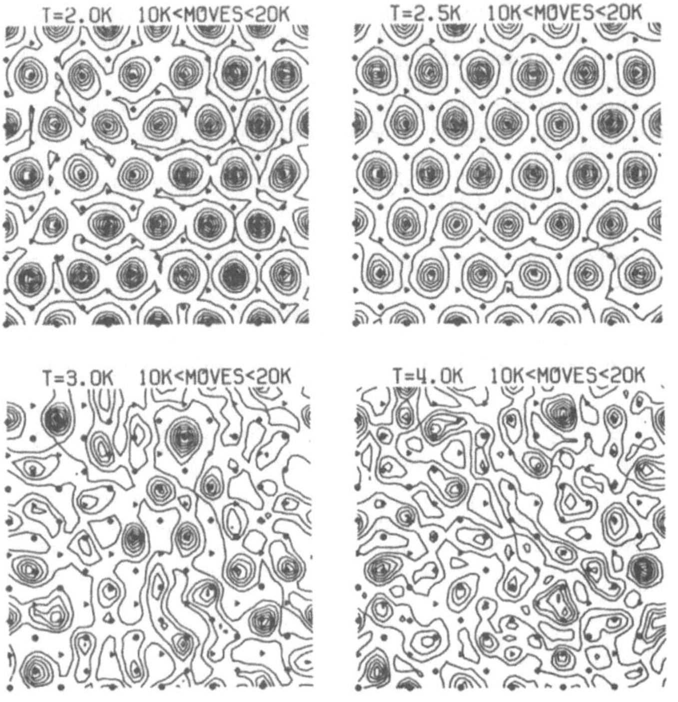

Fig. 6. The probability density contours for the helium beads averaged over 10,000 Monte Carlo moves. The coverage is unity, and the temperatures are 2.0K, 2.5K, 3.0K and 4.0K, respectively. The solid circles, triangles and diamonds denote the three-fold energetically degenerate adsorption sites on graphite for a commensurate solid.

of classical rare-gas atoms on graphite very well [50]. The graphite was modeled as a semi-infinite rigid solid.

By experimentation, we found that 96 beads are required to obtain energy convergence at 2.0K, and we used bead numbers of 84, 72 and 48 for temperatures of 2.5K, 3.0K and 4.0K, respectively. We also learned that fermion exchange is unimportant for temperatures equal to and above 2.0K [51] for the coverages of our study. In Fig. 5, the solid circles denote temperature-density locations where the Monte Carlo simulations were performed. In Fig. 6, the probability density contour plots for the helium beads are presented for a coverage of unity and for temperatures of 2.0K, 2.5K, 3.0K and 4.0K, respectively. The distribution was obtained by averaging over 10,000 Monte Carlo moves after at least 10,000 previous moves were made from the initialized triangular lattice. The averaging interval is given above each picture. As a consequence of the size of the helium atom, only one-third of the adsorption sites can be occupied by a commensurate solid, and there exist three energetically degenerate commensurate sublattices. The solid circles, triangles and diamonds denote these commensurate sublattices. We see that the commensurate solid state exists at 2.0K and at 2.5K and that the fluid state exists at 3.0K and 4.0K, in agreement with experiment (Fig. 5). In Fig. 7, the probability density contour plots for the helium beads are presented for the temperature of 2.0K and the coverages of 0.83, 1.00, 1.17 and 1.33, respectively. We see the low density fluid, commensurate solid, high-density reentrant fluid, and high-density incommensurate solid as we pass from low to high coverage, and these phases appear to be in agreement with the experimental phase diagram (Fig. 5). In particular, the microscopic structure of the reentrant fluid is observed to be a striped domain-wall liquid, in agreement with the experimental interpretation of MOTTELER [42] and the striped helical Potts model calculation of HALPIN-HEALY and KARDAR [43]. The shaded stripes pictorially denote the striped domain walls. The wall thickness is approximately one to two helium atomic diameters and is consistent with the prediction of theory [43].

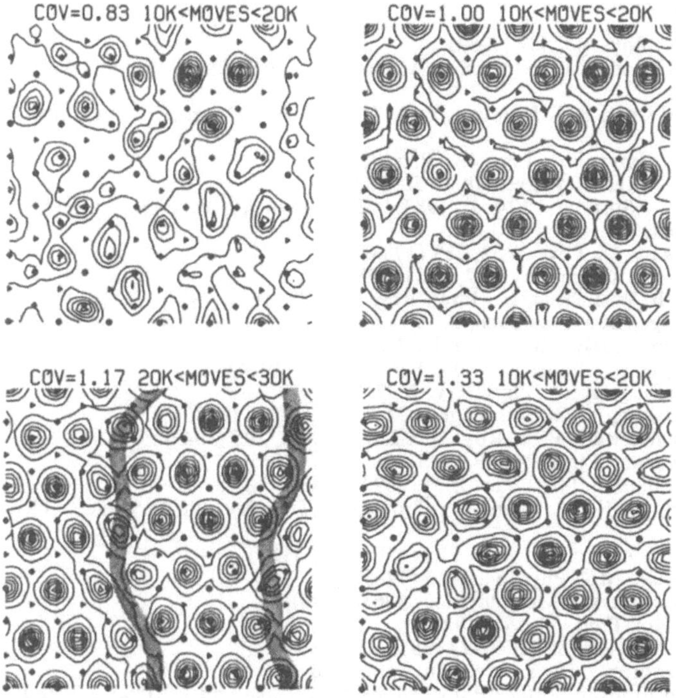

Fig. 7. The probability density contours for the helium beads averaged over 10,000 Monte Carlo moves. The temperature is 2.0K, and the coverages are 0.83, 1.00, 1.17 and 1.33, respectively. The shaded stripes pictorially denote the striped domain walls. See Fig. 6 for additional details.

It is essential to treat much larger size systems in order to achieve the needed detail required for a quantitative analysis. In particular, the microscopic structure of the domain-wall network is needed. Also, lower temperature simulations will allow a thorough examination of the phase diagram, e.g., the α-phase. We are presently implementing algorithms requiring fewer beads and amenable to parallel computation in hopes that we can approach systems of a thousand atoms using a special purpose array computer (50).

References

1. O. E. Vilches: Ann. Rev. Phys. Chem. 31, 463 (1980)
2. A. Thomy, X. Duval, J. Regnier: Surf. Sci. Rep. 1, 1 (1981)
3. M. Nielsen, J. Als-Nielsen, J. Bohr, J. P. McTague: Phys. Rev. Lett. 47, 582 (1981)
4. M. D. Chinn, S. C. Fain: Phys. Rev. Lett. 39, 146 (1977)
5. S. C. Fain, M. D. Chinn, R. D. Diehl: Phys. Rev. B 21, 4170 (1980)
6. P. W. Stephens, P. Heiney, R. J. Birgeneau, P. M. Horn: Phys. Rev. Lett. 43, 47 (1979)
7. P. W. Stephens, P. A. Heiney, R. J. Birgeneau, P. M. Horn: D. E. Moncton, G. S. Brown, Phys. Rev. B 29, 3512 (1984)
8. R. J. Birgeneau, E. M. Hammonds, P. Heiney, P. W. Stephens, P. M. Horn: In Ordering in Two Dimensions, ed. by S. K. Sinha (Plenum, New York 1980)
9. D. E. Moncton, P. W. Stephens, R. J. Birgeneau, P. M. Horn, G. S. Brown: Phys. Rev. Lett. 46, 1533 (1981)
10. P. Bak, D. Mukamel, J. Villain, K. Wentkowska: Phys. Rev. B 19, 1610 (1979)
11. H. Shiba: J. Phys. Soc. Jpn. 46, 1852 (1979)
12. J. Villain: In Ordering in Strongly Fluctuating Condensed Matter Systems, ed. by T. Riste (Plenum, New York 1980), p.221

13. J. Villain: In Ordering in Two Dimensions, ed. by S. K. Sinha (North-Holland New York 1980), p.123
14. P. Bak: Rep. Prog. Theor. Phys. 45, 587 (1982)
15. S. N. Coppersmith, D. S. Fisher, B. I. Halperin, P. A. Lee, W. F. Brinkman: Phys. Rev. Lett 46, 549 (1981)
16. S. N. Coppersmith, D. S. Fisher, B. I. Halperin, P. A. Lee, W. F. Brinkman: Phys. Rev. B 25, 349 (1982)
17. D. A. Huse, M. E. Fisher: Phys. Rev. Lett. 49, 793 (1982)
18. M. Kardar, A. N. Berker: Phys. Rev. Lett. 48, 1552 (1982)
19. J. Villain, M. B. Gordon: Surf. Sci. 124, 1 (1983). This gives an up-to-date review and comparison with experiment.
20. M. Schoebinger, S. W. Koch: Z. Phys. B 53, 233 (1983)
21. F.Hanson, J. P. McTague: J. Chem. Phys. 72, 6363 (1980)
22. F. F. Abraham, S. W. Koch, W. E. Rudge: Phys. Rev. Lett. 49, 1830 (1982)
23. F. F. Abraham, W. E. Rudge, D. J. Auerbach, S. W. Koch: Phys. Rev. Lett. 52, 445 (1984)
24. S. W. Koch, F. F. Abraham: Helv. Phys. Acta 56, 755 (1983)
25. S. W. Koch, W. E. Rudge, F. F. Abraham: Surf. Sci. 145, 329 (1984)
26. R. J. Gooding, B. Joos, B. Bergerson: Phys. Rev. B 27, 7669 (1983)
27. J. S. Whitehouse, D. Nicholson, N. G. Parsonage: Mol. Phys. 49, 829 (1983)
28. F. C. Frank, J. H. van der Merwe: Proc. Roy. Soc. London 198, 205 (1949)
29. J. P. McTague, A. D. Novaco: Phys. Rev. B 19, 5299 (1979)
30. M. Schoebinger, F. F. Abraham: Phys. Rev. B 31, 4590 (1985)
31. A. D. Migone, H. K. Kim, M. H. W. Chan, J. Talbot, D. J. Tildesley, W. A. Steele: Phys. Rev. Lett. 51, 192 (1983)
32. J. Talbot, D. J. Tildesley, W. A. Steele: Mol. Phys. 51, 1331 (1983)
33. S. Sokolowski, W. A. Steele: Mol. Phys. 54, 1453 (1985)
34. J. Piper, J. A. Morrison, C. Peters: Mol. Phys. 53, 1463 (1984)
35. C. Peters, M. L. Klein: Mol. Phys. 54, 895 (1985)
36. C. Peters, M. L. Klein: Phys. Rev. B 32, 6077 (1985)
37. F. F. Abraham, J. Q. Broughton: Phys. Rev. Lett. submitted.
38. M. Schick: In Phase Transitions in Surface Films, ed. by J. G. Dash and J. Ruvalds (Plenum, New York), p.68
39. B. J. Alder, D. M. Ceperley, E. L. Pollock: Acc. Chem. Res. 18, 268 (1985)
40. D. M. Ceperley, B. J. Alder: Science 231, 555 (1986)
41. D. M. Ceperley, E. L. Pollock: Phys. Rev. Lett. 56, 351 (1986)
42. F. C. Motteler: "A heat capacity study of p-H_2 monolayers on graphite," Ph.D. Dissertation (University of Washington, Seattle Washington 1986).
43. T. Halpin-Healy, M. Kardar: Phys. Rev. B34, 318 (1986).
44. S. C. Fain: private communication. One of the authors (FFA) is indebted to Professor Fain for pointing out that we were possibly simulating a domain wall liquid and for bringing references 42 and 43 to his attention.
45. M. Takahashi, M. Imada: J. Phys. Soc. Jpn. 53, 963 (1984)
46. M. Takahashi, M. Imada: J. Phys. Soc. Jpn. 53, 3765 (1984)
47. M. Imada, M. Takahashi: J. Phys. Soc. Jpn. 53, 3770 (1984)
48. In reference 45, it is stated that the determinate evaluation scales as the cube power of the number of particles. However, in the Monte Carlo procedure, only one row is changed for an attempted displacement, and an algorithm for the determinate evaluation has been devised that scales as the square of the number of particles (Nimrod Megiddo, IBM ARC, San Jose, California, private communication).
49. M. W. Cole, J. R. Klein: Surf. Science 124, 547 (1983)
50. F. F. Abraham: Adv. Phys. 35, Sec. 4.2, 1 (1986)
51. The details of convergence, exchange and energetics will be described in an expanded publication.

New Theory for Competing Interactions and Microstructures in Partially Ordered (Liquid-Crystalline) Phases

F. Dowell

Theoretical Division, Los Alamos National Laboratory,
University of California, Los Alamos, NM 87545, USA

1. SUMMARY

A summary of results from a unique statistical-physics theory to predict and explain competing interactions and resulting microstructures in some partially-ordered [in this case, liquid-crystalline (LC)] phases is presented. The static aspects of both partial orientational and partial positional ordering of the molecules into various microstructures in these phases (including the incommensurate smectic-Ad phase) can be understood in terms of various competing interactions (both entropic and energetic) involved in the packing together of the different molecular sub-units at given pressures and temperatures. These microstructures are predicted and explained (using no ad hoc or arbitrarily adjustable parameters) as a function of molecule chemical structure [including lengths and shapes (from bond lengths and angles), intramolecular rotations, site-site polarizabilities and pair potentials, dipole moments, etc.]. Theoretical results are presented for the nematic, re-entrant nematic, smectic-Ad, and smectic-A1 LC phases and the isotropic liquid phase.

2. INTRODUCTION

Liquid-crystalline (LC) phases are stable condensed phases in which the molecules pack together with order that is intermediate between the three-dimensional order of a crystalline solid and the disorder of an isotropic (ordinary) liquid. LC phases always have partial orientational order of the molecules, and some LC phases also have partial positional order of the molecules.

The partial molecular ordering that is characteristic of LCs occurs frequently in natural and synthetic materials. Thus, LCs are of considerable basic and applied interest [1].

LCs are formed by molecules with very anisotropic shapes (typically, with length-to-breadth ratios of three or more), with these shapes frequently changing as a function of temperature and density. From a basic viewpoint, the theoretical study of the partial orientational and positional ordering of such changing, highly anisotropic shapes in condensed phases is one of the most challenging problems in the statistical physics of many bodies.

From the standpoint of practical applications, LC ordering is the essential characteristic feature that determines the proper functioning of soaps and micelles (important in separation and extraction processes, such as enhanced oil recovery), LC polymers (important in their final solid state as stronger, lighter-weight replacements for metals in body armor, auto and airplane parts, and other structural applications), LC electro-optic devices (such as LC display devices--important, for example, in digital watches and calculators, because of their small energy requirements), and biomembranes and other biological structures. LC structures are also found between crystalline and amorphous layers in "semicrystalline" polymers (the common state of a very large number of solid polymers), in coals and other fossil energy systems, etc.

There are thousands of possible chemical structures in LCs. A reasonably typical example of a LC molecule structure is

$$\underbrace{H_3C\text{-}(CH_2)_y}_{\text{semiflexible tail-chain}}\text{-}\underbrace{\{\text{-O-}\phi\text{-N=N-}\phi\text{-O-}\}}_{\text{rigid core}}\text{-}\underbrace{(CH_2)_{y'}\text{-}CH_3}_{\text{semiflexible tail-chain}}$$

(ϕ indicates a para-substituted benzene ring. y, y' ~ 0 to 20. y and y' may be equal or unequal.)

The overlap of π orbitals in the aromatic, double, and triple bonds in the core section of a LC molecule leads to the rigidity of the core. In the isotropic (I) liquid phase and in the particular LC phases studied in this paper, there is essentially free rotation of the molecule about the core long axis, thereby giving an effective rodlike, cylinderical shape to the core.

The n-alkyl tail-chains are partially-flexible (semiflexible) since there are one trans and two gauche rotational energy minima for a carbon-carbon bond between methylene ($-CH_2-$) or methyl ($-CH_3$) units in a given chain section. There is an appreciable fraction of gauche states (the higher energy states) in n-alkyl tail-chains in LC and I phases. In this paper, each molecule has two tail-chains.

3. THEORY

In the theory of this paper, the chemical structure of each molecule is divided into a sequence of connected sites, where these sites correspond to small groups of atoms (such as benzene rings and methylene groups). We then use a localized mean-field (LMF) simple-cubic (SC) lattice theory to study the packing of the molecules in the system volume.

We use SC lattice theory since any orientation of a molecule or molecular part or bond can be decomposed into its x, y, and z components and mapped directly onto a SC lattice in a manner analogous to normal coordinate analysis in, for example, molecular spectroscopy. LMF means that there is a specific average neighborhood (of other molecular sites and empty space) in a given direction k around a given molecular site in a given local region in the system. These local regions are determined by the actual packing of the molecules in the system. This packing is done mathematically using lattice combinatorial statistics to determine the analytic partition function for the system. [The generalized combinatorics used in the theory of this paper have been found to be quite accurate when compared with Monte Carlo computer simulations[2] in at least one limiting case presently amenable to such simulations (see discussion in Ref. 3)]. Various continuum limits are taken in the theory of this paper.

The partition function and the resulting equations for static thermodynamic and molecular ordering properties are functions of the pressure P, temperature T, density ρ, lengths and shapes of the rigid core and the semiflexible tail-chains of a molecule, net energy difference E_g between trans and gauche states, dipole moments, site-site polarizabilities and Lennard-Jones (12,6) potentials, and orientational and one-dimensional (smectic-A) positional orderings of the different rigid and semiflexible parts of the molecules. The Lennard-Jones (LJ) potentials are used to calculate repulsions and London dispersion attractions between different molecular sites, and the dipole moments and polarizabilities are used to calculate dipole/dipole and dipole/induced dipole interactions between different molecular sites.

There are no ad hoc or arbitrarily adjustable parameters in this theory. All variables used in this theory are taken from experimental data for atoms or small groups of atoms or are calculated in the theory.

The theory used in this paper has been derived in detail elsewhere [4] and involves extension and refinement of earlier, very successful theories [3,5-7] for LCs. Due to length constraints on this paper, we note only the changes made in this paper to the equations of Refs. 3 and 5-7.

The equations in the theory of this paper are the same as the equations of Ref. 7, except for the following changes. (Variables not defined in this paper have been previously defined in Ref. 7.)

The treatment of the flexibility of a semiflexible tail-chain in Ref. 7 was significantly refined in Ref. 4: Thus, Eq. (14) of Ref. 7 is replaced in this paper by

$\nu = \langle(3\cos^2\psi - 1)\rangle/2 = 1 - 3u$, where f_γ is the number of semiflexible segments ($-CH_2-$ or $-CH_3$ groups) in tail-chain γ of the molecule,

$2u = (\Sigma_\gamma 2u_\gamma f_\gamma)/(\Sigma_\gamma f_\gamma)$, $f = \Sigma_\gamma f_\gamma$, $2u_\gamma = 2\zeta_2$ for $f_\gamma = 1$,

$2u_\gamma = \{(\Sigma_{j=1}^{2} Y_{1j}) + [(f_\gamma - 2)/2][\Sigma_{j=1}^{2} Y_{2j}]\}/f_\gamma$ for even $f_\gamma \geq 2$,

$2u_\gamma = \{(\Sigma_{j=1}^{3} Y_{3j}) + [(f_\gamma - 3)/2][\Sigma_{j=1}^{2} Y_{2j}]\}/f_\gamma$ for odd $f_\gamma \geq 2$,

$Y_{11} = 2(\zeta_1\zeta_2 + \zeta_2^2)/D_1$, $Y_{12} = 2(2\zeta_1\zeta_2 + \zeta_2^2)/D_1$,

$Y_{21} = Y_{22} = 2(\zeta_1^2 + 3\zeta_1\zeta_2 + 2\zeta_2^2)/D_2$, $Y_{31} = 2(\zeta_1^2\zeta_2 + 3\zeta_1\zeta_2^2 + \zeta_2^3)/D_3$,

$Y_{32} = 2(2\zeta_1^2\zeta_2 + 4\zeta_1\zeta_2^2 + \zeta_2^3)/D_3$, $Y_{33} = 6(\zeta_1^2\zeta_2 + \zeta_1\zeta_2^2)/D_3$,

$D_1 = \zeta_1^2 + 4\zeta_1\zeta_2 + 2\zeta_2^2$, $D_2 = 3\zeta_1^2 + 2(5\zeta_1\zeta_2 + 3\zeta_2^2)$,

$D_3 = \zeta_1^3 + 2(3\zeta_1^2\zeta_2 + 4\zeta_1\zeta_2^2 + \zeta_2^3)$,

$\zeta_1 = 1/(1 + 2\Lambda)$, $\zeta_2 = \Lambda/(1 + 2\Lambda)$, and $\Lambda = \exp[-E_g/(kT)]$.

Also, d_L in Eq. (16) in Ref. 7 has become $d_L = v_o^{1/3}\{r + f[(1 + 2\nu)/3]\} + (a - v_o^{1/3})$ in this paper.

4. RESULTS AND DISCUSSION

4.1 General Summary Remarks

This paper summaries theoretical results [4] calculated for the nematic (N), re-entrant N, smectic-Ad (SAd), and smectic-A1 (SA1) LC phases and the I liquid phase. In the LC phases, the molecules have partial orientational order, with the long axes of the rigid rodlike cores tending to align parallel to a preferred axis (axis z) in the system.

The molecules in the SA1 and SAd phases have total and partial one-dimensional positional order, respectively, such that oriented molecules position into layers with the planes of these layers perpendicular to the z-axis. [There is no positional order (regular spacing) along the x- and y-axes in these SA phases.] The positional order of oriented molecules along the z-axis in the SA1 phase is such that the rigid cores pack with other rigid cores, and the semiflexible tail-chains pack with other tail-chains. In the SAd phase, the cores and tails on oriented molecules each tend to pack with other like molecular parts. In the SA phases, L is the layer thickness, and d_L is the actual length of a molecule. In the SA1 phase, $L = d_L$. The SAd phase is a type of incommensurate phase since $d_L < L < 2d_L$. The flexibility of the tail-chains in smectic LC phases provides enough entropy (disorder) to keep the rigid cores from crystallizing totally, thus allowing the existence of partial positional order in smectic phases[4].

In the following theoretical results, the number of rigid segments (sites) r in the core of a molecule is 4. In each molecule, tail-chain 1 is $-(CH_2)_3-CH_3$; and tail-chain 2 is $-(CH_2)_y-CH_3$, with $y = f_2 - 1$.

In the particular calculations whose results are reported in this paper, these phase transitions were found to be second-order: SAd-N (except as noted) and SA1-SAd. In these calculations, these phase transitions were found to be weakly first-order: N-I, SAd-I, and SA1-I.

4.2 Hard Repulsions Generate I, N, Re-entrant N, SAd, and SA1 Phases

As seen in Table I, site-site hard [i.e., steric (infinitely large)] repulsions are sufficient [4] to generate N, low-T N (including re-entrant N), and multiple SA LC phases and the I liquid phase. (A re-entrant phase is a less-ordered phase that re-enters or re-appears at T below more-ordered phases.) Since there are no attractions, larger P and smaller T are needed for condensed-phase densities in Table I. (The transition between the high-T N phase and the SAd phase in column 1 of Table I is weakly first-order.)

Table I. Phases and transition T for $f_2 = 8$ and only hard repulsions

P =	149 atm	198 atm	149 atm	198 atm
E_g/k =	400 K	400 K	250 K	250 K
	I	I	I	I
	71.5 K	88.9 K	63.6 K	91.7 K
	N			
	70.1 K			
				SA1
				88.0 K
	SAd	SAd	SAd	SAd
	67.5 K	82.6 K	57.3 K	65.1 K
	N	N	N	N

In Table I, the general order of stable phases can be understood as follows. At higher T, the volume of the system is large enough that the molecules can pack randomly (thus forming the I phase). As T decreases, the volume of the system decreases, and the molecules are forced to order partially into N or SA LC phases, in order to pack the molecules into this decreasing volume. In the N phase, the molecules are forced to orient partially. In the SA phases, the molecules also have segregated (thus, more efficient) packing of oriented molecules, such that rigid cores pack with other cores and semiflexible tail-chains pack with other tail-chains; the tail-chains bend and twist well around each other, but do not pack as well with the oriented rigid cores. As T decreases even further, the tail-chains stiffen somewhat (become less flexible), thus decreasing the packing differences between cores and tails. The need for segregated packing of cores and tails decreases and is overcome by the entropy of unsegregated packing, thus giving a low-T N phase. (See Ref. 4 for actual density values in these phases and for more discussion.)

4.3 Effects of Pressure

As can be seen by comparing columns 1 and 2 and columns 3 and 4 in Table I, increasing P favors SA (first SAd, and then SA1) phases over N phases. As P increases, the volume decreases, thus leading to the segregated packing of the SA phases.

4.4 Effects of Tail-Chain Flexibility

As can be seen by comparing columns 1 and 3 and columns 2 and 4 in Table I, decreasing E_g/k (i.e., making the tail-chains more flexible) generally favors SA phases (especially SA1 phases) over N phases. As the tails become more flexible in a relatively dense system, the molecules are forced into the segregated packing of the SA phases. (See Ref. 4 for more discussion.)

4.5 Effects of Attractions and Soft Repulsions in Lennard-Jones Potentials

The addition[4] of LJ (12,6) potentials to the system of Table I adds London dispersion attractions and soft (finite-sized) repulsions to the system. As seen in Table II and elsewhere in this section, attractive forces pull the molecules closer together, thus allowing the condensed phases I, N, SAd, and SA1 to exist at larger T and smaller P than in Table I. (Specifically, P = 1 atm in Table II and elsewhere in this section.) In Table II and elsewhere in this section, the rigid core corresponds to $-CH_2-\phi-\phi-CH_2-$. From experimental data for atoms and small groups of atoms, the input variables (for definitions, see Ref. 7) were estimated (Ref. 4) after the manner of Ref. 7 to be $\epsilon_{cc}/k = 300$ K and $\epsilon_{tt}/k = 150$ K.

Table II. Transition T for a LJ system at P = 1 atm and $E_g/k = 250$ K

f_2	$T_{SA1-SAd}$ [K]	T_{SAd-N} [K]	T_{N-I} [K]
4	180.4	187.8	327.0
5	214.4	225.5	343.0
6	252.4	268.1	368.1
7	283.9	300.2	385.8
8	330.9	344.3	411.9
9	356.2	367.8	430.2

The effect of the soft repulsions (the net effect of the London dispersion attractions and the soft repulsions) in the LJ systems in Table II and elsewhere in this section is to push molecules further apart than in the system in Table I with only hard repulsions. This phenomenon allows a subtle competition between energy and entropy in determining the relative stabilities of the SA and N phases. In general, energy favors the SA phases, since the dispersion attractive energies are larger between two core sites (with their benzene rings) than between a core site and a tail site, thus favoring the segregated packing of the SA phases. Entropy favors the positional disorder of the N phase.

Increasing E_g/k from 250 K in Table II to 400 K leads [4] to the following phases and transition T (in K): For $f_2 = 8$: SA1 394.4 SAd 395.5 N 430.7 I. For $f_2 = 9$: N 416.6 SAd 417.1 SA1. Thus, increasing E_g/k to 400 K here leads to conditions for a re-entrant N phase between $f_2 = 8$ (with a high-T N phase) and f_2 = 9 (with a low-T N phase). While non-integer values of f_2 are not available experimentally in pure (single-component) LC systems, effective non-integer f_2 values are easily achieved experimentally in a mixture of LC components with different integer values of f_2.

In the steric system of Table I, increasing E_g/k decreases the density ρ. In contrast, in the LJ systems of this section, increasing E_g/k increases ρ slightly, thus leading to higher-T SA phases. As E_g/k increases in these LJ systems, the tail-chains get even stiffer at low T, thus leading to low-T N (including re-entrant N) phases. (See Ref. 4 for more discussion.)

4.6 Effects of Dipolar Forces

In this section, large dipolar forces (corresponding to those in a rigid core of -O-ϕ-ϕ-C=N-) have been added[4] to the system of Table II. The input variables for the calculations in this section are the same as in Table II, except for these additions (for definitions, see Ref. 7): μ_D = 5.2 D, $\alpha_c' = 24\times10^{-24}$ cm^3, and $\alpha_t = 2\times10^{-24}$ cm^3.

Adding dipolar forces in this section to the nondipolar LC system of Table II leads [4] to the following phases and transition T (in K): For f_2 = 4: N 115.2 SAd 115.7 SA1. For f_2 = 5: SA1 160.7 SAd 163.2 N 384.0 I. Thus, adding dipolar forces here leads to conditions for a re-entrant N phase between f_2 = 4 (with a low-T N phase) and f_2 = 5 (with a high-T N phase). Dipolar forces orient the cores more and pull the molecules closer to hard-repulsive separations, thus favoring SA and low-T N (including re-entrant N) phases. (See Ref. 4 for more discussion.)

5. CONCLUDING REMARKS

In this paper, we have shown theoretically how the following competing interactions in the packing of the molecules are important in determining the relative stabilities of the N, re-entrant N, SAd (incommensurate), and SA1 LC phases and the I liquid phase: tail-chain flexibility as a function of T, ρ as a function of T, hard repulsions, soft repulsions, London dispersion attractions, and dipolar forces. The theoretical results summarized in this paper have reasonable physical explanations and have also been found[4] to be in agreement with available experimental data. (See Ref. 4 for more discussion.)

6. ACKNOWLEDGMENT

This research was supported by the U. S. Department of Energy.

7. REFERENCES

1. For example, see the following general introductory references: P. G. de Gennes: The Physics of Liquid Crystals (Clarendon, Oxford, 1974); The Molecular Physics of Liquid Crystals, ed. by G. R. Luckhurst and G. W. Gray (Academic, London, 1979); Phys. Today, 35 (5), (1982).
2. F. L. McCrackin: J. Chem. Phys. 69, 5419 (1978).
3. F. Dowell: Phys. Rev. A 28, 3520 (1983).
4. F. Dowell: Phys. Rev. A 36, xxxx (1987) [scheduled for 1 Nov. 1987].
5. F. Dowell: Phys. Rev. A 28, 3526 (1983).
6. F. Dowell: Phys. Rev. A 31, 2464 (1985).
7. F. Dowell: Phys. Rev. A 31, 3214 (1985).

Part III

Dynamics

Defects, Hysteresis and Memory Effects in Modulated Systems

J.P. Jamet

Laboratoire de Physique des Solides
(associated to C.N.R.S.), Bâtiment 510,
F-91405 Orsay Cédex, France

Phase transitions in modulated systems are sensitive to the presence of defects, which may result in shifting of transition temperatures or changing critical behavior : defects are also responsible for specific hysteresis and memory effects. Some defects are intrinsic (discommensurations, phase vortices) some are extrinsinc (substitutional atoms, interstitials, irradiation defects, dislocations...). Frozen-in extrinsic defects deform the modulation phase and produce pinning of sliding modes, while mobile defects adjust to the modulation phase, giving rise to different properties. The memory effects which have been discovered in thiourea can be understood on the basis of mobile defects ordering in a defect density wave (DDW) with the modulation periodicity. This DDW traps in turn the modulation for the same wavevector. A variety of modulated structures has been shown recently to exhibit DDW condensation, both in insulators and charge density wave systems. In general, the nature of these mobile defects is unidentified. Irradiation defects give new problems : locked phases can be washed out, while new phases with arbitrary (irrational) wavevectors appear. Finally, the behavior of the memory effects in thiourea presents an interesting analogy with the properties of associative memories.

I. Introduction

Since the observation of an incommensurate structure by F. Reinitzer [1] in 1888, many modulated structures have been discovered. In solid state physics, an incommensurate spin modulation was evidenced in the metallic alloy $MnAu_2$ and later in CeSb. It was only in 1963 that the first structurally incommensurate phase appeared in the insulating $NaNO_2$ [2]. It is also valuable to mention the 1-d conductors where an electronic density modulation has been discovered : $K_2Pt(CN)_4Br_{0.3}$, xH_2O [3] and TTF-TCNQ [4]. The transition metal dichalcogenides have attracted a lot of interest due to their nonlinear conductivity properties. Finally, the so-called quasi-crystals can be considered as modulated systems in a high-dimensionality space [5].

The usual sequence of phases for decreasing temperature is disordered down to a second order phase transition T_i, then incommensurately modulated (with a decreasing modulation wavevector "q(T)"), down to T_c, where a first order phase transition occurs to a commensurate phase. Outside of this standard behavior, lock-ins for rational values of the modulation wavevector, hysteresis and metastability effects usually appear. Electric field and pressure effects have attracted a lot of interest among experimentalists and theoreticians [6-8, 54].

The origin of the incommensurability is believed to be a competition between short-range and more or less long-range interactions ; due to this delicate balance, the role of the defects, whether intrinsic or extrinsic, will be important. The more or less long-range character of the interaction will depend on the nature of the system (conductor or insulator) in magnetically modulated compounds for example. On the other hand, in 1-d conductors, the competition results from electron-phonon coupling. In ferroelectric or ferroelastic insating systems, the origin of the modulation is not clear : ab initio theoretical calculations have been done in K_2SeO_4 [9] and Rb_2ZnCl_4 [10] ; according to [10] in the last compound, incommensurability arises from the presence, above T_i, of a structure of imperfect and mechanically unstable helices of atoms or groups of atoms ; this mechanism could be very general in insulators.

Historically, the Frenkel-Kontorova discrete model [11] has been applied to the incommensurability problem by S. Aubry [12] following Frank and Van der Merwe [13]. On the other hand, a continuous model was elaborated by P. Bak and V. Emery [14].

Other models have been elaborated giving rise to specific phase diagrams [15]. All these models apply to pure systems but the presence of a thermal hysteresis of the modulation wavevector in many of these compounds has raised questions abouts its origin intrinsic or extrinsic.

II. The Defect Problem

The hysteresis effects appear on measurements of the modulation wave-vector q(T) as shown on Fig. 1, of the dielectric susceptibility, of the optical birefringence, etc. It is possible to relate the presence of this hysteresis to the discrete nature of the lattice which agrees with the predictions of the discrete microscopic theory ; nevertheless, extrinsic defects or impurities which are always present in real physical systems can help to understand hysteresis and metastability effects, but nucleation will also play a very important role.

1. Fixed impurities or defects

The presence of defects is known to have large effects in the vicinity of phase transitions because of the divergence of the correlation length : this has been demonstrated by specific heat measurements in $BaMnF_4$ [16] or sound attenuation in $RbH_3(SeO_3)_2$[17] ; these experimental results agree with defect theories [18]. In the Charge Density Wave (CDW) system : 2H-$TaSe_2$[19], the anomalous variation of the modulation wavevector q(T) agrees well with the defect-based calculation of Nakanishi [20].

On the other hand, in many insulating compounds, a global thermal hysteresis of the modulation wavevector is observed ; this hysteresis exists not only in the vicinity of the commensurate-incommensurate phase transition, but in the whole modulated phase, as shown on Fig. 1, in the case of thiourea. This

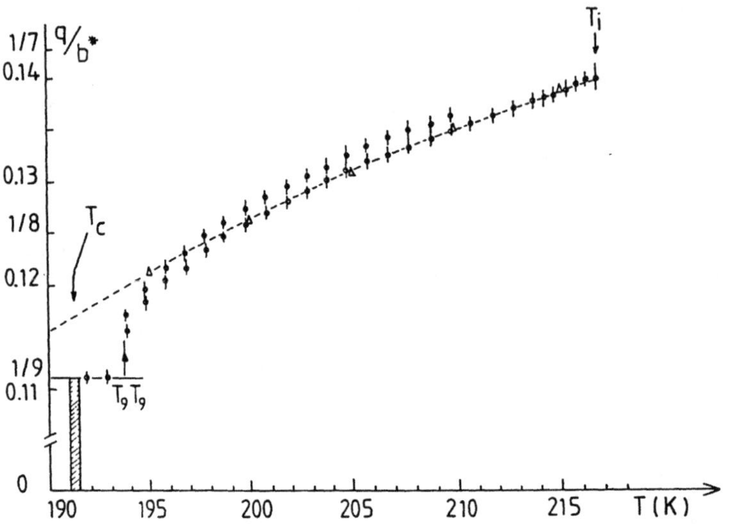

Fig. 1 : q-wavevector variation in $SC(ND_2)_2$ [21]. The thermal hysteresis on q is larger at higher temperature than at lower temperature (i. e. close to the incommensurate-commensurate phase transition to the locked phase with $q=b^*/9$). There is no locked phase for $q=b^*/8$, this phase being symmetry forbidden when there is no applied electric field (E = 0) [56]. The dotted line is the Landau Ginzburg prediction for the q(T) variation [38].

is due primarily to the nucleation of solitons inside the system by stripple nuclei [22], and not only to the pinning of these walls by defects [23]. Nevertheless, the situation is not very clear in the high temperature part of the modulated phase (in particular close to T_i) where there is no evidence for the presence of solitons : how does the system change the value of the modulation wavevector, in this case ? Are stripple nuclei still able to induce this change thanks to the local modulation distortion induced by some extrinsic defects ? To our knowledge, there is no answer to these questions at present time. Due to this nucleation property, the variation of q with T will depend on the time rate of variation of T, and this has been clearly seen in experiments on quartz [24] where for a rapid variation of T with time, q(T) appears to be monotonic, while for slow variations of T with time, q(T) appears like a staircase.

If, on the other hand, one introduces some disorder into the system, the hysteresis will be strongly increased ; this can be done by doping : experiments on $(Rb_{1-x}K_x)_2ZnCl_4$ display a large increase of the hysteresis of q(T) even for a x-value as low as 0.005 [25] ; similar observations can be done by X-ray irradiation [26] or electron bombardment [27].

2. Mobile impurities - memory effect

If a crystal is held for some minutes or more at $T = T^*$, within the modulated phase, a dielectric anomaly, a birefringence one, corresponding to a lock-in, will be observed at later times if T again runs through T^*. This q-memory effect involves mobile defects coupled to the modulation order parameter. This effect has been observed in various systems (CDW as well as insulators) since it was evidenced in thiourea [28, 29], in Rb_2ZnCl_4 [30] and in $Ba_2NaNb_5O_{15}$ [31]. Up to now the nature of the defects (impurities) responsible for these effects is unknown although one can think of chiral defects [32] or interstitial solvent molecules for example in as-grown samples. We shall go back to the experimental manifestations of this memory effect later on.

III. Extrinsic Defect Theories

The theoretical treatment of the defect influence on the properties of CDW systems has been extensively developed in the context of fixed impurities ; on the other hand, a similar framework can also be used for the case of the mobile impurities, i. e., of the memory effects.

1. Fixed impurities

According to [33], one can distinguish : the strong pinning limit (or very dilute limit) where the elastic energy is negligible and each impurity distorts the phase of the modulation ; the change in phase is $\pm \pi$ randomly from impurity site to impurity site; the weak pinning limit (or high concentration limit) where the elastic energy prevents the phase from adjusting at each impurity site ; the phase can then distort slightly over distances much larger than the average impurity-impurity distance.

2. Mobile impurities

If the preceding concepts have been applied to fixed impurities, it is also possible to apply them to mobile impurities, i. e. to the memory effect case which has been extensively studied in thiourea [29]. Then one starts with a random distribution of impurities, allow this density wave to distort and writes down a diffusion equation for the Defect Density Wave (DDW) in the periodic modulation potential. Two main cases can be considered :

a. The linear coupling case [34]

It seems to apply to systems where the order parameter and the disordered-modulated phase transition temperature T_i are time dependent ; this is observed in some blue bronzes and it appears in NMR experiments done on $RB_{0.3}MoO_3$ [35] ; birefringence experiments in $Ba_2NaNb_5O_{15}$ [36] display similar effects.

The problem can be treated by writing the coupling impurity modulation as :

$$V_{m-i} = - V_o c_q \rho_q$$

where ρ_q is the equilibrium distribution of impurities, c_q is the amplitude of the DDW and V_o is chosen positive and identical for all the impurities.

The free energy of the pure modulated system is (if we neglect gradient terms) :

$$F_m = \frac{T - T_i}{2} \rho_q^2 + \frac{T_i}{4} \rho_q^4$$

On the other hand, the energy of the impurities is :

$$F_i = T \int [c \, \ell n \, c + (1 - c) \, \ell n(1 - c)] dr$$

with $c = c_o - c_q \cos qr$.

One sees easily that at equilibrium (i. e. at t infinite) the order parameter is increased :

$$\rho_q^2 = \frac{T_i - T}{T_i} + \frac{2c_o V_o^2}{T_i T}$$

It is the same situation for the modulated-disordered transition temperature which is increased :

$$T_i' = T_i \frac{1 + \sqrt{1 + \frac{8c_o V_o^2}{T_i^2}}}{2}$$

This simple model gives $T_i' - T_i \simeq 50$ K for $V_o = 0.2$ eV, $T_i = 200$ K, and $c_o = 10^{-3}$; this is in agreement with classical experimental situations in 3d- systems. Time evolutions of the order parameter ρ_q as a function of temperature have been given in [29, 34].

b. The quadratic coupling case [29]

It has been elaborated for the case of thiourea and uses a standard Landau-Ginzburg functional. The basic model (without impurities) involves the polarization along the ferroelectric axis $P_z(x)$ [55] so that the free energy is :

$$F = \int \left\{ \frac{1}{2} A_o P_x^2 + \frac{1}{4} B P_x^4 + \frac{1}{2} \alpha \left[\frac{\partial P_x}{\partial z}\right]^2 + \frac{1}{4} \delta \left[\frac{\partial^2 P_x}{\partial z^2}\right]^2 + \eta \left[P_x \frac{\partial P_x}{\partial z}\right]^2 - E P_x \right\} dz$$

E is the applied electric field, $A_o = T - T_o/C$, B and δ are positive constants and α is negative. η is very important because it allows for the q(T) curvature.

Let us now introduce a quadratic coupling between the modulation order parameter and the impurity :

$$V(r) = V_o P_q^2(r) \, \delta(r)$$

with V_o identical for all the impurities and $V_o P_q^2 << k_B T$ (i. e. we assume the weak coupling limit). The diffusion equation for the impurities is supposed to involve a unique diffusion constant D. If c(r) is the impurity concentration, then :

$$\frac{\partial c}{\partial t} = D \Delta c + \frac{D}{k_B T} \nabla (c \nabla V)$$

The impurities are going to diffuse in the potential V(r) with wavevector $q = q^*(T^*)$ if the temperature is stabilized at $T = T^*$ within the modulated phase ; after a time Δt^* a periodic component of the DDW will appear with an amplitude :

$$b_o'(t) = \frac{V_o P_q^2}{k_B T^*}[1 - \exp(- 4Dq^{*2} \Delta t^*)]$$

As a result this DDW will be able to generate a pinning potential at $q = q^*$ with the amplitude :

$$V_p = V_o c_o b_o' P_q^2 \delta(q, q^*)$$

If now, the trajectory of the system again crosses the $q = q^*$ line by varying T, for example, we shall observe a lock-in on a temperature interval :

$$\Delta T = \frac{2}{\pi} \zeta_m V_o \left[\frac{c_o}{k_B \; T \gamma q^{*4}}\right]^{1/2} P_q^*[1-\exp(-4Dq^{*2}\Delta t^*)]^{1/2}$$

This increase of ΔT with the various parameters q^* and Δt^* has been studied experimentally in thiourea and there is a very good agreement with our theoretical model. Contrary to the linear coupling case, the metastability effects are very small : there is no variation of the transition temperature and no increase of the order parameter (ΔT_i <10mK), but the memory effects have ΔT of the order of 2 to 3 K in thiourea.

Other couplings are possible and have been suggested recently [37] ; they can contribute as well as the basic linear or quadratic couplings to the DDW creation process.

Another model has been elaborated starting with a d-dimensional ANNNI Hamiltonian [50] with an exchange coupling between the spins of the matrix and the impurity spins. The impurities are allowed to come to thermal equilibrium which is relevant for the experiments (time scales long as compared to the impurity relaxation times) ; furthermore, even if they do not perturb the wavevector of the underlying spin structure, they nevertheless order with this wavevector.

However, in all these models, the impurities are independent ; this does not necessarily reflect the experimental situation : in the case of high impurity concentration one can then expect qualitative changes in the phase diagram.

IV. Experiments on Memory Effects in Thiourea

The memory effects have been discovered recently [28, 30, 31] and studied extensively in thiourea from the theoretical [29] and experimental points of view. In thiourea, the quadratic coupling mechanism between the impurities and the modulation order parameter, explains the main properties of the effect with a very good quantitative agreement. Experimental manifestations of this effect are illustrated on Figures 2 and 3. The basic experiment to get a memory effect is a two-step procedure.

1. Writing of a memory effect

a. Creation of a defect density wave (DDW)

By leaving the temperature T of the system to stabilize at $T = T^*$ within the modulated phase for a time Δt^*, we allow the mobile impurities to order by coupling with the modulation order parameter (the electric modulated polarization P_q in thiourea) : this gives rise to a spatial ordering of the impurities with the wavevector $q = 2q(t^*)$ in the quadratic coupling case or $q = q(T^*)$ in the linear coupling case (i. e. a DDW). This first step is the writing of the memory effect.

b. Observation of a defect-induced locked incommensurate phase (DILI Phase)

If after some temperature variations, we allow the temperature to go again through T^*, the DDW will create a periodic potential and lock the modulation in a way similar to a classical Umklapp term.

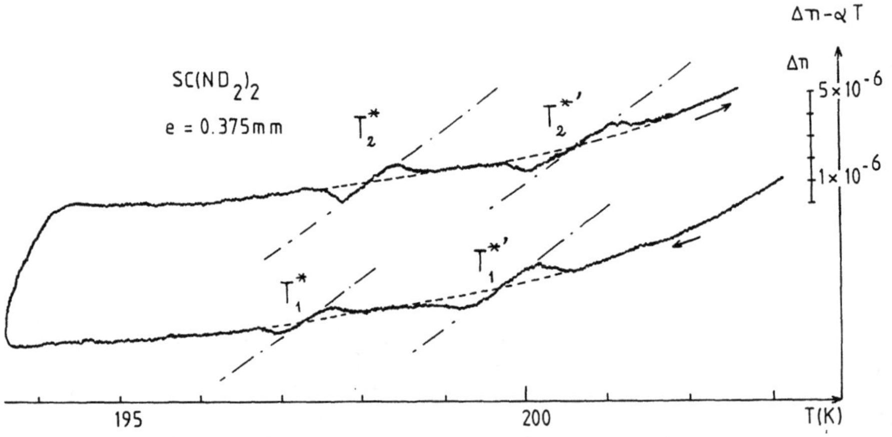

Fig. 2. Two memory effects appear in a birefringence measurement after two temperature stabilizations during an earlier cooling experiment with Δt = 3 hours each time. The different T-positions of the memory effects between the heating and cooling runs correspond to the thermal hysteresis on q(T) as shown on Fig. 1. The central part of the memory effects indicated by dashed-dotted lines corresponds to q = cste birefringence variations.

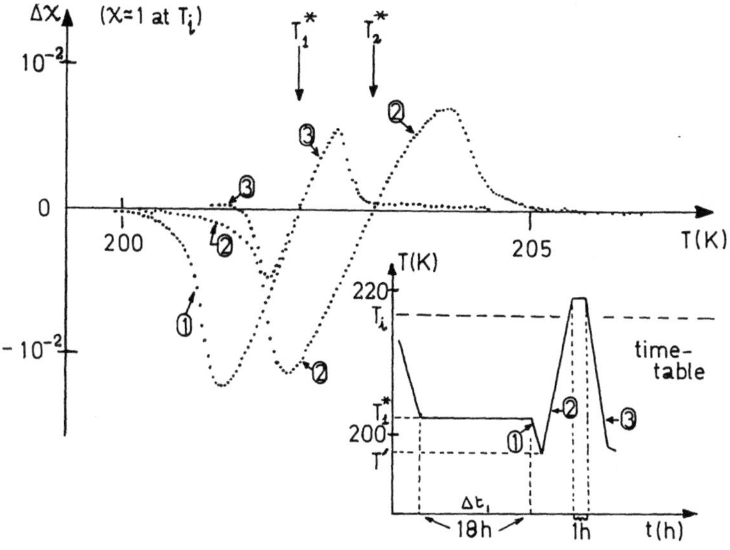

Fig.3 : Dielectric susceptibility measurements of a memory effect in thiourea with its timetable, after 18 hours stabilization time at T_1^* ; the points are the differences between experimental measurements with and without memory effects. Curve 1 is the direct reading (no relaxation of the DDW has yet occurred : it is somehow a virgin effect), while curves 2 and 3 are delayed readings occurring after some relaxation. Curve 3 has been got after the system has been heated 1K above T_i. The difference between T_1 and T_2^* is related to the global hysteresis on q(T) between cooling and heating. Cooling and heating rates : 3 mK/sec.

This second step is the reading of the memory effect. The anomaly which is observed at $T = T^*$ is a defect-induced locked incommensurate (DILI) phase.

A direct reading can be done just after the writing time, and it is illustrated on Fig. 4 where we have reproduced the differences between an experiment without the memory effect and other ones with the memory effect but for various writing times ; it is clear that the temperature width and the amplitude of the effects increase with the writing time Δt^*, but a saturation is observed for $\Delta t^* \sim 100$ hours. One can associate this saturation value with a diffusion constant $D \simeq 10^{-19} cm^{-2}$ [Fig. 5].

This value does not vary much if the stabilization temperature is changed ; in particular, stabilization of the temperature at a value where q is a rational ($q = b^*/8$ in thiourea) does not seem to critically change the memory effect ; this is partly in contradiction with a recent paper [39] which predicted a diffusion constant D(q) identical for all irrational q but a q-dependent D(q) for all rational q.

Without entering into too many details which are outside of the scope of this paper, one can mention another important

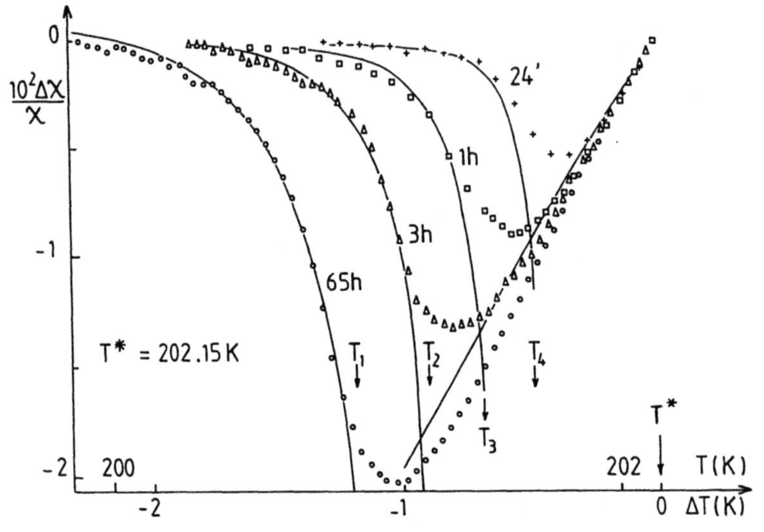

Fig.4 : Direct readings of memory effects obtained after several writing times Δt^* at the same temperature T^*=202.15 K. All the experimental curves have the same tangent at T^*, which means that they are locked phases : the slightly curved solid line corresponds to the mean field calculation of $\Delta\chi/\chi$ for $q=q^*\simeq$cste, while the other solid lines correspond to the critical regime in the vicinity of the DILI phase [38].

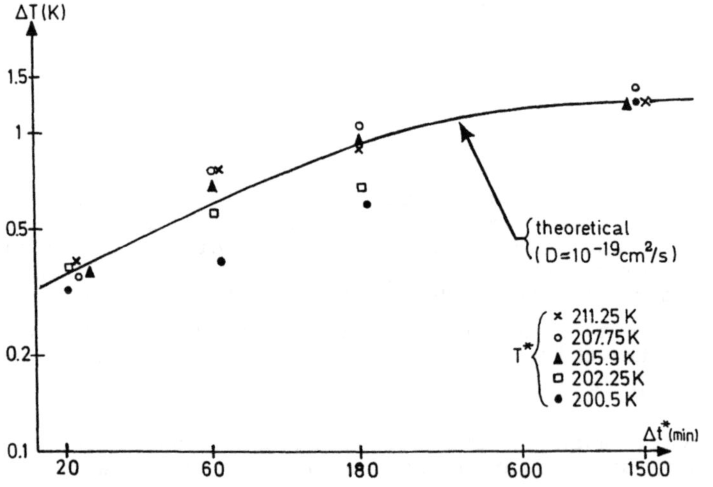

Fig. 5. Saturation of the direct reading for various conditions in thiourea. The solid line corresponds to the ΔT law obtained from the theoretical calculation with the quadratic coupling hypothesis. The dispersion of the points is due to the use of different samples. N.B. The direct reading ΔT represents the half-width of the locked phase.

characteristic of this memory effect : it is possible to write several memory effects at different temperatures and then to read them, which means that several DDW patterns with different q-wavevectors can be stored at the same time in the same crystal.

2. Erasing of a memory effect [Fig. 6 and 6b]

To erase a memory effect, one can just stabilize the temperature of the system at $T > T_i$ for a sufficient amount of time [38]. This raises the question of the nature of the defects : if the defects had an intrinsic nature, the memory effect would be instantaneously destroyed as soon as the temperature is larger or equal to T_i, (i. e. when the order parameter P_q is zero) ; this is not the case in thiourea, where, depending upon the writing time, a memory effect will not be completely erased even after a long stabilization time in the disordered phase (for example, 1 hour at T_i + 10 K). This indicates that the defects responsible for the memory effect have presumably an intrinsic origin : they can be solvent interstitial molecules for example. Indeed, if one sublimates thiourea, the crystals obtained by this method do not possess any memory property, contrary to the crystals where the memory effect is observed and which were grown in methanol ; nevertheless, in both cases, the thermal hysteresis measured on q(T) are equivalent : this means that the mechanism responsible for the change of q with T (or the nature of the defects responsible for this hysteresis) is the same one.

V. Memory Effects : Experiments in Systems Different From Thiourea

1. As-grown samples

a. Insulators

In thiourea, the metastability of the system appears only through the memory effect ; in particular during the stabilization time, in the modulated phase, which can be up to 100 hours in our experiments, apparently no relative variations of the dielectric susceptibility X or of the optical birefringence are observed to, respectively, more than 2×10^{-4} and 10^{-2} which are our long-term experimental accuracies. In other systems, it is not usually the case : in particular, in $Ba_2NaNb_5O_{15}$ where the properties of the system depend strongly on the annealing temperature T_a inside or outside the modulated phase [36]. Nevertheless, the dynamics of the order parameter in this system is very complicated due to the coexistence of two phases [40] which was confirmed recently by Transmission Electron Microscopy.

In quartz, where experiments are delicate in the modulated phase due to its small temperature extent, the hysteresis has been shown to depend strongly on the time variation of the temperature $|\Delta T/\Delta t|$: for very slow cooling or heating rates, the optical birefringence displays a regular succession of steps [24] ; they are probably related to q(T) steps through the gradient-amplitude coupling term present in the free energy of this system ; the mechanism responsible for these steps has not been discussed but it can be related to successive approaches to equilibrium through nucleation processes of stripple nuclei, atlernating with metastable states.

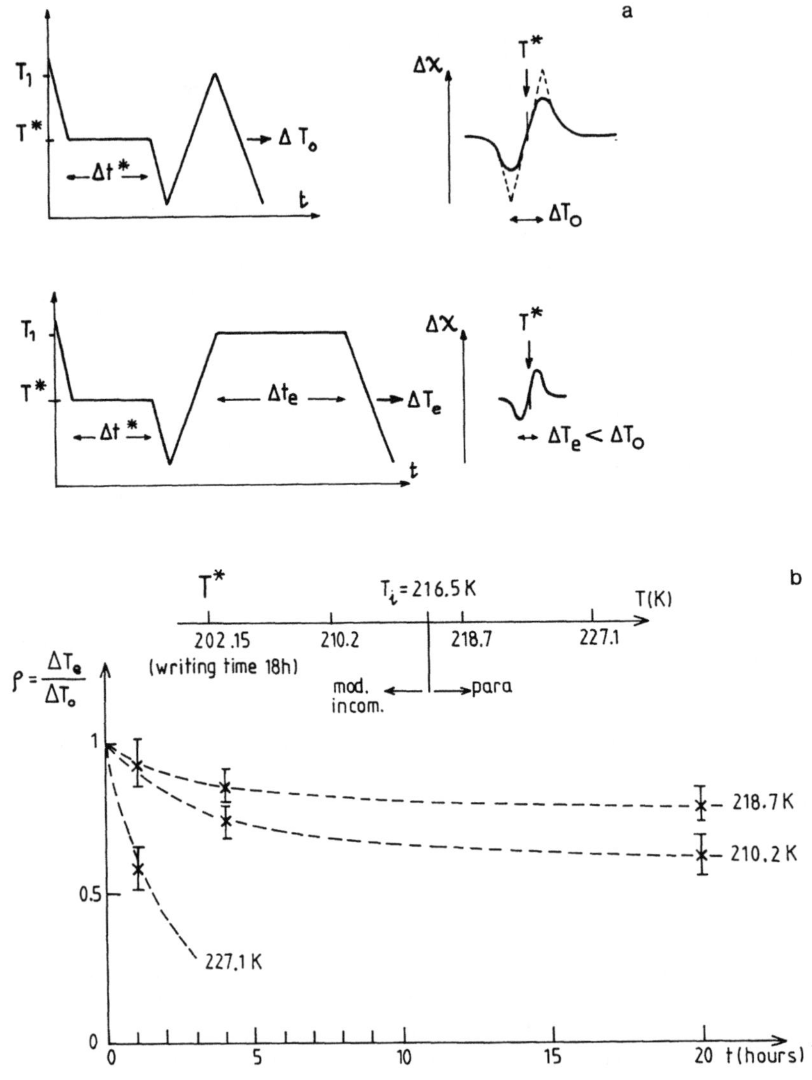

Fig. 6 : Erasing of the memory effects due to the return of the DDW to randomness for T_1 in the disordered phase or to another q-value for $T_1 \neq T$, but in the modulated phase. As shown in Fig.6a, each data point has been got thanks to two experiments : in a first experiment, we have written a memory effect at T during Δt , then gone to the erasing temperature T_1 and gone back again through T ; this gives the width ΔT_0 ; in the second experiment, the process was identical but at T_1 the temperature has been stabilized during Δt_e ; so we deduced from the data $\rho = \Delta T_e / \Delta T_0$ which expresses the memory erasing. Fig.6b shows the ratio $\rho = \Delta T_e / \Delta T_0$ for the same writing conditions. One sees clearly that the memory effect is not completely erased even after 1 h at T_i+10K which presumably means that the impurities have an extrinsic origin.

These mechanisms have been described by H. G. Unruh [30] and discussed recently by T. Nattermann [41] in the context of the commensurate-incommensurate transition.

In quartz [24] it has been proven that the memory effect, which we thought to be a lock-in effect of the modulation by the DDW [28], is indeed this effect ; the memory effect has been observed simultaneously by optical birefringence and Y-ray ; these techniques give access respectively to the order parameter and to the modulation wave vector q(T).

Other modulated systems display memory effects in particular in the A_2BX_4 family [42] ; finally, one can also mention the observation of memory effects in NH_4HSeO_4 [43] or $AgAsS_3$ [44], but their interpretation is complicated as it is in $Ba_2NaNb_5O_{15}$.

b. Charge density wave systems

In these systems, the memory effect is rather well described by the linear coupling theory, i. e. when the temperature of the system is stabilized, the modulation order parameter slowly increases ; furthermore, the modulated-disordered transition temperature is shifted upwards. This has been demonstrated by NMR in $Rb_{0.3}MoO_3$ [35].

2. Irradiated samples (see Fig. 7)

The irradiation by X-rays or electrons has strong effects on the modu- lated systems, due to the creation of defects inside the structure. Experiments have been done on insulating and CDW systems. Heavy irradiation by γ-rays on $SC(ND_2)_2$ [26] has been reported to wash out the normal locked phases for large irradiation doses (100 Mrad), but to induce new quasi-locked phases on irrational q-values close to the modulated-disordered transition temperature T_i, and to increase significantly the thermal hysteresis on q(T). Similar effects are observed in even slightly doped crystals (see for example the properties of $(Rb_xK_{1-x})_2ZnCl_4$ in [25]). These

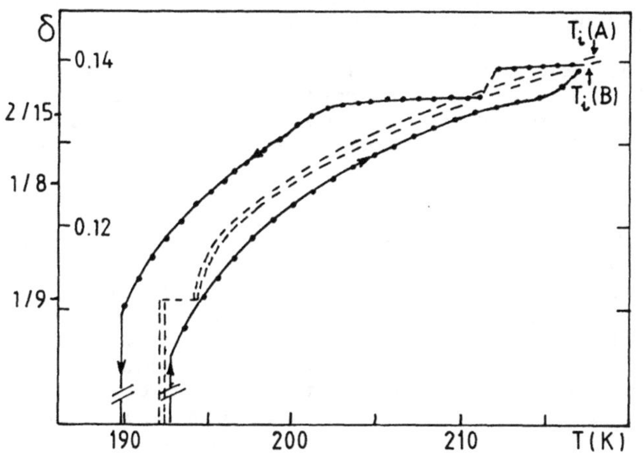

Fig. 7 : Modulation wave vector evolution in thiourea after heavy γ-ray irradiation (100 Mrad) ; the dotted lines are the evolutions before irradiation [26] ; plateaus appear close to T_i, but the normal lock-in ($q = b^*/9$) has disappeared

effects have been discussed in the context of the incommensurate-commensurate transitions [44-45] ; it is known that quenched random impurities destroy the long-range order in the incommensurate phase below d = 4 dimensions ; this can easily explain the disappearance of normal locked phases. Nevertheless, the appearance of quasi-plateaus on the variation of q(T) close to T_i is more difficult to explain : in this part of the phase diagram the memory effect with a quadratic coupling impurity-modulation can be efficient only if we take account of the existence of correlated regions in the system with a finite P_q and a well defined q at T_i ; then there is a possibility of a quasi lock-in even at T_i ; on the other hand, a linear coupling will allow such a memory effect and it cannot be ruled out due to the presence of a shift of T_i. To clarify these possibilities, time-dependent effects must be studied.

Irradiation effects have been also studied in CDW systems [46] showing time dependent effects and clear memory effects observed by resistivity measurements. Here the irradiation has been introduced by fast electrons with doses up to 400 mC/cm^2, and the hysteresis is strongly increased as in irradiated thiourea. In [46] according to the authors' interpretation, the memory effect comes from the pinning of q and from a conduction mechanism which happens in a band separated from the Peierls distorted bands. The strong time dependent effects argue in favor of the presence of a linear coupling impurity-modulation.

In these various systems, the memory effects have various aspects ; even if their dynamics are usually slow and memory effects appear, the coupling mechanism between the defects and the modulation can be different from one system to the others ; it is important, from this point of view, to determine whether the defects come from the growing-conditions of the crystals (extrinsic character) or from the topology of the modulation itself (intrinsic character) - if we exclude the special case of the extrinsic irradiation defects.

VI. Memory Effects in Modulated Systems and Other Systems with Competing Interactions.

Many systems display memory effects, from the neural networks in the brain to spin-glasses or photorefractive materials for example.

There is actually a great deal of interest in these systems not only because of their intrinsic physical properties, but because theoretical models have been developed, in particular for spin-glasses, which exhibit fascinating properties like learning, memory or pattern recognition [47, 48].

In these systems, as in spin-glasses, the competition between the exchange interactions on the same site are at the origin of the dynamics of the system, furthermore, no impurity is necessary to get the spin-glass behavior which is an intrinsic property of the spin lattice. In the case of the memory effect observed in thiourea for example, the memory, in contrast to the spin-glass case, is due to the presence of extrinsic defects or impurities which are coupled to the modulation and, to first order, does not affect its properties. In spin-glasses, there is no long-range order but the spin-glass (SG) state is a cooperative phenomenon due to the presence of the exchange coupling between spins : it is observed even for a very low concentration of spins (some ppm of Mn^{2+} ions) in the case of the canonical CuMn [49] : this is only due to the oscillating long-range RKKY interaction. The very large distribution of time constants observed in these systems is due to the competition between interactions at the same site.

Can we do a comparison with the impurity lattice which is present in thiourea ? Indeed, the time constant necessary to obtain the

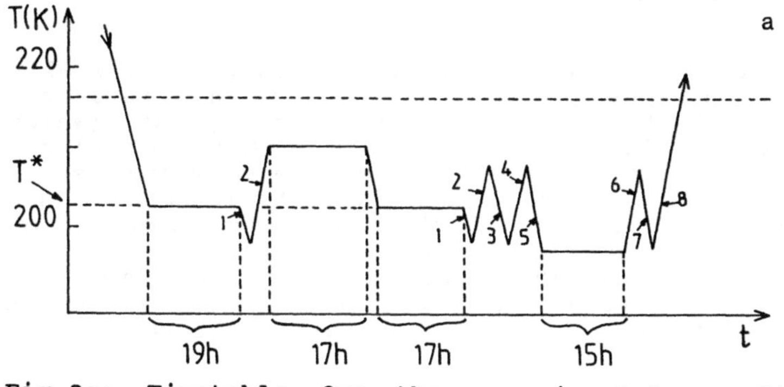

Fig.8a: Timetable for the experimental results of Fig.8b. The stabilization at T^* has been reproduced after a long stabilization at 210 K. The 2 direct readings (1) are indistinguishable as well as the 2 delayed readings (2) ; readings (2) have not been reproduced.

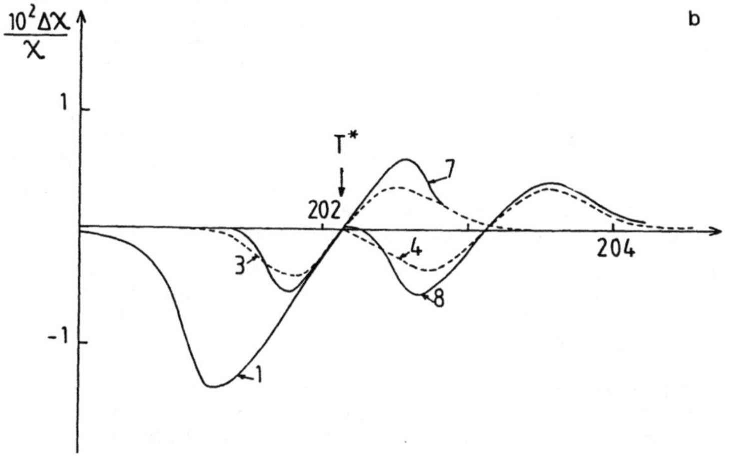

Fig. 8b : Only (1) has been reported, but (3) and (4) have anomalous shapes and amplitudes : - when there is not the second writing at T^*, (3) and (4) are respectively identical to (7) and (8) - they don't correspond to a normal locked phase : the long-range ordering is "broken". But curiously in (7) and (8) the normal regime is restored : the system has retrieved its original pattern [38]. This behaviour is typical of associative memories [48] and memory effects in spin-glasses.

maximum memory effects is rather long (100 hours) ; this means that the impurities are probably linked together by a 3d long-range coupling interaction : this coupling is ignored in the theoretical treatment [29, 50] which have been elaborated and only take account of the coupling between the impurities and the modulation. The effect of a uniform external magnetic field on the spins in the SG state can be compared to the modulated electrical polarization P_q present in the modulated phase of thiourea which acts on the defect lattice : in both cases, a long relaxation time to a metastable state is observed ; another difference between the two systems is the presence in the second one of a spatial periodicity in one direction which can be adjusted by choosing the appropriate temperature.

Is thiourea able to learn several patterns ? Thiourea is really able to learn several patterns of the modulation, if one stabilizes the temperature several times in the modulated phase (which corresponds to several q values) : 10 times can be easy to realize. Furthermore, when two patterns are written at the same wavevector but at different times, the system is also able to reject the new pattern and retrieve the old one ; this property, typical of associative memories, has been demonstrated in spin-glasses with field-reverse sequences [51]. This superposition experiment is reported on Fig. 8a and 8b and it shows that the system can reject the noise i. e. the difference between the learnt pattern and the second one [8, 38].

Conclusion

Hysteresis and memory in thiourea are still not fully understood. Many experiments can be done in these modulated systems to try to understand them but some conditions must be fulfilled and first of all the chemistry or the growing conditions of these crystals ; it is clear that the modulated systems are very sensitive to defects, so that mastering the impurity content as well as their nature is essential to determine whether the hysteresis effect is intrinsic or partly extrinsic. The same comment can be made about the memory effect origin : in thiourea it is presumably extrinsic. The interest of the study of the memory effect in modulated systems is also to be related to the actual studies on neural networks and associative memories. Other memory effects have been found in a CDW system : $K_{0.3}MoO_3$ associated with short electric pulses [58] ; this new memory effect which is related to metastable states of a moving CDW is a pulse-duration memory effect ; it is on a time scale very different from ours and involves different mechanisms : this means nevertheless that these systems are very rich.

At last, visualization experiments which have not been reported here, are beginning in the case of insulating systems : Some recent studies show clearly the presence of stripples in $Ba_2NaNb_5O_{15}$ [40] and Rb_2ZnCl_4 [52]. On the other hand, X-ray topography has shown displacements of entities in $TMAZnCl_4$ [53]. Nevertheless, these techniques must be used with care in some of the insulating systems because they can create irradiation defects and modify the normal behavior of these systems [26].

Acknowledgements

The author is indebted to P. Lederer and G. Montambaux for their collaboration and for many helpful discussions as well as M. Chauvin for the experiments. It is a pleasure to thank F. Denoyer, D. Durand, J. Ferré, for valuable discussions ; lastly, I am pleased to acknowledge P. E. Littlewood and R. M. Fleming for enlightening discussions on the memory effects while visiting them at Bell Laboratories.

References

1. F. Reinitzer, Montash Chem. 9, 421 (1888)
2. Y. Yamada, I. Shibuya and S. Hoshino, J. Phys. Soc. Japan 18, 1594 (1963)
3. R. Comès, M. Lambert, H. Launois and H. R. Zeller, Phys. Rev. B 8, 571 (1973)
4. F. Denoyer, R. Comès, A. F. Garito and A. J. Heeger, Phys. Rev. Lett. 35, 445 (1975)
5. F. Denoyer, private communication
6. Among review papers, see for example : P. Bak, Rep. Prog. Phys. 45, 587 (1982)
7. Incommensurate Phases in Dielectrics, vol. 14.1 and 14.2, R. Blinc and A. P. Levanyuk editors in Modern Problems in Condensed Matter Sciences, North Holland Physics Publishing (1986)
8. Review Papers on "Phase transitions in the presence of small concentration of defects", J. C. Toledano ed. to appear in a special issue of Phases Transitions, Gordon and Breach Publishers (1988)
9. M. S. Haque and J. R. Hardy, Phys. Rev. B 21, 245 (1980)
10. V. Katkanant, P. J. Edwardson, J. R. Hardy and L. L. Boyer, Phys. Rev. Lett. 57, 2033 (1986)
11. Y. I. Frenkel and T. Kontorova, Zh. Eksp. Teor. Fiz. 8, 1340 (1938)
12. S. Aubry in Solitons and Condensed Matter Physics, A. R. Bishop and T. Schneider Ed. Springer Verlag Publishers, p. 264 (1979) and J. Physique (Paris) 44, 147 (1983)
13. F. C. Frank and J. H. Van Der Merwe, Proc. Roy. Soc. 198, 205 (1949)
14. P. Bak and V. J. Emery, Phys. Rev. Lett. 36, 978 (1976)
15. See T. Jansen, p. 67 in vol. 14.1 of Ref. [7]
16. J. F. Scott, Ferroelectrics 36, 375 (1981) and ibid. 66, 11 (1986)
17. S. Kh. Esayan, V. V. Lemanov, N. Mamatkulov and L. A. Shuvalov, Sov. Phys. Crystallogr. 26, 619 (1981)
18. A. P. Levanyuk, A. S. Sigov, V. V. Osipov and A. A. Sobyanin, Sov. Phys. JETP 49, 176 (1979)
19. D. E. Moncton, J. D. Axe and F. J. Disalvo, Phys. Rev. Lett. 34, 734 (1975)
20. K. Nakanishi, J. Phys. Soc. Jap. 46, 1434 (1979)
21. A. H. Moudden, Thesis, Orsay (France) 1980, unpublished ; F. Denoyer, A. H. Moudden, R. Currat, C. Vettier, A. Bellamy and M. Lambert, Phys. Rev. B 25, 1697 (1982) ; F. Denoyer and R. Currat, p. 129 in Vol. 14.2 of ref. [7]
22. W. L. Mac Millan, Phys. Rev. B 14, 1496 (1976) ; V. Janovec, Phys. Lett. 99A, 384 (1983)
23. T. Nattermann, J. Phys. C : Solid State Phys. 18, 5683 (1985)
24. G. Dolino, Jap. J. of Applied Phys., 24, Suppl. 24-2, 153 (1985)

25. K. Hamano, K. Ema and S. Hirotsu, Ferroelectrics 36, 343 (1981) and K. Hamano in [8] ; H. Mashiyama, S. Tanisaki and K. Hamano : J. Phys. Soc. Jap. 51, 2538 (1982)
26. G. André, D. Durand, F. Denoyer, R. Currat and F. Moussa, Phys. Rev. B 35, 2909 (1987)
27. S. Barre, H. Mutka, C. Roucau and G. Errandonea, Phase Transitions 9, 225 (1987)
28. J. P. Jamet and P. Lederer, J. Phys. Lettres (Paris) 44, L-257 (1983) and Ferroelectric Lett. 1, 139 (1984)
29. P. Lederer, G. Montambaux, J. P. Jamet and M. Chauvin, J. de Phys. Lettres (Paris) 48, L-627 (1984) ; P. Lederer, J. P. Jamet, G. Montambaux, Ferroelectrics 66, 25 (1986) ; M. Chauvin, Thèse Orsay (France) (1985) unpublished
30. H. G. Unruh, J. Phys. C : Solid State Phys. 16, 3245 (1983)
31. G. Errandonea et al., J. de Phys. Lettres (Paris) 45, L-329 (1984)
32. J. F. Scott, Ferroelectrics 66, 11 (1986)
33. H. Fukuyama and P. A. Lee, Phys. Rev. B 17, 535 (1978) ; P. A. Lee and T. M. Rice, Phys. Rev. B 19, 3970 (1979)
34. P. Lederer, G. Montambaux and J. P. Jamet, Mol. Cryst. and Liq. Cryst. 121, 99 (1985)
35. P. Butaud, P. Segransan, C. Berthier, J. Dumas and C. Schlenker, Phys. Rev. Lett. 55, 253 (1985)
36. J. C. Toledano, G. Errandonea, J. Schneck, A. Litzler, H. Savary, F. Bonnouvrier and M. L. Esteoule, Jap. J. Appl. Physics, 24, Supp. 24-2, 290 (1985)
37. V. S. Vikhnin, Sov. Phys. Crystallogr. 31, 374 (1986)
38. M. Chauvin, G. Montambaux, J. P. Jamet to appear and M. Chauvin, Thèse Orsay (France) (1985) unpublished
39. K. Golden, S. Goldsein and J. L. Lebowitz, Phys. Rev. Lett. 55, 2629 (1985)
40. C. Manolikas, J. Schneck, J. C. Toledano, J. M. Kiat and G. Calvarin, Phys. Rev. B 35, 8884 (1987)
41. T. Nattermann, J. Phys. C : Solid State Phys. 18, 5683 (1985)
42. O. G. Vlokh, B. V. Kaminskii, A. V. Kityk, I. I. Polovinko and S. A. Sveleba, Sov. Phys. Sol. State 28, 1226 (1985)
43. I. P. Aleksandrova, Yu. N. Mosvitch, O. V. Rozanov, A. F. Sadreev, I. V. Seryukova and A. A. Sukhovsky, Jap. J. of Appl. 24, Suppl. 24-2 (1985) and Ferroelectrics 67, 63 (1986)
44. J. Villain, J. de Phys. (Paris) Lettres 43, L-551 (1982) ; P. Prelovsek and R. Blinc, J. Phys. C : Solid State Phys. 17, 577 (1984) and P. Prelovsek in [8] ; R. Blinc, P. Prelovsek, V. Rutar, J. Seliger and S. Zumer in [7].
45. T. Nattermann, J. Phys. C : Solid State Phys. 16, 6407 (1983) and ibid, 18, 5683 (1985)
46. H. Mutka, F. Rullier-Albenque and S. Bouffard, J. de Phys. (Paris) 48, 425 (1987)
47. J. J. Hopfield : Proc. Natl. Acad. Sci. (U.S.A.) 79, 2554 (1982), J. J. Hopfield, D. I. Feinstein and R. G. Palmer, Nature 304, 158 (1983)
48. J. P. Nadal, G. Toulouse, J. P. Changeux and S. Dehaene, Europhysics Lett. 1, 535 (1986)
49. E. C. Hirschoff, O. G. Symko and J. C. Wheatley, Phys. Lett. 33A, 19 (1970)
50. H. Roeder and J. Yeomans, J. Phys. C : Solid State Phys. 18, L-163 (1985)
51. L. Lundgren, P. Nordblad and L. Sandlund, Europhysics Lett. 1, 529 (1986)
52. H. Bestgen, Solid State Comm. 58, 197 (1986)
53. M. Ribet, Ferroelectrics 66, 259 (1986) and J. de Phys. (Paris) Lettres 44, L-963 (1983)
54. M. Barreto, J. P. Jamet and P. Lederer, Phys. Rev. B 28, 3994 (1983)

55. Y. Ishibashi and H. Shiba, J. Phys. Soc. Jap. 45, 409 (1978)
56. J. P. Jamet, P. Lederer and A. H. Moudden, Phys. Rev. Lett. 48, 442 (1982)
57. W. Kinzel, Z. Phys. B 60, 205 (1985)
58. R. M. Fleming and L. F. Schneemeyer, Phys. Rev. B 33, 2930 (1986)
59. S. B. Coppersmith and P. B. Littlewood, preprint.

Competing Interactions: Charge Density Waves and Impurities

G. Grüner

Department of Physics and Solid State Sciences, University of California, Los Angeles, CA 90024, USA, and
Los Alamos National Laboratory, Los Alamos, NM 87545, USA

ABSTRACT

Electron-phonon interactions lead to the formation of charge-density-waves (CDW), which have been observed in several inorganic linear chain materials. Because of the absence of a gap in the CDW excitation spectrum, impurities have a profound influence on the statics and dynamics of charge-density-waves. They lead to the absence of long range order, to smearing of the phase transitions and pinning of the collective mode. These effects are discussed by presenting experimental results on several model systems.

1. CHARGE DENSITY WAVES

Consider a one-dimensional metal. In the absence of an interaction with the lattice, the ground state is as shown in Fig. 1(a), where the electron states are filled up to the Fermi level and the underlying lattice is that of a periodic array of atoms with lattice constant a. As first pointed out by PEIERLS [1], this state is not stable for a coupled electron-phonon system. In the presence of an interaction (of any strength), it is energetically more favorable to distort the lattice periodically with period λ related to the Fermi wave vector k_F

$$\lambda = \frac{\pi}{k_F} \quad . \tag{1}$$

A lattice distortion with this period opens up a gap at the Fermi level, as shown in Fig. 1(b) where the situation appropriate for a half-filled band is drawn. As states only up to $\pm k_F$ are occupied, the opening of the gap leads to the lowering of the electronic energy. The lattice distortion leads to an increase of the elastic energy, but in 1D the total energy is lower than that of the undistorted metal (this is the consequence of the divergent Linhard Function at $q = 2k_F$ in 1D). Consequently a distorted state is stable at T=0 K. The state has a gap in the single particle excitation spectrum, and this gap opening also leads to the modification of the electron density, much in the same way as in the nearly free electron theory of metals. The density $\rho = |\psi|^2$ will be a periodic function of the position x with the period given by (1), i.e., determined by the band filling. Thus, for an arbitrary band filling, the period of this modulated charge density [and the accompanying periodic lattice distortion, see Fig. 1(b)] will be incommensurate with the underlying lattice. At finite temperatures, normal electrons excited across the single particle gap Δ screen the electron-phonon interaction. This in turn leads to a reduction of the gap and eventually to a (second order) phase transition. Such semiconductor to metal transition is called the Peierls transition.

Calculations which lead to the Peierls transition are performed by using the 1D electron-phonon Hamiltonian [2]

$$H = \sum_k \varepsilon_k C_k^+ C_k + \sum_q \omega_q^o b_q^+ b_q + \sum_{k,q} g\, C^+_{k+q} C_k (b_q + b_{-q}^+) \;, \tag{2}$$

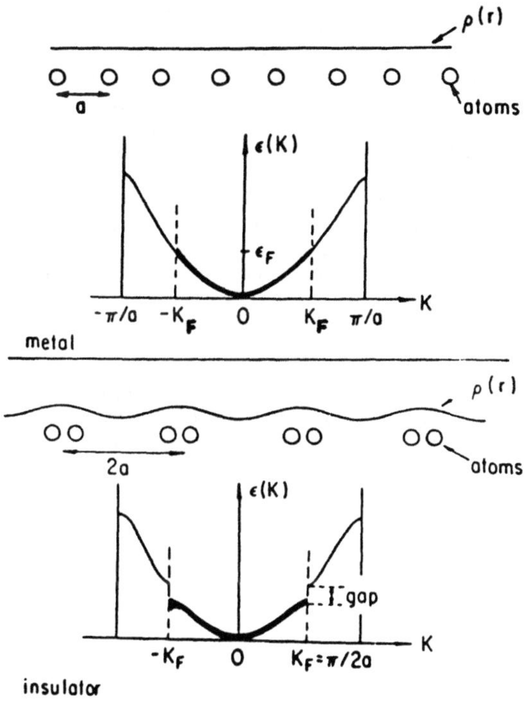

Fig. 1. A one-dimensional half filled band without (1a) and with (1b) Peierls distortion. $\rho(r)$ refers to the electron density, and the single particle gap is also indicated on the figure.

where C_k, C_k^+ and b_q, b_q^+ are the electron and phonon creation and annihilation operators with moments k and q, ε_k and ω_q^0 are the electron and phonon dispersions and g is the electron-phonon coupling constant. In (2), spin is omitted.

Defining a complex order parameter

$$\Delta e^{i\phi} = g\langle b_{2k_F} + b^{\dagger}_{-2k_F}\rangle, \tag{3}$$

where Δ and ϕ are real, the displacement field of the ions is

$$\langle b_{2k_F} + b^{\dagger}_{-2k_F}\rangle \, e^{2ik_Fx} + \text{H.C.} = \frac{2\Delta}{g}\cos(2k_Fx + \phi) \ . \tag{4}$$

One can diagonalize the electronic part of the Hamiltonian by setting up a self-consistent equation for Δ by replacing b_{2k_F} by $\langle b_{2k_F}\rangle$. Such mean field approximation leads to a BCS gap equation

$$\Delta = 2D \exp\left(- \frac{1}{\lambda} \right) , \tag{5}$$

where $\lambda = D(\varepsilon_F)g^2/(M\omega^2_{2k_F})$ is the dimensionless electron-phonon coupling constant and D is the electronic bandwidth. The temperature dependence of Δ also has the characteristic BCS form, and the transition temperature is given by $kT_p = \Delta/1.76$. The spatially dependent electron density can also be evaluated, and

$$\rho(x) = \rho_o + \frac{\rho_o\Delta}{\lambda^2\varepsilon_F}\cos(2k_Fx + \phi) = \rho_o + \rho_1\cos(2k_Fx + \phi) \ , \tag{6}$$

where ρ_o is the electron density in the absence of electron-phonon interactions. The second term on the right-hand side describes the CDW with period π/k_F and amplitude $\rho_o\Delta/\lambda^2\varepsilon_F$.

The fact that the CDW ground state has always a lower energy than the undistorted metallic state is the consequence of the logarithmically divergent one dimensional Linhard function at $q = 2k_F$. It is not surprising, therefore, that materials which undergo Peierls transitions have predominantly linear chain structure. The most well known examples [3] (aside from many organic linear chain compounds) include transition metal trichalcogenides, MX_3, where M = Nb or Ta and X = S or Se, tetrachalcogens, such as $(TaSe_4)_2I$ or bronzes, like $K_{0.3}MoO_3$; all have a chain structure. These materials undergo a Peierls transition at temperature somewhat below room temperature [note that the cut-off energy is the bandwidth and not the phonon frequency in (5) and consequently the gap is much larger than the superconducting gap] as evidenced, for example, by the dc electrical conductivity $\sigma(T)$. This is shown in Fig. 2 where $\sigma(T)$ measured on various materials is displayed. The arrows represent the temperature T_p where a phase transition to the Peierls state occurs. In the above cases the gap opens up over the entire Fermi surface, leading to a metal-insulator transition. $NbSe_3$ is an exception; here two transitions occur, one at $T_1 = 145$ K, the other at $T_2 = 59$ K, both removing only part of the Fermi surface.

In all of the materials mentioned above, the periodic lattice distortion has been observed by structural studies. Also, in all cases, the CDW is incommensurate with the underlying lattice, and the measured scattering intensity as the function of temperature indicates a BCS-like order parameter. The dynamics of the collective mode is characterized by an amplitude and by a phase mode, the former has a gap, the latter is gapless.

Treating the phase as a classical field, the Lagrangian density associated with $\phi(x,t)$ is given by

$$L = n\left[\frac{m^*}{2} \left[\frac{1}{2k_F} \frac{d\phi}{dt} \right]^2 - \frac{\kappa}{2} \left[\frac{d\phi}{dx} \right]^2 \right] , \tag{7}$$

where the first term is the kinetic energy of a line of mass m^*n per unit length. The potential energy is the second term, with a phenomenological elastic constant κ. The dispersion relation corresponding to wave-like excitations of the form $\exp[i(\omega t-kx)]$ is given by

$$\omega^2 = \left[\frac{\kappa}{m^*} \right] \left[2k_F q \right]^2 , \tag{8}$$

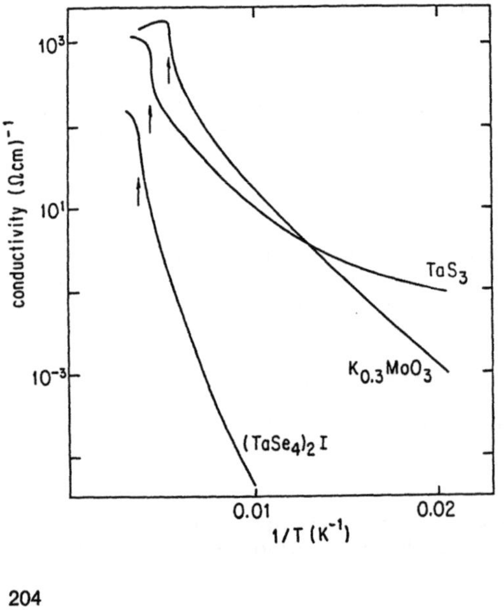

Fig. 2. Temperature dependence of the dc resistivity in linear chain inorganic compounds with CDW transitions, indicated by arrows in the figure.

and comparison with the microscopic theory gives

$$k = m \left[\frac{v_F}{2k_F}\right]^2 . \tag{9}$$

The excitations, which correspond to local modulation of the phase ϕ with a period q, are called phasons.

2. THE EFFECTS OF IMPURITIES

The fact that the phason mode is gapless has important consequences as far as the effect of impurities is concerned. FUKUYAMA and LEE, and EFETOV and LARKIN [4] considered the problem of impurity pinning in detail. The Hamiltonian in one dimension

$$H = \frac{\hbar v_F}{4\pi} \int dx\left(\nabla\phi(x)\right)^2 - V_o\rho_1 \sum_i \cos\left[2k_F x_i + \phi(x_i)\right] , \tag{10}$$

where the first term represents the elastic energy of the deformable CDW, and the second term describes the interaction between the CDW and the impurities [with potential $V(x-x_i) = V_o\delta(x-x_i)$] distributed at random at positions i. In less than four dimensions, (10) leads to a finite phase-phase correlation length and also to a finite pinning enegy in the thermodynamic limit. Also, strictly speaking, no phase transition occurs in the thermodynamic limit.

The first term in (10) favors a uniform phase, the second favors local distortions around impurities. A dimensionless parameter

$$\varepsilon = \frac{V_o\rho_1}{\hbar v_F n_i} ,$$

tells us which of these is more important [4] [5]. For $\varepsilon \gg 1$ (strong pinning) the phase of the CDW is fully adjusted to every impurity site to obtain a maximum potential energy gain, the cost in elastic energy for doing this is negligible. For $\varepsilon \ll 1$ (weak pinning) the phase cannot be adjusted fully at every impurity site but only over a length scale L_o longer than $1/n_i$, the average distance between the impurities.

For strong impurity pinning, where the elastic energy of the CDW is neglected, the pinning energy is trivially given by

$$E_{PIN} = \Delta E_{POT} = - V_o\rho_1 n_i , \tag{11}$$

where n_i is the number of impurities in a unit volume. As the phase is fully adjusted at every impurity site, the phase-phase correlation length

$$L_o \sim \frac{1}{n_i} . \tag{12}$$

The case of weak impurity pinning is more interesting. In 3D, assuming a volume L_o^3 over which the phase is constant but is adjusted to the impurity fluctuations, one obtains a potential energy gain

$$\Delta E_{POT} = - V_o\rho_1 \left(\frac{n_i}{L_o^3}\right)^{1/2} . \tag{13}$$

The elastic energy, from (10)

$$E_{EL} = \frac{3\pi}{4} \frac{\hbar v_F}{L_o^2} . \tag{14}$$

Minimizing the total energy with respect to L_o leads to a finite phase-phase correlation length

$$L_o = \left[\frac{\pi\hbar v_F}{V_o\rho_1}\right]^2 \frac{1}{n_i} , \tag{15}$$

and

$$E_{PIN} = -\frac{1}{4}\frac{V_o^4\rho_1^4}{\pi^3\hbar^3 V_F^3} n_i^2 , \qquad (16)$$

i.e., the pinning energy is proportional to the square of the impurity concentration.

3. THE EFFECTS OF IMPURITIES ON THE PHASE TRANSITIONS

In contrast to superconductors, or ferromagnets where the transition to the ordered state remains sharp and long range order is maintained even in the presence of considerable disorder, arguments advanced before suggest that this is not the case for charge density waves. The consequences are twofold: 1) smearing of the x-ray lines due to the finite phase-phase correlation length, and 2) smearing of the phase transition. The first has not been addressed to date, but several recent studies suggest that impurity induced smearing of the phase transition is significant. The temperature dependence of the derivative dR/dT is shown [6] in several materials in Figs. 3 and 4. While for the nominally pure compounds dR/dT is sharply peaked, it is progressively smeared with increasing impurity concentrations. Data, such as shown in Figs. 3 and 4, can qualitatively be interpreted as the ef-

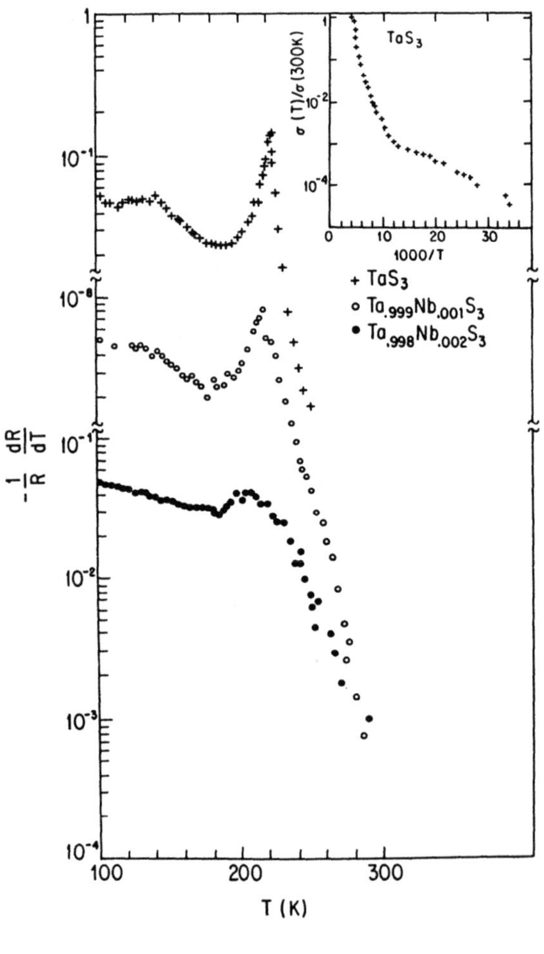

Fig. 3. Temperature derivative of the dc resistivity dR/dT in TaS_3 alloys.

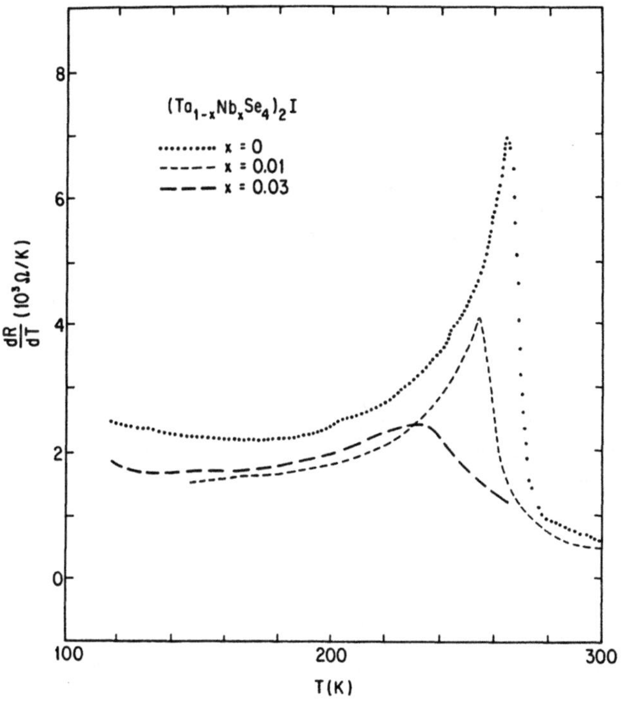

Fig. 4. Temperature derivative of the resistivity dR/dT in $(TaSe_4)_2I$ alloys.

fect of impurity limited correlation length. For pure materials the temperature dependent correlation length ξ is given by

$$\xi = \frac{a}{(T-T_o)^{\nu}} , \tag{17}$$

with the initial exponent $\nu = 0.5$ for a mean field transition. The resistivity depends also on the available momentum states for scattering and

$$\frac{dR}{dT} \sim \frac{1}{(T-T_o)^{\nu+\alpha}} , \tag{18}$$

where α depends on the dimensionality of the system. Impurities do not have an effect on the resistivity, as long as $\xi < L_o$ with L_o given by (12) and (15). This defines a crossover temperature where impurity effects become dominant. From $a(T-T_o)^{-\nu} = b/n_i$, one obtains

$$T_{co} = T_o + \left(\frac{an_i}{b} \right)^{-\nu} . \tag{19}$$

Experiments such as displayed in Figs. 3 and 4 while showing the qualitative features suggested by this argument do not have the sufficient precision which would enable the evaluation of the critical exponent ν, and by performing careful experiments on pure specimens, and comparing the measured dR/dT to (18) also α.

4. THE EFFECTS OF IMPURITIES ON THE DYNAMICS OF THE COLLECTIVE MODE

Impurities also have a profound influence on the dynamics of the collective mode. In the absence of impurities, the equation of motion with an applied electric field E is

$$\frac{d^2\phi}{dt^2} + v_F^2 \frac{m}{m^*} \frac{d^2\phi}{dx^2} = \frac{2k_F eE}{m^*} , \tag{20}$$

and the frequency dependent conductivity is

$$\sigma(\omega) = \frac{j(\omega)}{E(\omega)} = \frac{m}{m^*} \quad \frac{i\omega_p^2}{4\pi(\omega + i\delta)} , \tag{21}$$

$$\mathrm{Re}\sigma(\omega) = \frac{m}{4m^*} \omega_p^2 \, \delta(\omega) . \tag{22}$$

$\mathrm{Re}\sigma(\omega)$ then has a $\delta(\omega)$ peak at zero frequency. This feature, together with the gap Δ in the single particle excitation spectrum is reminiscent of superconductivity.

Impurities with pinning energies given by (11) or (16) act as restoring forces, shifting the oscillator strength from zero to finite frequency. The equation of motion then is that of a harmonic oscillator

$$\frac{d^2x}{dt^2} + \Gamma \frac{dx}{dt} + \omega_0^2 \, x = \frac{eE}{m^*} e^{i\omega t} , \tag{23}$$

where m* is the effective mass of the collective mode, Γ is a phenomenological damping constant, x refers to deviation from the equilibrium position x_0 of the pinned collective mode (with x_0 depending on the impurity positions). In Fig. 5, $\mathrm{Re}\sigma(\omega)$ is displayed for various materials [7]. The gap is clearly evident and optical experiments are shown as full lines with the dotted line showing the collective mode response, with (in general) $\omega_0 \sim 10^{14}$ sec^{-1}, $\omega_0\Gamma \sim 1$ and $m^* \sim 10^3$–10^4. Writing the overall pinning energy as

$$E_{PIN} = - \kappa(x-x_0)^2 = - k\frac{\lambda^2}{2} , \tag{24}$$

where the displacement from equilibrium of a half wavelength gives E_{PIN}, suggests that $\omega_0^2 = + k/m^*$ should increase with increasing impurity concentration, for both weak and strong impurity pinning $\omega_0 \sim n_i$. $\sigma(\omega)$ measured in TaS_3[8] and $(TaSe_4)_2I$[9] alloys is displayed in Figs. 6 and 7. In both cases, ω_0 increases dramatically with increasing impurity concentration, and is roughly proportional to $n_i^{1/2}$, as expected for $\omega_0 = (k/m^*)^{1/2}$ with k given by (24). The magnitudes are also in agreement with estimations based on (24) and (11), with $V_0 \sim 10^{-2}$ eV and $\rho_1 \sim 0.1\rho_0$.

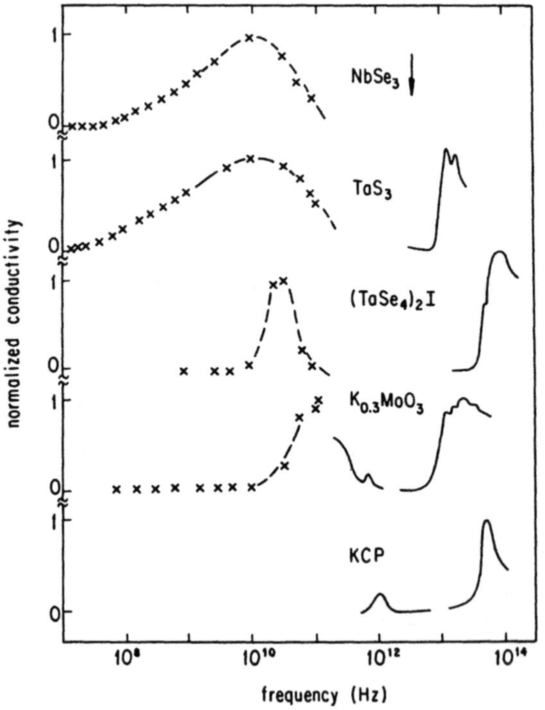

Fig. 5. Frequency dependence of the conductivity in the CDW state of several nominally pure materials.

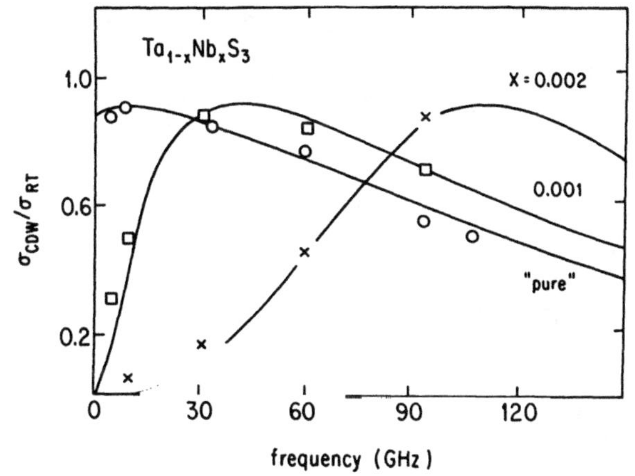

Fig. 6. Frequency dependence of the conductivity in TaS_3 alloys.

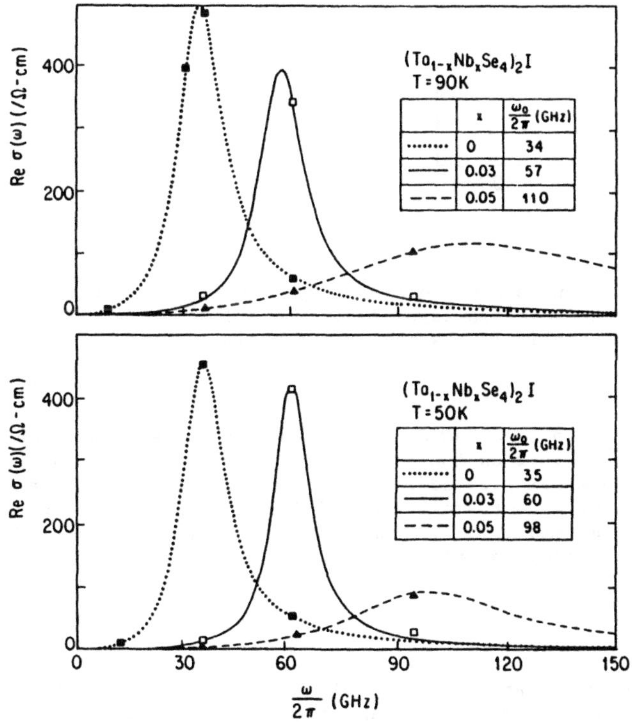

Fig. 7. Frequency dependence of the conductivity in $(TaSe_4)$ alloys. The pinning frequencies are indicated on the Figure.

Random impurity distributions also lead to pronounced effects concerning the details of the dynamics, and, in particular at low frequencies effects similar to those observed in glasses and spin glasses are observed [10]. The simplest (and probably not correct) approach to describe these is to assume that because of random impurity distributions, ω_0 has a distribution, and the frequency dependent conductivity is given by

$$\sigma(\omega) = \int_{-\infty}^{\infty} P(\omega-\omega')\ \sigma(\omega')d\omega' \ , \tag{25}$$

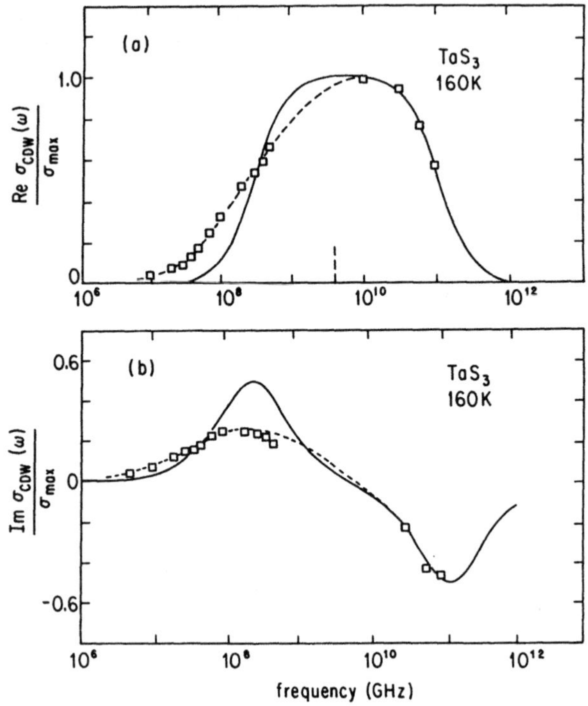

Fig. 8. $\mathrm{Re}\sigma(\omega)$ and $\mathrm{Im}\sigma(\omega)$ in nominally pure TaS_3. The full line is calculated on the basis of Eq. (25).

with $\sigma(\omega')$ the conductivity with a single pinning frequency. In Fig. 8, $\mathrm{Re}\sigma(\omega)$ and $\mathrm{Im}\sigma(\omega)$ measured in nominally pure TaS_3 is displayed [7]. The full line is a fit to a harmonic oscillator, the dotted line is (25) with a distribution

$$P(\omega-\omega_c) = \begin{cases} 1 & \omega > \omega_c \\ 0 & \omega < \omega_c \end{cases} , \tag{26}$$

where ω_c is a cutoff frequency. While both $\mathrm{Re}\sigma(\omega)$ and $\mathrm{Im}\sigma(\omega)$ can be well represented by such a linear response with a distribution of resonant frequencies, the actual state of affairs is probably more complicated with also pronounced nonlinear and hysteresis effects observed even for rather small applied fields. They become progressively more important when the amplitude of the driving field is increased. Such effects are discussed, for example, in Reference 11.

5. CONCLUSIONS

Initial progress in the area suggests that charge density waves can be regarded as good models for the study of both statics and dynamics of random systems. The qualitative features of the broadening of the phase transitions are well established, but more careful experiments on materials with different anisotropies would be needed to clarify the details of the smearing mechanism. The small amplitude dynamics of the collective mode is well understood, but further work is required to explore the various aspects of the nonlinear ac dynamics of these systems.

6. ACKNOWLEDGMENTS

The experiments reported here were performed by L. Mihály, D. Reagor, Tae Wan Kim, and T. Chen. This research was supported by the National Science Foundation grant DMR 86-12011.

REFERENCES

1. R.E. Peierls: *Quantum Theory of Solids* (Oxford University Press, 1955).
2. See for example, J. Berlinski: Rep. Progr. Phys. 42, 1243 (1979) and references cited therein.
3. See for example, P. Monceau: In *Electronic Properties of Low Dimensional Materials*, ed. by P. Monceau (Reidel Co., 1985).
4. H. Fukuyama and P.A. Lee: Phys. Rev. B17, 535 (1978); K.B. Efetov and A.I. Larkin: Zh. Eksp. Theor. Fiz. 72, 2350 (1977).
5. R.A. Lee and T.M. Rice: Phys. Rev. B19, 3970 (1979).
6. N.P. Ong, J. W. Brill, J. C. Eckert, J. W. Savage, S. K. Khanna, and R. B. Somoano: Phys. Rev. Lett. 42, 811 (1979).
7. D. Reagor, S. Sridhar and G. Grüner: Phys. Rev. B34, 2212 (1986); S. Sridhar, D. Reagor and G. Grüner: Phys. Rev. B34, 2223 (1986).
8. D. Reagor and G. Grüner: Phys. Rev. Lett. 56, 659 (1986).
9. Tae Wan Kim and G. Grüner (to be published).
10. See for example, *Charge Density Waves in Solids*, Lecture Notes in Physics, ed. by Gy. Hutiray and J. Solyom, Vol. 217 (Springer-Verlag, 1985).
11. L. Mihaly and G. Grüner: In *Nonlinearity in Condensed Matter*, ed. by A. R. Bishop, D.K. Campbell, P. Kumar, and S.E. Trullinger, Springer Series in Solid State Sciences Vol. 69 (Springer-Verlag, Berlin, 1987), p. 308.

Dynamical Excitations of Site-Diluted Magnets

R. Orbach, A. Aharony, S. Alexander, and O. Entin-Wohlman

Department of Physics, University of California,
Los Angeles, CA 90024, USA

Abstract

Site-diluted magnets are realizations of percolating networks with "scaler" interactions controlling the elementary magnetic excitations at low temperatures. We are able to map the solution to the classical diffusion problem on a fractal onto the dynamical excitations of such systems. Controlling are the fractal dimension, D, which determines the site density on the infinite network (containing the magnon excitations), and the fracton dimension, $\bar{\bar{d}}$, which determines the magnetic excitation density of states and, together with D governs the dispersion law for "fracton" excitations. A cross-over frequency, ω_c, exists such that, for $\omega < \omega_c$, the excitations are magnon-like, while for $\omega > \omega_c$, the excitations are fracton-like. The cross-over frequency ω_c is related to the characteristic length in the problem, the pair-connectedness length ξ_p: $\omega_c \propto \xi_p^{-D/\bar{\bar{d}}}$.

We have derived the structure factor for site-diluted magnets both within the effective-medium-approximation, and analytically for asymptotic behavior . We find a (broadened) asymetric peak at the magnon dispersion law for momentum transfers $q < \xi_p^{-1}$, together with broad structure at ω_c. For large momentum transfers, $q > \xi_p^{-1}$, the structure factor is a broad peak centered at the fracton dispersion frequency, with a width equal to the fracton frequency. The fracton structure factor increases as ω until the peak, then falls off as ω^3.

This behavior has been observed in very recent neutron diffraction experiments by Y. J. Uemura and R. J. Birgeneau on $Mn_{0.5}Zn_{0.5}F_2$. Their latest data finds a value for ξ_p from elastic diffuse scattering measurements in accord with a value for ξ_p generated from inelastic scattering data, interpreted in terms of fracton excitations. Structure at ω_c for $q < \xi_p^{-1}$ is observed through a deconvolution which finds a fracton width $1/\tau \sim \omega_c$ for $q < \xi_p^{-1}$, and $1/\tau \sim \omega_{fr}$ for $q > \xi_p^{-1}$ [Here, ω_{fr} is the fracton frequency proportional to $q^{D/\bar{\bar{d}}}$]. These results can be understood within the terms of a recent scattering calculation where it is argued that the fractons experience scattering at the Ioffe-Regel limit ($\omega_{fr}\tau \sim 1$).

I. Introduction

The approach we take towards the solution of the magnetic excitation problem for site-diluted magnets follows our philosophy for other random materials [1]. One formulates the excitations of such materials in the short length scale regime starting from a presumption that the excitations arise from a network in the material which does not possess translation invariance, but rather dilation invariance [2]. That is, the structure (or network) responsible for the relevant excitation spectrum possesses self- similar geometry [3] for length scales larger than a lower crossover length, a (of the order of the size of the fundamental building block for the structure), but less than a characteristic upper crossover length, ξ. Such invariance requires that the structure "looks the same" for length scales between a and ξ. We refer to this regime as fractal. Conversely, examining the structure in this spatial range does not produce a length scale. For length scales larger than ξ, the structure is presumed to appear uniform. We refer to this regime as Euclidean. Thus, the structure crosses over from fractal to Euclidean as the length scale increases through ξ.

For site-diluted magnets, a will be of the order of an atomic length, while for aggregates (e.g. smoke particle aggregates of small silica particles [4]) a can be of the order of 40 Å, while for sinters (e.g. sintered copper powders [5]) a can vary from 5,000 Å to 100,000 Å. The upper crossover length will depend upon the material parameters. We have estimated [1] ξ to lie between 20 - 40 Å for epoxy-resin and for some glasses, so that ξ/a ranges from approximately 4 to 10. For the site-diluted antiferromagnet $Mn_{0.5}Zn_{0.5}F_2$, UEMURA and BIRGENEAU /6/ find $\xi^{-1} = 0.15 \pm 0.01$ rlu, so that $\xi/a \sim 3$. For sinters and aggregates, ξ/a is roughly an order of magnitude. One would wish the range over which self-similar geometry exhibits itself to be larger. Nevertheless, we feel it sufficient to prove the adequacy of our approach.

It is important to recognize that the average density of the material may be quite substantial, even though the excitations arise from a network exhibiting self-similar geometry. This is clear for the percolating network. For a sample with concentration p, the number of sites on the infinite network scales [7] as $(p - p_c)^\beta$. For a site-diluted antiferromagnet, magnetic order is expected only on those sites which lie on the infinite network. That number can be much less than p for p close to p_c. Likewise, for epoxy-resin and most glasses, the density is order unity. However, the network which supplies the elastic rigidity may be mapped onto the infinite percolation network, and thus involve only a small fraction of the atoms making up the glass [8]. While the latter interpretation is not without controversy, we emphasize that an overall non-critical density (e.g., not a function of $p - p_c$ in a percolating network) does not mean that critical networks do not exist [e.g., the infinite cluster in a percolating network whose density increases as $(p - p_c)^\beta$].

II. Fractal and Fracton Dimensionalities

We shall make use of self-similar geometry for length scales less than ξ to define two important dimensionalities:

$\underline{D}$, fractal dimension [3]; (1a)
$\bar{\bar{d}}$, fracton dimension [9]. (1b)

The former relates the mass (or number) density to the length scale r. In particular, the number of sites within a radius r, n(r), increases as,

$$n(r) \propto r^D \quad , \tag{2}$$

where for the infinite cluster of a percolating network embedded in a three-dimensional Euclidean space (d = 3),

$$D \simeq 2.5 \quad , \quad r < \xi \quad ; \tag{3a}$$
$$D = d = 3 \quad , \quad r > \xi \quad . \tag{3b}$$

The density of sites on the infinite network thus varies as $\rho(r) \simeq r^{D-d}$, $r < \xi$, leading to a very sparse structure for large ξ.

III. Dynamics on a Fractal Network

The fracton dimensionality is important for the dynamics of a fractal network. It was developed from the GEFEN, AHARONY, and ALEXANDER /10/ solution of the DE GENNES /11/ "ant problem" which can be formulated as follows. Suppose an ant parachutes down onto a site contained on the infinite cluser of a percolating network. The ant executes random walk between connected sites on the infinite network at a constant step rate. What is the mean square distance the ant travels in a time t after landing on the network?

In any actual calculation, the random character of the network means that we must perform an ensemble average (i.e. the ant must parachute down over and over again, so that we can take the average of all the mean square distances the ant travels over after landing on all available sites). GEFEN et al /10/ write the mean-square distance travelled in time t in the form of the usual diffusion equation:

$$\langle r^2(t)\rangle = D(r)t \quad , \tag{4}$$

but with a length-dependent diffusion "constant,"

$$D(r) \propto r^{-\theta} \quad , \quad r < \xi \quad ; \tag{5a}$$
$$= \text{constant} \quad , \quad r > \xi \quad . \tag{5b}$$

For percolating networks in d = 2, $\theta \simeq 0.8$; while for d = 3, $\theta \simeq 1.5$. The finiteness of θ leads to a slowing-down of diffusion. For example, the ratio of mean square distances the ant travels in d = 2 after times t_1 and t_2 [assuming that $\langle r^2(t)\rangle < \xi^2$] is,

$$\left[\langle r^2(t_2)\rangle/\langle r^2(t_1)\rangle\right] = \left(t_2/t_1\right)^{1/[1+(\theta/2)]} \quad ,$$
$$\simeq (t_2/t_1)^{5/7} \quad ; \tag{6}$$

whereas the exponent for a Euclidean walk would of course be unity. ALEXANDER and ORBACH /9/, for reasons we shall see below, introduced the fracton dimensionality,

$$\bar{\bar{d}} = 2D/(2+\theta) \tag{7}$$

At the time of their work, they noted that "experimental" values for D and θ (i.e. machine simulations) gave values for $\bar{\bar{d}}$ roughly equal to 4/3 for percolating networks in $d \geq 2$. They suggested that $\bar{\bar{d}}$ might equal 4/3 exactly (as it does in $d \geq 6$, the mean field for percolation networks) for all $d \geq 2$. Although this "Alexander-Orbach conjecture" appears to break down [10] very weakly in $d = 6-\varepsilon$ $\{\bar{\bar{d}} = (4/3)[1-(\varepsilon/126)+O(\varepsilon^2)]\}$, it remains controversial at d = 2. Numerical simulations suggest $\bar{\bar{d}} \simeq 1.32$ [12], but a more recent series expansion of ESSAM and BHATTI /13/ finds $\bar{\bar{d}} = 1.334 \pm 0.007$. The precise value of $\bar{\bar{d}}$ is important because $\bar{\bar{d}}$ will play a crucial role for the scattering linewidth of excitations on a fractal. Departures from $\bar{\bar{d}} = 4/3$ will therefore have experimental consequences.

Given Eq. (7), we can rewrite Eq. (6) as,

$$\langle r^2(t)\rangle \simeq t^{\bar{\bar{d}}/D} \tag{6'}$$

[as we shall see, this is equivalent to a dispersion law]. Equation (6') enables us to obtain the diffusion equation spectral density of states (and thereby, via an appropriate mapping, the vibrational and magnetic excitation energy density of states).

We have already seen that the number of sites inside of sphere of radius r scales as r^D (2). Hence, in time t, the ant has visited $S(t) \propto \langle r^2(t)\rangle^{D/2}$ distinct sites. Using (6'),

$$S(t) \propto t^{\bar{\bar{d}}/2} \quad . \tag{8}$$

For compact diffusion ($\bar{\bar{d}} < 2$ satisfies this condition [14]), the autocorrelation function, $P_0(t)$, (the probability after starting from the origin of remaining at the origin at time t) is proportional to $S^{-1}(t)$, or,

$$P_0(t) \propto t^{-\bar{\bar{d}}/2} \quad . \tag{9}$$

The spectral density of states, $N(\varepsilon)$, is just,

$$N(\varepsilon) = -(1/\pi)\ \mathrm{Im}\ \langle\tilde{P}_0(-\varepsilon+i0^+)\rangle \quad , \tag{10}$$

where $\tilde{P}_0$ is the Laplace transform of $P_0(t)$. Inserting (9),

$$N(\varepsilon) \propto \varepsilon^{(\bar{\bar{d}}/2)-1} \quad . \tag{11}$$

The importance of this result is immediately evident if we substitute a mass M for every occupied site on the infinite cluster, and where bonds exist, replace them by springs of force constant K. Then, the equation of motion for the mass M at the origin is,

$$d^2\vec{u}_0/dt^2 = \Sigma_n(K_{0\leftrightarrow n}/M)(\vec{u}_n - \vec{u}_0) \quad . \tag{12a}$$

Comparing with the diffusion equation,

$$dP_0(t)/dt = \Sigma_n W_{0\leftrightarrow n}(P_n - P_0) \quad , \tag{12b}$$

we see that the right-hand sides are identical (component by component) if we set $W_{0\leftrightarrow n} = K_{0\leftrightarrow n}/M$. Taking the Fourier transform of Eq. (12a) (spectral variable ω) and the Laplace transform of Eq. (12b) (spectral variable ε), the solutions of the latter map onto the former if we analytically continue from ε to $-\omega^2$. This trick (originally due to MONTROLL /15/) solves the vibrational excitation problem for scalar elasticity. We find the vibrational density of states to equal,

$$N(\omega) \propto \omega^{\bar{\bar{d}}-1} \quad . \tag{13}$$

This was first derived by ALEXANDER and ORBACH /9/, and is the reason for their definition of $\bar{\bar{d}}$ as a dimension: the Euclidean density of states is just $N(\omega) \propto \omega^{d-1}$.

The vibrational dispersion law in the fractal regime follows immediately from (6'). Letting $t \sim \varepsilon^{-1}$, we have upon analytically continuing,

$$\lambda(\omega) \propto \omega^{-\bar{\bar{d}}/D} \quad , \tag{14}$$

where $\lambda(\omega)$ is the length scale appropriate to the eigenfrequency ω.

We are similarly able to map the (linearized) equation of motion for an antiferromagnet onto the diffusion equation. This was first carried out by YU and ORBACH /16/ in the absence of anisotropy. The antiferromagnetic dispersion law then maps directly onto that for lattice vibrations, leading to an excitation law of the form of (14).

We refer to the excitations described by (13) and (14) as fractons. These are high energy excitations, appropriate to length scales $\lambda(\omega) < \xi$. Length scales larger than ξ refer to the Euclidean regime where the relevant excitations are either phonons (vibrational excitations) or magnons (magnetic excitations). It is clear that the length scale, ξ, separates the two regimes. We define a crossover frequency between the two regimes by setting $\lambda(\omega_c) = \xi$. Then from (14),

$$\omega_c \propto \xi^{-D/\bar{\bar{d}}} \quad . \tag{15}$$

For $\omega < \omega_c$, the relevant excitations are phonons/magnons, while for $\omega > \omega_c$, the relevant excitations are fractons.

There is more than just a name change involved at ω_c. The dispersion law for long length scales $[\lambda(\omega) > \xi]$ is [17],

$$\omega = \xi^{-\theta/2}\ \lambda^{-1} \quad . \tag{16}$$

[It is easy to show that (16) is equivalent to (15) at $\lambda = \xi$.] These long length scale excitations are extended at low frequencies. They may be Anderson localized [18] at a frequency $\omega_\ell < \omega_c$, depending on the strength of the scattering.

IV. Excitation Linewidths

At $\omega = \omega_c$, the crossover to fractons leads to Ioffe-Regel localization [19], considerably stronger than Anderson localization. This can be seen by calculation of, say, the phonon and fracton lifetimes. For the former, one can show [19] that,

$$1/\tau_{ph} \sim \omega^{d+1}\ \omega_c^{-d} \quad . \tag{17}$$

This implies that $1/\tau = \omega$ at $\omega = \omega_c$. But ω_c is just the crossover frequency between phonon and fracton excitations. A calculation of the lifetime of the latter [19] leads to,

$$1/\tau_{fr} \sim \omega^{5-3\bar{\bar{d}}} \quad . \tag{18}$$

For $\bar{\bar{d}} = 4/3$, the two results connect smoothly at a single crossover frequency ω_c. There are problems if $\bar{\bar{d}} \neq 4/3$. For example, if $\bar{\bar{d}} > 4/3$, the Ioffe-Regel limit $\omega\tau = 1$ is achieved at ω_c, but then the scattering diminishes from this limit at higher frequencies, presumably an unlikely eventuality. For $\bar{\bar{d}} < 4/3$, a second length scale must be introduced, and the analysis becomes much more complex [19]. A plot of (17) and (18), with $\bar{\bar{d}} = 4/3$, takes the form,

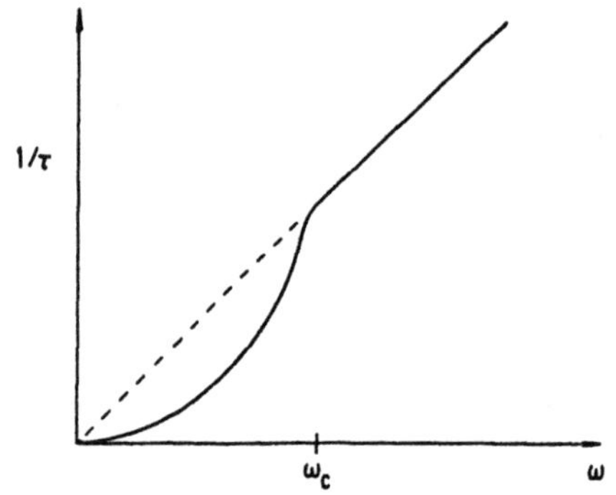

Figure 1. Schematic plot of $1/\tau$, crossing over from (17) for $\omega < \omega_c$ to the Ioffe-Regel limit (18) for $\omega > \omega_c$

V. Application to Site-Diluted Antiferromagnets: The Scattering Structure Factor

Neutron diffraction experiments of UEMURA and BIRGENEAU /6/ on the site-diluted antiferromagnet $(Mn_{0.5}Zn_{0.5})F_2$ appear to be consistent with these ideas. Here, the site dilution implies that the ordered magnetic structure occupies the infinite cluster of a percolation network. The concentration, $p = 0.5$, is rather far from $p_c \simeq 0.25$, so that ξ_p will not be large (see below). Both static (diffuse) and dynamic (inelastic) scattering measurements were performed near the Bragg peak at [0,0,1]. The former yielded the value for the percolation correlation length, $\xi_p = 0.15 \pm 0.01$ rlu. The Brillouin zone boundary occurs at 0.5 rlu, so that the ratio ξ_p/a is only ~ 3. Nevertheless, there turns out to be sufficient room in the Brillouin zone for the experiments to determine that the excitations can be usefully expressed in terms of fractons. Dynamically, an analysis of the data yielded a crossover frequency ω_c which corresponded to the percolation correlation length ξ_p. Thus, both static and dynamic determinations of the crossover length coincide, giving confidence in the consistency of the picture.

The experiments measure the scattering structure factor $I(\vec{q},\omega)$. Such a quantity has only been computed fully for antiferromagnets within the effective medium approximation [20]. Nevertheless, one can make some qualitative arguments which can help to understand the data. The magnons in the long length scale limit are sharp in energy (see Fig. 1), and their contribution to $I(\vec{q},\omega)$ will be a sharp structure at the magnon dispersion law. The contribution of the fractons to $I(\vec{q},\omega)$ is more complex. First, the fractons are localized, resulting in a large spread in $\vec{q}$ space, so much so that they will contribute to $I(\vec{q},\omega)$ even for $\xi q < 1$.

A straight forward calculation [21] yields the following scaling form for the fracton contribution to $I(\vec{q},\omega)$,

$$I_{fr}(\vec{q},\omega) \sim \omega^{-2} f(\omega/\omega_x) \quad . \tag{19}$$

The characteristic frequency ω_x depends upon the neutron momentum transfer length scale, $1/q$. Thus,

$$\omega_x = \omega_c \quad , \quad \xi q < 1 \quad ; \tag{20a}$$
$$\omega_x = \omega(q) = q^{D/\bar{\bar{d}}} \quad , \quad \xi q > 1 \quad ; \tag{20b}$$

while the scaling function,

$$f(z) \sim z^3 \quad , \quad z \ll 1 \quad ; \tag{21a}$$
$$\sim 1/z \quad , \quad z \gg 1 \quad . \tag{21b}$$

Hence, in the magnon regime, $q\xi < 1$, where the magnon contribution to $I(\vec{q},\omega)$ is a sharp structure at the magnon dispersion law $\omega = \omega_{mag}(q)$, the fracton contribution to $I(\vec{q},\omega)$ is,

$$I_{fr}(\vec{q},\omega) \sim \omega/\omega_c^3 \quad , \quad \omega \ll \omega_c \quad ; \tag{22a}$$
$$\sim \omega_c/\omega^3 \quad , \quad \omega \gg \omega_c \quad . \tag{22b}$$

Thus, one would expect the following structure in an ω scan at fixed q for $\xi q < 1$: a sharp peak at the magnon frequency $\omega_{mag}(q)$, and a second (but broad) peak centered at ω_c, rising as ω for $\omega \ll \omega_c$, and falling off as ω^{-3} for $\omega \gg \omega_c$. This double peaked structure has been observed by UEMURA and BIRGENEAU /6/ for $Mn_{0.5}Zn_{0.5}F_2$ at $q = 0.125$ rlu, near but below the crossover wavevector $q = 0.15$ rlu.

For q in the fracton regime, the magnons cannot contribute to $I(\vec{q},\omega)$, for otherwise self-similar geometry would be violated (the dispersion law for magnons contains ξ in the stiffness constant, and ξ cannot be present in any property in the fracton regime). Here, $\omega_x = \omega_{fr}(q) = q^{D/\bar{\bar{d}}}$ so that (19) generates,

$$I(q,\omega) \sim \omega/\omega_{fr}^3(q) \quad ; \quad \omega \ll \omega_{fr}(q) \quad ; \tag{23a}$$
$$\sim \omega_{fr}(q)/\omega^3 \quad ; \quad \omega \gg \omega_{fr}(q) \quad . \tag{23b}$$

This can be represented as a very broad line, centered at the fracton dispersion frequency, with a width comparable to the peak frequency (consistent with the Ioffe-Regel condition for the fracton lifetime--see Fig. 1).

The asymptotic limits of the scaling function are the same as the damped harmonic oscillator introduced by UEMURA and BIRGENEAU /6/ in their fit to $I_{fr}(\vec{q},\omega)$:

$$I_{fr}(\vec{q},\omega) \sim \frac{\omega\omega_{fr}(q)\Gamma_{fr}(q)}{[\omega^2 - \omega_{fr}^2(q)]^2 + \omega^2\Gamma_{fr}^2(q)} \quad , \tag{24}$$

where $\omega_{fr}(q)$ is the fracton frequency at momentum transfer q, and $\Gamma_{fr}(q)$ is the fracton linewidth.

VI. Experimental Results for $(Mn_{0.5}Zn_{0.5})F_2$

We reproduce the measured scattering structure factor of UEMURA AND BIRGENEAU /6/ for $(Mn_{0.5}Zn_{0.5})F_2$ at momentum transfer 0.125 rlu as a function of energy transfer ω in Fig. 2. Remarkably, a double peaked structure, as predicted [20], is

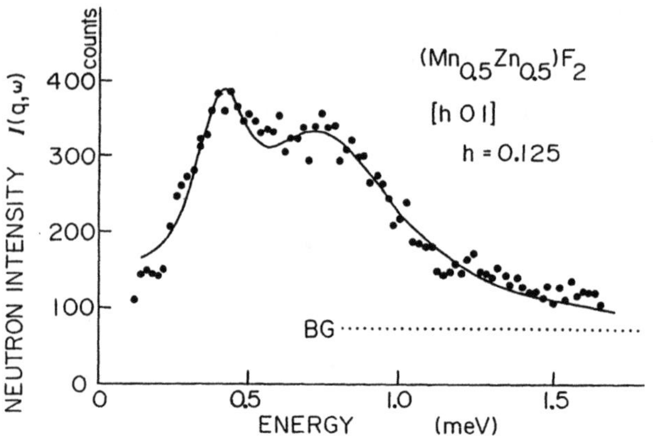

Figure 2. Energy spectra from $(Mn_{0.5}Zn_{0.5})F_2$ observed at wavevector h = 0.125 rlu in the [h 0 1] direction at T = 5 K. The solid line represents the fit to the sum of (24) and (25); the dotted line shows the background level. A double-peak feature characteristic of the wavevector around h_c (the crossover wavevector) ~ 0.15 rlu is demonstrated

observed. The first peak corresponds to the magnon excitation contribution to $I(\vec{q},\omega)$, represented by the solid line fitted to a sharp Gaussian shape:

$$I_{mag}(\vec{q},\omega) \sim \exp\left[-\left\{[\omega-\omega_{mag}(q)]^2/2\Delta_{mag}^2\right\}\right] \tag{25}$$

where $\omega_{mag}(q)$ is the magnon frequency at momentum transfer q, and Δ_{mag} is the magnon linewidth. The solid line through the second peak is a plot of (24).

The two features are clearly separate: a sharp peak at the magnon energy superposed upon a broad feature centered at the fracton energy. As the momentum transfer is increased, the magnitude of the magnon contribution to $I(\vec{q},\omega)$ diminishes, while the magnitude of the fracton contribution increases. At the crossover momentum transfer [corresponding to ω_c of (15)], the magnon contribution to $I(\vec{q},\omega)$ falls to zero, and only the fracton contribution remains. This requirement of self-similarity (one must not be able to sense ξ in the fracton regime, so that no magnon contribution can be present) is clearly evident in the experimental work of UEMURA and BIRGENEAU /6/.

The fitting parameters in (24) and (25); the peak frequencies $\omega_{fr}(q)$ and $\omega_{mag}(q)$, and the fracton linewidth $\Gamma_{fr}(q)$, are plotted in Fig. 3 below. If the Ioffe-Regel limit exhibited in Fig. 1 is to apply, $\Gamma_{fr}(q) \approx \omega_{fr}(q)$. Note that for $\omega < \omega_c$, $I_{fr}(\vec{q},\omega)$ from (19) is centered at ω_c, so that both $\Gamma_{fr}(q)$ and $\omega_{fr}(q)$ asymptotically approach ω_c as $q \to 0$. On Fig. 3, $\omega_{mag}(q)$ remains finite as $q \to 0$ because of the finite anisotropy.

It can be seen that indeed the Ioffe-Regel limit is approximately obeyed for excitations in the fracton regime. The curvature in Fig. 3 for h beyond about 0.3 rlu is probably caused by the proximity of the zone boundary at h = 0.5 rlu.

These experimental results are the clearest examples known for the relevance of fractal concepts to the dynamics of random structures. It is significant that the theory has proceeded to the point that one can now actually fit scattering structure factors using specific predictions derived from fracton excitations.

VII. Conclusions and Acknowledgements

Though this paper has focused on the scattering structure factor for site-diluted antiferromagnets, it is in fact considerably more general in its application. For example, the formalism should apply to structures which exhibit a Euclidean to fractal crossover in their vibrational excitation spectrum. The precise values for D and $\bar{d}$ may be different from those appropriate to percolation networks which apply to site-diluted antiferromagnets, but our overall interpretation will still apply, and the scaling laws should continue to hold.

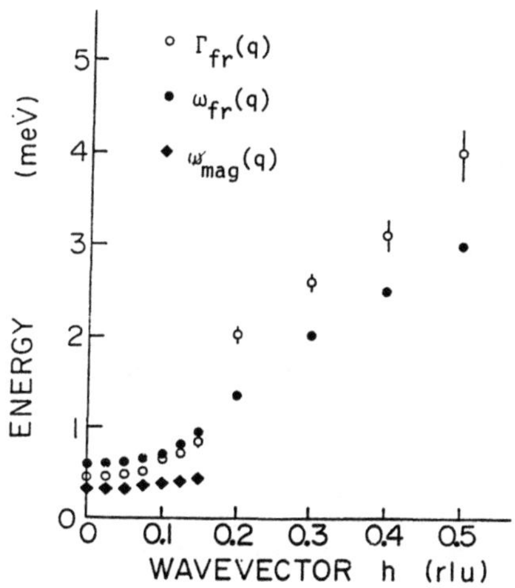

Figure 3. The best fit values of the peak energy $\omega_{fr}(q)$ [$\omega_{mag}(q)$] and the energy width $\Gamma_{fr}(q)$ for the damped-harmonic-oscillator [Gaussian] part of the energy spectra from $(Mn_{0.5}Zn_{0.5})F_2$ at T = 5 K. The data at h $\leqslant$ 0.15 rlu are fitted to the sum of the two parts, while those at h > 0.2 rlu are fitted to the damped-harmonic-oscilator shape alone.

Thus, the long length scale (phonon) excitations should contribute to $I(\vec{q},\omega)$ in the manner of (25), while the short length scale (fracton) excitations should contribute in the manner of (24). Both contributions will appear for small momentum transfers ($q\xi < 1$), while only the latter will be present for larger momentum transfers ($q\xi > 1$). The contribution of the fracton excitations will follow the scaling law (19), with asymptotic forms (22).

Epoxy resin appears to be one of the vibrational systems which follows the predictions of the phonon/fracton picture most closely [1]. We feel that the recent inelastic neutron scattering measurements on epoxy resin of ARAI and JØRGENSEN /22/ are consistent with the form for $I(\vec{q},\omega)$ outlined in this paper. It is now time, in our view, to directly test the ideas contained in our proposed view of the vibrational properties of random structures by ascertaining whether the measured scattering structure factor is or is not consistent with our detailed predictions.

This research was supported by the National Science Foundation through grant DMR 84-12898. The author acknowledges important conversations with Prof. R.J. Birgeneau and Dr. Y.J. Uemura, and thanks them for permission to reproduce their figures prior to publication.

VIII. References

1. S. Alexander, C. Laermans, R. Orbach, and H.M. Rosenberg: Phys. Rev. B28, 4615 (1983)
2. R. Rammal and G. Toulouse: J. Phys. (Paris) Lett. 44, L-13 (1983)
3. B.B. Mandelbrot: in The Fractal Geometry of Nature (Freeman, San Francisco, 1982)
4. T. Freltoft, J. Kjems, and D. Richter: Phys. Rev. Lett., submitted for publication, (1987)
5. J.H. Page and R.D. McCulloch: Phys. Rev. Lett. 57, 1324 (1986)
6. Y.J. Uemura and R.J. Birgeneau: Phys. Rev. Lett. 57, 1947 (1986); and Phys. Rev., submitted for publication, (1987)
7. D. Stauffer: Phys. Rep. 54, 2 (1979)
8. A. Aharony, S. Alexander, O. Entin-Wohlman, and R. Orbach: Phys. Rev. Lett. 58, 132 (1987)
9. S. Alexander and R. Orbach: J. Phys. (Paris) Lett. 43, L-625 (1982)
10. Y. Gefen, A. Aharony, and S. Alexander: Phys. Rev. Lett. 50, 77 (1983)

11. P.G. de Gennes: Recherche 7, 919 (1976)
12. D.C. Hong, S. Havlin, H.J. Herrmann, and H.E. Stanley: Phys. Rev. B30, 4083 (1984); J.G. Zabolitzky: Phys. Rev. B30, 4077 (1984); H.J. Herrmann, B. Derrida, and J. Vannimenus: Phys. Rev. B30, 4080 (1984); and C.J. Lobb and D.J. Frank: Phys. Rev. B30, 4090 (1984)
13. J.W. Essam and F.M. Bhatti: J. Phys. A18, 3577 (1985)
14. P.G. de Gennes: C. R. Seances Acad. Sci. II 296, 881 (1983)
15. E. Montroll: Proceedings of the Third Berkeley Symposium on Mathematical Statistics and Probability (U. Calif. Press, Berkeley, 1955) p.209
16. K.-W. Yu and R. Orbach: Phys. Rev. B30, 2760 (1984)
17. A. Aharony, S. Alexander, O. Entin-Wohlman, and R. Orbach: Phys. Rev. B31, 2565 (1985)
18. E. Abrahams, P.W. Anderson, D.C. Licciardello, and T.V. Ramakrishnan: Phys. Rev. lett. 42, 673 (1979)
19. A. Aharony, S. Alexander, O. Entin-Wohlman, and R. Orbach: Phys. Rev. Lett. 58, 132 (1987)
20. R. Orbach and K.-W. Yu: J. Appl. Phys. 61, 3689 (1987)
21. A. Aharony, O. Entin-Wohlman, and R. Orbach: to be published (1987)
22. M. Arai and J.-E. Jørgensen: Phys. Rev. Lett., submitted for publication, (1987)

Random Field Effects in Dilute Antiferromagnets

D.P. Belanger

Department of Physics, University of California,
Santa Cruz, CA 95064, USA

Experiments in d = 2 and d = 3 random field Ising systems are briefly reviewed. Random fields destroy the phase transition for the d = 2 system $Rb_2Co_xMg_{1-x}F_4$. The d = 3 $Fe_xZn_{1-x}F_4$ system shows a new phase transition with unusual dynamics. Extreme critical slowing down plays an essential role in the d = 3 measurements.

1. Introduction

IMRY and MA [1] introduced a simple but elegant statistical argument that random fields would destroy the phase transition for dimensions $d \leq 2$. Much of the subsequent theory and experiments focused on whether the lower critical dimension d_l , below which no phase transition occurs, is equal to 2, the lower bound suggested by Imry and Ma, or larger, perhaps 3. It is now widely accepted from experimental and theoretical studies that $d_l = 2$. Present efforts are largely concerned with the characterization of the static critical behavior and the unusual critical dynamics associated with the d = 3 phase transition.

The random field Ising model remained outside the realm of experimental study for several years, simply because no mechanism was known for easily generating ordering fields with uncorrelated randomness on the scale of a crystal lattice spacing. The situation changed when FISHMAN and AHARONY [2] proposed that an external uniform field applied to an anisotropic antiferromagnet with random exchange (RE) couplings will effectively generate random fields (RF). Among their predictions are the following: the random field strength h is proportional to the applied field H; the crossover to this behavior is governed by a function of the variable $|t|^{\phi} h^2$; and, consequently, if a new transition occurs it will be at a temperature

$$T_c(h) = T_N - bh^2 - ch^{2/\phi}, \tag{1}$$

where bh^2 is a small mean field temperature shift with $b \sim 1/x$. These predictions, along with the seemingly simple problem of determining d_l, were the motivation of the early experimental work. Cardy [3] generalized the theory by demonstrating the equivalence of the site dilute anisotropic antiferromagnet and the pure Ising RF model in the sense that both belong to the same static critical behavior universality class.

$Rb_2Co_xMg_{1-x}F_4$ and $K_2Co_xMg_{1-x}F_4$ are ideal examples of d = 2 Ising antiferromagnets and $Fe_xZn_{1-x}F_2$ is an ideal d = 3 Ising antiferromagnet. RE behavior is observed in these systems for H = 0 and RF behavior occurs for $H \neq 0$. We will concentrate on the above antiferromagnetic crystals in our dicussions of the RE and RF model since they are the best candidates for studies of dilute Ising spin systems at the present time. As we shall see, in almost every respect the d = 2 RF Ising system behaves differently from the d = 3 one. This is a direct consequence of the destruction of the phase transition by the random field for d = 2, but not for d = 3.

2. Concentration Gradients

The physical realizations of the RF and RE models must have quenched randomness which is uncorrelated on length scales of no more than a few lattice spacings. The quenched randomness, e.g. magnetic dilution in the case of $Fe_xZn_{1-x}F_2$, is introduced during the crystal growth process. Some degree of macroscopic inhomogeneity is inevitable. However, if the concentration gradients can be well characterized, the best part of a high quality crystal can be selected for use in experiments. The effects of a concentration gradient on specific heat and scattering experiments have been studied in some detail [4]. An appreciable linear gradient may cause the specific heat peak not to appear at the average T_c and, in scattering experiments, κ may appear not to reduce to zero at T_c.

The scattering and pulsed specific heat experiments require large crystals, typically on the order of 50 mm^3 in volume. In such cases it is of utmost importance to use crystals of extremely high homogeneity. It is precisely in the experiments most sensitive to concentration gradients, particularly the neutron scattering studies, that very discrepant interpretations have been proposed for data obtained in similar systems.

An excellent single crystal of $Fe_{0.46}Zn_{0.54}F_2$ has been produced in the Materials Preparation Laboratory at the University of California Santa Barbara. It is a disk of 4 mm radius and 2 mm thickness. With a room temperature birefringence technique, the variation of x has been measured to be about 2×10^{-4} [5] and no effects of the gradient on critical behavior measurements will appear for $|t| > 4 \times 10^{-4}$. Such high quality samples are essential to further experimental progress.

3. Specific heat results

If a phase transition occurs, the asymptotic specific heat may be described as $C_m = A^{\pm}|t|^{-\alpha} + B$ where A^+ (A^-) is the amplitude for $T > T_c$ ($T < T_c$), α is the specific heat critical exponent and B includes the noncritical contributions to the specific heat. For the case $\alpha = 0$, the form $C_m = A\log|t| + B$ applies.

The optical linear birefringence technique for determining the specific heat critical parameters has been established theoretically [6,7] and experimentally [8] to give the specific heat behavior in the Ising antiferromagnet FeF_2. Analysis of the data gives the critical parameters $\alpha = 0.11 + 0.005$ and $A^+/A^- = 0.53 \pm 0.01$. These values are in superb agreement with the theoretical values $\alpha = 0.110 \pm 0.005$ [9] and $A^+/A^- = 0.5$ [10].

The birefringence technique was successfully employed to determine the specific heat of the dilute system $Fe_{0.6}Zn_{0.4}F_2$ with $H = 0$ [11]. The quality of available samples prevented at the time a comparable study using conventional specific heat techniques which require large samples. Fig. 1 shows $d(\Delta n)/dT$ vs. $\log|t|$. It is apparent that for small $|t|$ the curvature is negative, a behavior characteristic of $\alpha < 0$ (in contrast with

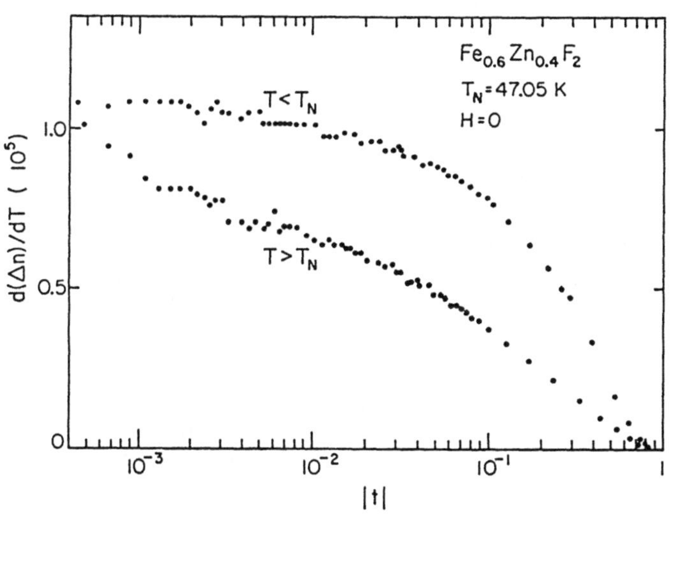

Figure 1. $d(\Delta n)/dT$ vs. $\log|t|$ for $Fe_{0.6}Zn_{0.4}F_2$ showing the d = 3 RE critical behavior of the magnetic specific heat near T_N. The data show the behavior of an asymmetric cusp ($\alpha < 0$).

pure FeF_2 which shows positive curvature with $\alpha > 0$ throughout the critical region). Analysis of the data gives $\alpha = -0.09 \pm 0.03$ and $A^+/A^- = 1.6 \pm 0.3$. This behavior is in agreement with the theoretical values $\alpha = -0.09$ [12] and $\alpha = -0.04$ [13] for the RE Ising model.

Using the identical sample $Fe_{0.6}Zn_{0.4}F_2$, the d = 3 RF Ising transition was observed to be sharp for small h ($H \leq 2.0T$) on the scale of experimental resolution, i.e. $|t| \approx 10^{-3}$. The sharpness of the transition, apparent in the $d(\Delta n)/dT$ vs. $\log|t|$ plot of Fig. 2, is strong evidence that the transition is not destroyed by the random fields, contrary to some earlier suggestions based upon theory and experiment.

The correspondence between the magnetic specific heat and $d(\Delta n)/dT$ in an applied field has been established experimentally [14,15,16] when the samples are first cooled in zero field, the field is applied and the sample is heated through the transition (ZFC). When the uniform field H is applied, the shape of the peak is changed markedly [17]. For small $|t|$ the behavior is nearly a symmetric logarithmic divergence for $|t| < 10^{-2}$. Analysis of the data gives $\alpha = 0.00 \pm 0.03$ and $A^+/A^- = 1.1 \pm 0.1$. Hence, the new behavior, which approximates a logarithmic divergence, is in this sense *sharper* than the H = 0 peak over the reduced temperature range accessible in the experiment.

In the $Fe_{0.6}Zn_{0.4}F_2$ specific heat behavior, no hysteretic behavior was observed for $H \leq 2.0T$. Hysteresis is observed in the $Fe_xZn_{1-x}F_2$ system using other techniques such as neutron scattering and capacitance. More recent experiments on the $Fe_{0.6}Zn_{0.4}F_2$ crystal at high field and on the

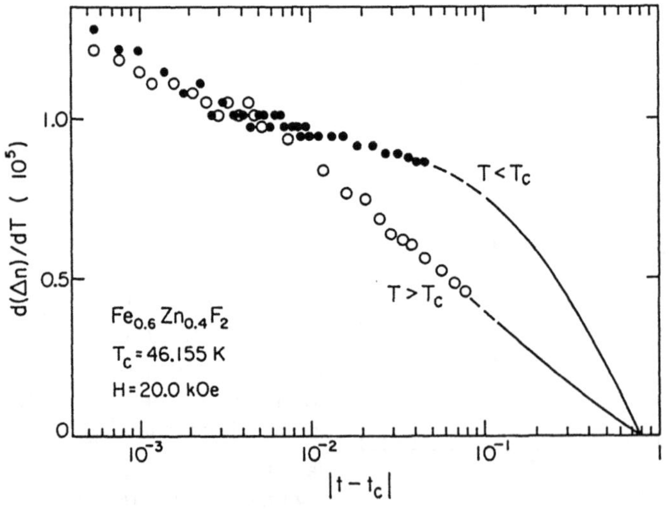

Figure 2. $d(\Delta n)/dT$ vs. $\log|t|$ for $Fe_{0.6}Zn_{0.4}F_2$ showing the d = 3 RF critical behavior of the magnetic specific heat near $T_c(H)$. The data show the approximate behavior of a symmetric logarithmic divergence ($\alpha \approx 0$).

more dilute $Fe_{0.46}Zn_{0.54}F_2$ sample at low field do reveal hysteresis. This is evident in that when the sample is FC a very rounded peak is produced, whereas the ZFC peak is very sharp. This is observed for the latter system for both the d(Δn)/dT measurements and the pulsed specific heat measurements [15,16].

The specific heat behavior of the d = 2 RF Ising model has also been obtained using the optical linear birefringence technique in $Rb_2Co_xMg_{1-x}F_4$. The pure system Rb_2CoF_4 is a well characterized one and the behavior of the magnetic specific heat has been determined optically [18]. Close to T_c the observed symmetric logarithmic peak is in agreement with the exact theory [19]. The dilute sample $Rb_2Co_{0.85}Mg_{.15}F_4$ sample has been studied in a similar manner by FERREIRA, et al. [20]. A symmetric logarithmic divergence is still a good description of the critical behavior. The theoretical prediction [21] is that the behavior should crossover to log(log|t|) as |t| approaches zero, but that this may happen very slowly and would be difficult to distinguish experimentally from log|t| behavior.

Random fields destroy the transition for d = 2. Fig. 3 shows d(Δn)/dT vs. T for H = 0 and for H = 2.0T. The peak is unmistakably rounded for H = 2.0T. The degree of rounding increases with H as described by a general scaling function of the variable $|t|^{\phi} H^2$ with $\phi = 1.7 \pm 0.2$.

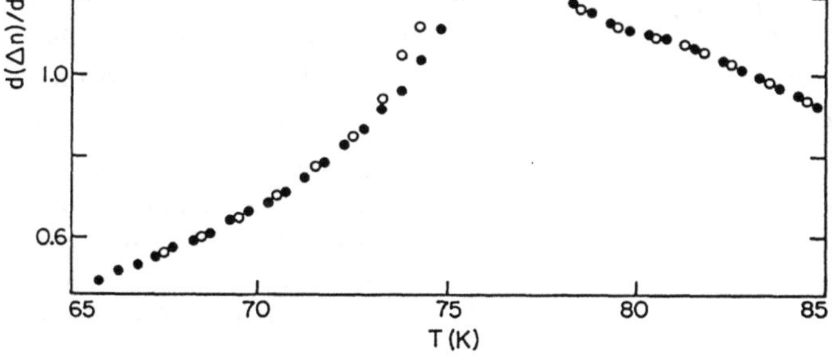

Figure 3. d(Δn)/dT vs. T for the d = 2 crystal $Rb_2Co_{0.85}Mg_{0.15}F_4$ for H = 0 and H = 2.0T. The small random fields generated by the applied field are sufficient to destroy the transition. The rounded peak is in a temperature range for which hysteresis is not observed.

4. AC Susceptibility

Recently, using a high quality $Fe_xZn_{1-x}F_2$ crystal, the critical dynamics of the random field Ising system have been characterized with an ac susceptibility technique (MYDOSH, KING and JACCARINO [22]). Extreme critical slowing down is observed for $H \neq 0$, in keeping with the very long time scales associated with nonequilibrium effects [23].

For the RE case of $H = 0$, the uniform ac susceptibility showed a frequency-independent cusp at $T = T_N$ as predicted [2]. For $H \neq 0$, the RF peak appears symmetric with no hysteresis observed upon ZFC and FC. The peak is rounded, however, and both the amplitude $[\chi_c'(\omega)]_p$ and width $t^*(\omega)$ are frequency dependent. Significant rounding is observed for ω as low as 10Hz (Fig. 4). The peak shape can be fit to a conventional dynamic scaling theory with $t^*(\omega) \sim \omega^{1/z\nu}$ and $[\chi_c'(\omega)]_p \sim |\log(\omega)|$, yielding $z\nu \approx 14$, indicating a very large z value since ν is of order unity. It can equally well be fit to activated dynamic scaling [24] with $t^*(\omega) \sim |\log(\omega)|^{-1}$ and $[\chi_c'(\omega)]_p \sim \log(\log(\omega))$.

Not only is the ac susceptibility characterization of the critical dynamics essential to the understanding of the d = 3 RF Ising phase

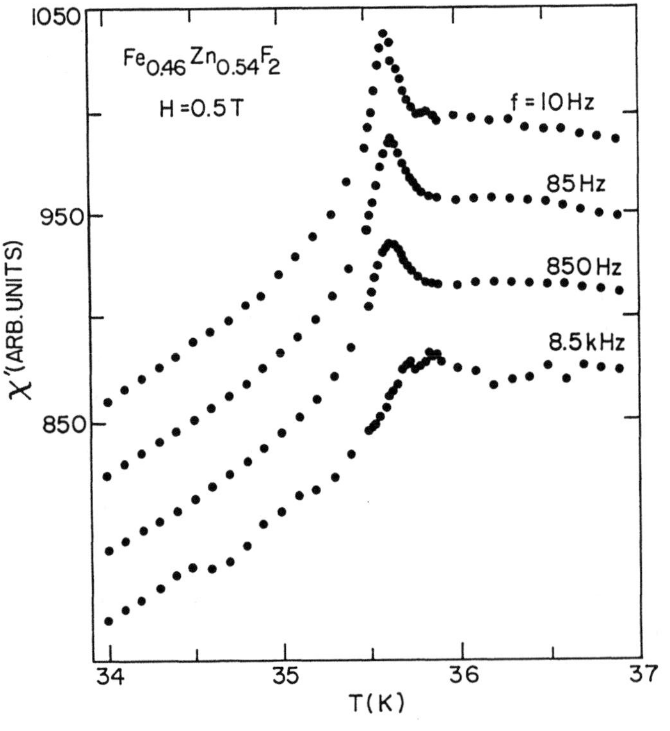

Figure 4. The ac susceptibility vs. T near $T_c(H)$ for $Fe_{0.46}Zn_{0.54}F_2$ with H = 0.5T for frequencies f = 10, 85, 850 and 8500Hz. The peak width increases and the height decreases as f increases, indicating extreme critical slowing down. The scans are displaced vertically for clarity.

transition, it also has important implications regarding other experimental methods which are normally considered to be static in nature. Techniques such as Faraday rotation, birefringence and neutron scattering, where time scales of 10^2-10^3 seconds are typical for each measurement, may be significantly affected by extreme critical slowing of fluctuations close to the transition [14]. Indeed, recent neutron scattering studies using this same crystal have clearly shown dynamical effects for $H \neq 0$, as will be discussed.

5. Neutron Scattering

In the quasielastic approximation [25,26] the scattering intensity $I(\mathbf{q})$ is proportional to $S(\mathbf{q}) = \overline{\langle s_{\mathbf{q}} \cdot s_{-\mathbf{q}} \rangle}$, the Fourier transform of the two spin correlation function $\overline{\langle \mathbf{s}(0) \cdot \mathbf{s}(\mathbf{r}) \rangle}$, provided that the instrumental resolution is properly accounted for. Here < > denotes a thermal average and — represents a configurational average. Using the $\mathbf{q}$ dependent susceptibility $\chi(\mathbf{q}) = \overline{\langle (s_{\mathbf{q}} - \langle s_{\mathbf{q}} \rangle) \cdot (s_{-\mathbf{q}} - \langle s_{-\mathbf{q}} \rangle) \rangle}$, the correlation function can be written as $S(\mathbf{q}) = \chi(\mathbf{q}) + \overline{\langle s_{\mathbf{q}} \rangle} \cdot \overline{\langle s_{-\mathbf{q}} \rangle}$. To separate out the $\mathbf{q}$ dependent fluctuations, a new function is introduced as

$$C(\mathbf{q}) = \overline{\langle s_{\mathbf{q}} \rangle \cdot \langle s_{-\mathbf{q}} \rangle} - m_s^2 \delta(\mathbf{q}) \tag{2}$$

giving us the convenient form for the correlation function

$$S(\mathbf{q}) = \chi(\mathbf{q}) + C(\mathbf{q}) + m_s^2 \delta(\mathbf{q}) , \tag{3}$$

where m_s is the staggered magnetization for an antiferromagnet and obeys the power law $m_s = m_0 |t|^\beta$ for $t < 0$ and is zero otherwise. The last term gives rise to magnetic Bragg scattering which is a measure of long range order. $C(q) = 0$ for all $\mathbf{q}$ and temperatures for a pure, translationally invariant sample.

$\chi(\mathbf{q})$ is known to varying degrees for different models. The most simple form is that obtained from the mean field approximation

$$\chi(\mathbf{q}) = A_0^{\pm} \kappa^{\eta} / (q^2 + \kappa^2) \tag{4}$$

where the scaling requirement $\chi(0) = A_0^{\pm} \kappa^{-2+\eta} = \chi_0^{\pm} |t|^{-\gamma}$ is satisfied with $\gamma = \nu(2-\eta)$. At $d = 2$, $\eta = 1/4$ and for $d = 3$, $\eta \approx 0.03$. This Lorentzian expression is widely used to analyze neutron scattering profiles because it is simple and is often the only theoretical lineshape available.

For the pure d = 2 and d = 3 Ising models the lineshapes are well understood and one need not resort to the mean field form (4). For both d = 2 and 3, expressions which are believed to be accurate for experiments have been developed [27,28,29]. Details of the expressions need not be given here, but two important general aspects are noted. First, the deviation from the Lorentzian behavior is greater for d = 2. Second, the deviation is much greater for both d = 2 and 3 for $T < T_c$.

The theory is much less developed for the situation in which translational invariance does not apply. $S(\mathbf{q})$ is largely unknown for either the RE or RF Ising models. In addition, $C(\mathbf{q})$ is expected to be nonzero for both cases, qualitatively changing the lineshapes observed in experiments. PELCOVITS and AHARONY [30] predict $C(\mathbf{q})$ to contribute considerable scattering intensity at small $\mathbf{q}$ for $T < T_c$ in the RE Ising model. In particular, they predict that when the Lorentzian form alone is used, the ratio χ_0^+/χ_0^- will be overestimated.

It can be argued [31,32] that the random field induces a $\mathbf{q}$ dependent magnetization approximately proportional to the susceptibility. Hence,

$$C(\mathbf{q}) = B_0^{\pm}\kappa^{-\tilde{\eta}}/(q^2+\kappa^2)^2 \tag{5}$$

where B_0 is proportional to h^2. The interpretation of neutron scattering data is seriously hindered by the lack of a well developed theory for $S(\mathbf{q})$. It may also be that the observed lineshapes are influenced by dynamics near the RF phase transition. Although necessary, resorting to mean field approximations could introduce serious systematic errors. Much can be learned by first examining the scattering behavior in the well understood d = 2 and d = 3 pure Ising systems.

The scattering lineshapes for the d = 2 Ising model are well understood. In addition, the Lorentzian expression (4) can be used to analyze the data to test whether the mean field approximation introduces significant systematic errors into the determination of critical scattering behavior. Such systematic errors should be more significant for d = 2 than for the d = 3 case. Nevertheless, in the experiments [33] it was difficult to observe, at any given temperature, a significant decrease in the quality of the fit to the scattering intensity data when the mean field approximation was employed rather than the more accurate theoretical lineshapes. However, when the temperature dependences of κ and $\chi(0)$, as determined from the such fits, are analyzed using the power law expressions $\kappa = \kappa_0^{\pm}|t|^{\nu}$ and $\chi(0) = \chi_0^{\pm}|t|^{\gamma}$, serious disagreement is found. The experimentally determined amplitude ratios $\kappa^+/\kappa^- = 5.25 \pm 0.50$ and $\chi_0^+/\chi_0^- = 53.4 \pm 5.5$ do not compare well with the theoretical values $\kappa^+/\kappa^- = 2.0$ and $\chi_0^+/\chi_0^- = 37.33$. When the more accurate lineshapes are utilized, the ratios $\kappa^+/\kappa^- = 1.85 \pm 0.2$ and $\chi_0^+/\chi_0^- = 32.6 \pm 3.0$ are obtained, in good agreement with theory. The significant differences are largely attributable to the fits for $T < T_N$, as one would expect. The exponent value obtained from the data, $\nu = 1.1 \pm 0.1$, compares well with the exact value $\nu = 1.0$ from theory. The behavior of $\chi(0)$ yields the value $\gamma = 1.9 \pm .2$ which agrees with the exact theoretical value $\gamma = 1.75$. The critical behavior of the staggered magnetization is determined from the Bragg intensity, yielding the exponent $\beta = 0.15 \pm 0.02$, which is in reasonable agreement with the exact value $\beta = 0.125$.

In the d = 3 Ising system FeF_2 [34], at any given temperature, the lineshape fit is only slightly worse when the mean field Lorentzian form is used rather than the approximants. The most detrimental effect of using the Lorentzian is observed in the values obtained for κ and $\chi(0)$ for $|t| < 10^{-3}$, especially for $T < T_N$. The fitted values for $|t| < 10^{-3}$ show clear rounding and κ does not appear to approach zero. In contrast, when the more accurate forms for $S(\mathbf{q})$ are used in the data analysis, power law behavior is observed for κ at all temperatures in the range $10^{-4} < |t| < 10^{-2}$. From a fit to the data the exponent $\nu = 0.64 \pm 0.01$ and amplitude ratio $\kappa^+/\kappa^- = 0.53 \pm 0.01$ are obtained. This is in superb agreement with the theoretical value $\nu = 0.63$ (see [35] for references) and with the theoretical ratios $\kappa^+/\kappa^- = 0.524$ [10] and 0.510 ± 0.008 [28]. Similarly, from the behavior of $\chi(0)$ over the temperature range $10^{-4} < |t| < 10^{-2}$, the exponent $\gamma = 1.25 \pm 0.02$ and ratio $\chi_0^+/\chi_0^- = 4.6 \pm 0.2$ are found. These are in good agreement with theories predicting $\gamma = 1.24$ [35] and $\chi_0^+/\chi_0^- = 4.8$ [10] and 5.10 ± 0.05 [29].

Unfortunately, the critical behavior of the staggered magnetization could not be determined in this FeF_2 sample. Since the sample is of very high quality and the Bragg scattering is so intense, the beam is essentially depleted in the first part of the sample and the observed Bragg scattering intensity quickly saturates. Extinction has also prevented the determination of β for the d = 3 RF Ising model.

The mean field approximation must be resorted to for the dilute systems until the theory is more adequately developed. A recent study using a very homogeneous sample of $Fe_{0.46}Zn_{0.54}F_2$ [36] covered the range $10^{-3} \leq$

$|t| \leq 10^{-1}$ and was limited experimentally only by spectrometer resolution as the concentration gradient effects should not be observed in the sample for $|t| > 4 \times 10^{-4}$. No evidence from the data fits directly indicates non-Lorentzian scattering contributions. Over the range of $|t|$ investigated, power law behavior for κ and $\chi(0)$ is observed. A fit to this behavior yields the exponents $\nu=0.69 \pm 0.01$ and $\gamma = 1.31 \pm 0.03$ and amplitude ratios $\kappa^+/\kappa^- = 0.69 \pm 0.02$ and $\chi_0{}^+/\chi_0{}^- = 2.8 \pm 0.2$. The exponents are in agreement with RE theory where $\nu= 0.70$ and $\gamma= 1.39$ [12] and $\nu= 0.68$ and $\gamma= 1.34$ [13]. The amplitudes $\kappa^+/\kappa^- = 0.83$ and $\chi_0{}^+/\chi_0{}^- = 1.7$ [20] are not as accurately known.

In the d = 2 $Rb_2Co_xMg_{1-x}F_4$ system long range order is not achieved when the crystal is FC [38] as manifested by the lack of a Bragg scattering peak at low T. Long range order can be frozen in, however, by ZFC [39]. After ZFC, the intensity I(0) remains essentially unchanged as H is increased, as can be seen from Fig. 5. As T is subsequently increased at constant H, a fairly narrow region of temperature is reached in which I(0) rapidly decreases with T, signifying the decay of long range order. No hysteresis is observed at higher temperatures. The temperatures for which the birefringence measurements show the peak of the rounded transition are in the range for which no hysteresis is observed, so the rounding occurs in equilibrium. The inflection points $T_F(H)$ of the ZFC I(0) vs. T curves scale as $T_N(0)-T_F(H) \sim H^{2/\phi}$ after mean-field corrections. At $T_F(H)$, the ZFC I(t) is, to a good approximation, found to relax logarithmically in time.

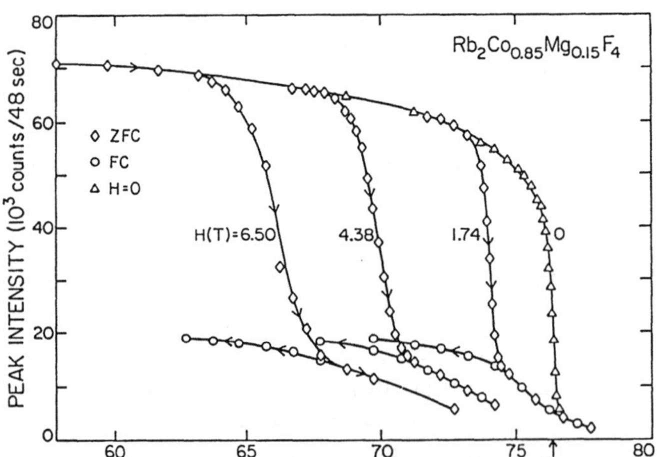

Figure 5. Bragg peak scattering intensity vs. T for $Rb_2Co_{0.85}Mg_{0.15}F_4$ with H = 0, 1.74, 4.38 and 6.50T for the FC and ZFC procedures. The broadened Bragg peak observed in FC samples can be fit rather well by a squared Lorentzian lineshape.

YOSHIZAWA, et al. [38] observed that in the d = 3 $Co_{0.3}Zn_{0.7}F_2$ system which, like FeF_2, is very anisotropic, long range order is not established when the sample is FC, as evident from the lack of a resolution limited Bragg scattering peak. The original interpretation of the experimental result was that the phase transition was destroyed by the random field, just as for d = 2. It is now widely accepted that, for d = 3, the interpretation of the data as indicating a destroyed phase transition is inconsistent with other experiments and theory.

In the similar, very anistropic $Fe_xZn_{1-x}F_2$ system, the FC domain state was shown to be metastable in neutron scattering experiments [40]. This is consistent with the rigorous finding by IMBRIE [41] that the ground state at T = 0 for d = 3 is the antiferromagnetically ordered state. From the results of the capacitance measurements (KING, et al. [23]) using the $Fe_xZn_{1-x}F_2$ systems, it is clear that the d = 3 system does not develop long range order upon FC because it enters a region between $T_{eq}(H)$ and $T_c(H)$ where complete equilibrium is lost, in a way not yet well understood.

The broadened FC Bragg peak can be fit rather well by a squared Lorentzian of the form

$$I(\mathbf{q}) = C/(q^2+\kappa^{*2})^2 \tag{6}$$

where κ^* is some measure of the typical domain structure length scale [42]. It must be kept clearly in mind that this FC lineshape, observed below T_c, replaces the δ function (3) and can not be identified with $C(\mathbf{q})$, which may also have a squared Lorentzian form (5).

The lineshapes and critical scattering behavior in the sample $Fe_{0.6}Zn_{0.4}F_2$ have been investigated [43] for temperatures $T > T_{eq}(H)$, where equilibrium behavior is observed. In contrast to the H = 0 case, the scattering data are poorly described unless a squared Lorentzian is allowed in the fit. The Lorentzian plus squared Lorentzian line profile is presumably merely an approximate form suggested by the mean-field argument and extremely accurate measurements would be required to detect further details experimentally. All analysis of the critical scattering behavior is subject to the limitations of this approximation until a suitable theory is developed. From fits of the scattering q scans for $T > T_{eq}(H)$ to the sum of the Lorentzian and squared Lorentzian with appropriate resolution corrections, κ vs. T is obtained. The linear dependence of κ on T suggests a correlation exponent near unity. It is estimated to be $\nu = 1.0 \pm 0.15$.

The scattering lineshapes near to and below the transition are not well understood but show strong dynamical effects [44]. Using the ZFC procedure at fields up to 3.0T q scans were obtained for $Fe_{0.46}Zn_{0.54}F_2$ below and near to the phase boundary. Everywhere below $T_c(H)$ a well defined Bragg peak is observed. However, the scattering just beyond the Bragg tails is very narrow, very weak and poorly fit by a Lorentzian plus squared Lorentzian, even with κ set to zero.

In Fig. 6 the scattering intensities at the Bragg point (1 0 0) and at (1 -0.004 0), the latter being outside the q range for which Bragg scattering can be observed, is shown versus T for $H \neq 0$. The asymmetry in the scattering from fluctuations is readily apparent, being much smaller for $T < T_c$ (H) and rising to a much higher value above $T_c(H)$. The rounding from the gradient in concentration is indicated in the figure and is seen to be negligibly small.

The effects of critical slowing down on fluctuation scattering intensities were observed using the following ZFC procedure. The crystal was cooled

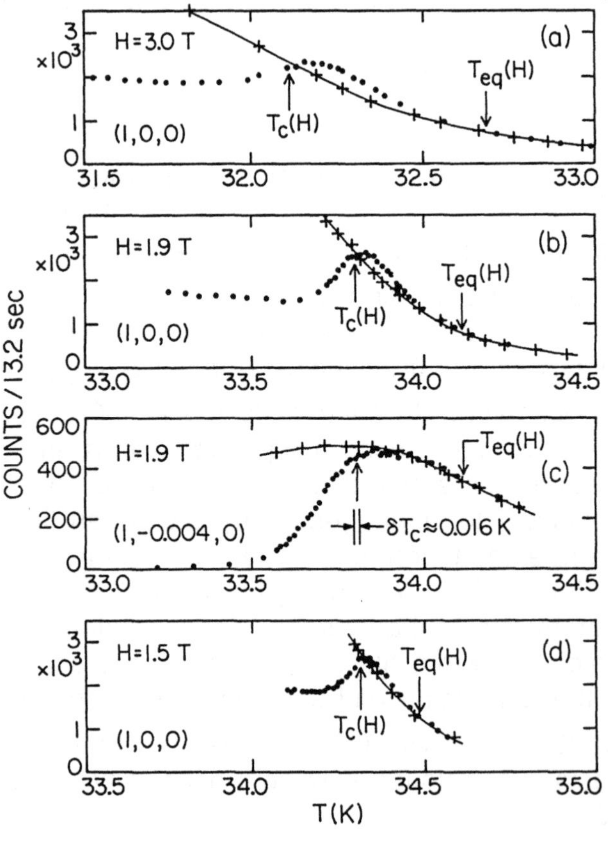

Figure 6. The scattering intensity at (1 0 0) for $Fe_{0.46}Zn_{0.54}F_2$ vs. T for H = 1.5 (a), 1.9 (b) and 3.0T (d) and at (1 -.004 0) for H = 1.9T (c). Each figure shows the ZFC (dots) and FC (crosses) behavior. The smearing of T_c, $\delta T \approx 0.0016$K as determined with a room temperature birefringence technique, is indicated in figure c. T_c and T_{eq} are determined from capacitance data. The scattering data in figure c are from fluctuations alone, whereas the other data include Bragg contributions below T_c.

in zero field to a temperature well below $T_N(0)$. The magnetic field was subsequently increased to a point near the phase boundary $T_c(H)$. The intensity was then measured at $q = 0$ for times up to about 10^4 seconds. Except for initial variations for $\tau < 30$ seconds, possibly attributable to instrumental equilibration, approximate logarithmic dependence on time is observed for $I(\tau)$ over two decades in time τ for points near $T_c(H)$ for H=1.9 and 3.0T. No time dependent effects are observed well below $T_c(H)$ or above $T_{eq}(H)$ or for the FC procedure.

6. The exponent ϕ for crossover from RE to RF critical behavior

In the original treatment of the random field effects in dilute antiferromagnets [2], it was shown that the exponent ϕ for crossover from *pure* to RF critical behavior is equal to the staggered susceptibility exponent, i. e. $\phi = \gamma$. At the time it was believed that RE critical behavior would be difficult to observe at experimentally accessible reduced temperatures. Hence, it was anticipated that the experimental measurements of ϕ would yield the *pure* $d = 3$ Ising value $\phi = \gamma = 1.25$. The RE behavior is observed in $Fe_xZn_{1-x}F_2$, however, and, as discussed earlier, the neutron scattering measurements on the $Fe_{0.46}Zn_{0.54}F_2$ crystal yield $\gamma = 1.31 \pm 0.03$. From various determinations of the shift in T_c as a function of H (1) in the $Fe_xZn_{1-x}F_2$ system, the best value for the crossover exponent is $\phi = 1.42 \pm 0.03$. This gives a ratio $\phi/\gamma = 1.08 \pm 0.05$, which is in good agreement with the prediction of AHARONY [45] that for the case of crossover from RE to RF, the *inequality* $\phi > \gamma$ holds, with an estimated ratio $1.05 < \phi/\gamma < 1.1$.

7. Conclusions and Future Directions

For the $d = 2$ RF Ising model the phase transition is clearly destroyed by a relatively weak random field. Neutron scattering experiments have shown this rounding to occur in a region free of nonequilibrium effects on experimental time scales. Hence the theoretical arguments that $d_l \geq 2$ are well confirmed. At lower temperatures logarithmic time dependence is observed if nonequilibrium long-range order is introduced.

Quite contrasting behavior is observed for $d = 3$. Experiments show static behavior consistent with a new second-order phase transition which is different from the pure and RE Ising behaviors for $d = 3$, although intriguingly similar to that of the d = 2 Ising model. The scattering lineshapes are not yet well understood but are manifestly different from either the pure or RE systems. It is clearly of interest to obtain measurements of the staggered magnetization exponent β and efforts toward this end are being made.

Much of the current experimental effort is directed at characterizing the dynamical behavior of the d = 3 RF Ising model. The dynamical effects are being studied using ac susceptibility, NMR [46], Mossbauer and neutron spin-echo techniques using the $Fe_xZn_{1-x}F_2$ system.

I wish to acknowledge the many people with whom I have had the pleasure of collaborating on random field problems. I particularly thank Allan King and Vince Jaccarino for many collaborations and for their generous support of my research efforts. I also thank Hideki Yoshizawa (now at the ISSP, Tokyo) and Bob Nicklow for fruitful collaborations at the Brookhaven and Oak Ridge National Laboratories, respectively. Support from the UCSC Committee on Research and a UCSC Junior Faculty Fellowship is gratefully acknowledged.

References

1. Y. Imry and S. Ma : Phys. Rev. Lett. **35**, 1399 (1975)
2. S. Fishman and A. Aharony : J. Phys. C **12**, L729 (1979)
3. J. L. Cardy : Phys. Rev. B **29**, 505 (1984)
4. D. P. Belanger, D. P., A. R. King, I. B. Ferreira and V. Jaccarino : Submitted to Phys. Rev. B (1987)
5. A. R. King, I. B. Ferreira, V. Jaccarino and D. P. Belanger : Submitted to Phys. Rev. B (1987)
6. G. A. Gehring : J. Phys. C **10**, 531 (1977)
7. J. Ferre and G. A. Gehring : Rep. Prog. Phys. **47**, 513 (1984)
8. D. P. Belanger, D. P., P. Nordblad, A. R. King, V. Jaccarino, L. Lundgren, and O. Beckman : J. Magn. Magn. Mater. **31-34**, 1095 (1983)
9. J. C. Le Guillou and J. Zinn-Justin : Phys. Rev. B **21**, 3976 (1980)
10. E. Brezin, J. C. LeGuillon, and J. Zinn-Justin: Phys. Lett. **47A**, 285 (1974)
11. R. J. Birgeneau, R. A. Cowley, G. Shirane, H. Yoshizawa, D. P. Belanger, A. R. King and V. Jaccarino : Phys. Rev. B **27**, 6747 (1983)
12. K. E. Newman and E. K. Riedel : Phys. Rev. B **25**, 264 (1982)
13. G. Jug : Phys. Rev. B **27**, 609 (1983)
14. W. Kleemann, A. R. King, and V. Jaccarino : Phys. Rev. B **34**, 479 (1986)
15. K. E. Dow and D. P. Belanger : Unpublished (1987)
16. I. B. Ferreira, A. R. King and V. Jaccarino : Unpublished (1987)
17. D. P. Belanger, A. R. King, V. Jaccarino and J. L. Cardy : Phys. Rev. B **28**, 2522 (1983)

18. P. Nordblad, D. P. Belanger, A. R. King, V. Jaccarino, and H. Ikeda : Phys. Rev. B **28**, 278 (1983)
19. L. Onsager : Phys. Rev. **65**, 117 (1944)
20. I. B. Ferreira, A. R. King, V. Jaccarino, J. L. Cardy, and H. J. Guggenheim : Phys. Rev. B **28**, 5192 (1983)
21. V. S. Dotsenko and V. S. Dotsenko : J. Phys. C **15**, 495 (1982)
22. J. Mydosh, A. R. King and V. Jaccarino : J. Magn. Magn. Mater. **54-57**, 47 (1986) ; A. R. King, J. A. Mydosh, and V. Jaccarino : Phys. Rev. Lett. **56**, 2525 (1986)
23. A. R. King, V. Jaccarino, D. P. Belanger, and S. M. Rezende : Phys. Rev. B **32**, 503 (1985)
24. D. S. Fisher : Phys. Rev. Lett. **56**, 416 (1986)
25. J. Als-Nielsen : In Phase transitions and critical phenomena, Vol. 5a , ed. by C. Domb and M. S. Green (Academic Press, New York 1976)
26. S. W. Lovesey : Theory of neutron scattering from condensed matter, (Clarendon Press, Oxford 1984)
27. C. A. Tracy and B. McCoy : Phys. Rev. B **12**, 368 (1975)
28. M. E. Fisher and R. J. Burford : Phys. Rev. **156**, 583 (1967)
29. H. B. Tarko and M. E. Fisher : Phys. Rev. B **11**, 1217 (1975)
30. R. A. Pelcovits and A. Aharony : Phys. Rev. B **31**, 350 (1985)
31. S. W. Lovesey : J. Phys. C **17**, 6213 (1984)
32. P. M. Richards : *Phys. Rev. B* **30**, 2955 (1984)
33. R. A. Cowley, M. Hagen, and D. P. Belanger : J. Phys. C **17**, 3763 (1984)
34. D. P. Belanger and H. Yoshizawa : Phys. Rev. B **35**, 4823 (1987)
35. K. E. Newman and E. K. Riedel : Phys. Rev. B **30**, 6615 (1984)
36. D. P. Belanger, A. R. King and V. Jaccarino : Phys Rev. B **34**, 452 (1986)
37. S. A. Newlove : J. Phys. C **16**, L423 (1983)
38. H. Yoshizawa, R. A. Cowley, G. Shirane, R. J. Birgeneau, H. J. Guggenheim, and H. Ikeda : Phys. Rev. Lett. **48**, 438 (1982)
39. D. P. Belanger, A. R. King and V. Jaccarino : Phys Rev. Lett. **54**, 577 (1985)
40. D. P. Belanger, A. R. King and V. Jaccarino : Sol. St. Comm. **54**, 79 (1985)
41. J. Z. Imbrie : Phys. Rev. Lett. **53**, 1747 (1984)
42. P. Debye, H. R. Anderson, and H. J. Brumberger : J. Appl. Phys. **28**, 679 (1957)
43. D. P. Belanger, A. R. King and V. Jaccarino : Phys. Rev. **31**, 4538 (1985)
44. D. P. Belanger, A. R. King, V. Jaccarino and R. M. Nicklow : Submitted to Phys. Rev. Lett. (1987)
45. A. Aharony : Europhys. Lett. **1**, 617 (1986)
46. C. Magon, J. Sartorelli, A. R. King, V. Jaccarino, M. Itoh, H. Yasuoka and P. Heller : J. Magn. Magn. Mater. **54-57**, 49 (1986)

Spin Glasses

A.P. Young

Physics Department, University of California,
Santa Cruz, CA 95064, USA

This talk reviews the ingredients necessary for a system to be a spin glass and describes some spin glass materials. Some key experiments on these systems are discussed which lead to the idea of a phase transition. It will be emphasized that there is divergent correlation length at this transition, contrary to what is often supposed. Results of numerical calculations on simplified models will be compared with experiment and the status of our current understanding of the low temperature state will be discussed. Finally, some remarks on "reentrant" spin glass systems will be given.

1. What is a spin glass?

There has been an enormous amount of activity on spin glasses in the last 12 years or so. This talk will only touch on a very few aspects; much more complete discussions can be found in Refs. 1 and 2. One might ask what is the cause of this interest. One reason seems to be that the characteristic phenomena are fairly universal, occurring in metals and insulators, crystalline materials and amorphous structures, and systems with electric dipoles or quadrupoles rather than magnetic moments. Additionally, the theory has proved to be difficult, and hence a challenge, and ultimately rather novel and different in several key aspects from phase transitions in other systems. Also, insight from spin glasses has been very useful in other areas such as optimization [3] and neural networks [4].

The name *spin glass* would suggest that we are dealing with (a) magnetic systems, and (b) a disordered structure like window glass. In fact, although most measurements have been done on magnetic systems we have already discussed that the phenomena are more general than this. Furthermore, it is not clear whether the analogy with real glasses is very deep or just expresses the obvious fact that both have slowly decaying metastable states at low temperatures.

To be a spin glass, it is necessary that the system be disordered and have competition or "frustration" in the interactions, see eg. Ref. 5. It is then clearly nontrivial to find the true ground state, let alone understand properties at finite temperature T. One also finds that there are many local minima which are nearly ground states, but which are widely separated in configuration space. At low temperatures fluctuations between these minima take place by activation over the intervening barriers, and hence are extremely slow. Finding a deep local minimum is therefore a complicated optimization problem. One technique for doing this is to perform a Monte-Carlo simulation starting at high T and to gradually reduce T to zero. This is called "simulated annealing" and has been successfully applied [3] to other optimization problems as well.

2. Spin Glass Systems.

The most completely studied spin glass systems are dilute metallic alloys such as $CuMn$ with concentration of the magnetic species, Mn in this example, varying typically from a fraction of a percent to several percent. The magnetic atoms interact with the Ruderman-Kittel-Kasuya-Yosida (RKKY) interaction, which, at large distances, varies as

$$J_{ij} \sim \frac{\cos(2k_F R_{ij})}{R_{ij}^3} \tag{1}$$

where R_{ij} is the distance between moments i and j, and k_F is the Fermi wavevector. Concentrated insulators can also be spin glasses, for example [6] $Eu_xSr_{1-x}S$, in which the Eu atoms lie on a face centered cubic lattice with a ferromagnetic first neighbor coupling and anti-ferromagnetic second neighbor interaction. The pure material, $x = 1$, is ferromagnetic but it becomes a spin glass for $0.2 < x < 0.5$. The quenched disorder necessary to produce a spin glass may also be due to noncrystallinity, eg. [7] in $CoO \cdot Al_2O_3 \cdot SiO_2$. Analogous phenomena are also found in dielectric relaxation measurements on disordered ferroelectrics, such as [8] $KTaO_3$ diluted with Li, which is interpreted as an electrical dipole glass, or in diluted molecular crystals such as [9] $K(CN)_xBr_{1-x}$ which is a "quadrupolar" or "orientational" glass.

3. Spin Glass Transition.

We shall see in Section 4 below that there is evidence for a sharp second order phase transition. It must be emphasized that, as always at a second order transition, there are correlations extending over large distances and there is a correlation length, ξ_{SG}, which diverges at the transition temperature T_c. To understand this, consider first a ferromagnetic system where typically one has

$$< S_iS_j>_T \sim \exp(-R_{ij}/\xi_F) \tag{2}$$

$$N^{-1}\sum_{i,j}< S_iS_j>_T = T\chi, \tag{3}$$

where $< \cdots >_T$ denotes a thermal average, S_i is the spin on site i, ξ_F is the ferromagnetic correlation length and χ is the uniform susceptibility. In a spin glass, correlations extend over large distances but have a random sign, so if we define ξ_F in (2) by averaging over all pairs of a given relative separation, then it stays small because of cancellations in signs. Suppose instead we square the averages in (2), i.e.

$$[< S_iS_j>_T^2]_{av} \sim \exp(-R_{ij}/\xi_{SG}) \tag{4}$$

$$N^{-1}\sum_{i,j}[< S_iS_j>_T^2]_{av} = \chi_{SG} \tag{5}$$

where $[...]_{av}$ denotes an average over the ensemble of random interactions or equivalently an average over pairs of sites with fixed separation for a given realization of the interactions. The spin glass correlation length ξ_{SG} and spin glass susceptibility χ_{SG} can be much larger than their ferromagnetic counterparts and indeed diverge at the spin glass transition.

Fortunately, a quantity very like χ_{SG} can be measured experimentally. If we expand the magnetization, m, in powers of the field, h, i.e.

$$m = \chi h - \chi_{nl} h^3 + \cdots \tag{6}$$

then the nonlinear susceptibility, χ_{nl}, is very like χ_{SG} and is expected to diverge in the same way.

4. Experiments

Measurements of the a. c. susceptibility $\chi(\omega)$ by Canella and Mydosh [10] sparked the wave of interest in the subject. They, and subsequently many other people, found a sharp cusp at a freezing temperature, T_c, which depends only weakly on frequency. Below T_c one observes hysteresis effects and very slow decay of the remanent magnetization. To a first approximation this varies linearly with the logarithm of the time. Recently, though, it has been argued [11] that a better fit is obtained from a "stretched exponential" decay, $\exp-(t/\tau)^x$ where $0 < x < 1$. There is currently no satisfactory theoretical explanation for the long time dynamics below T_c, though qualitatively it is clearly related to the activation over barriers.

As mentioned above, it is very valuable to look at the nonlinear susceptibility because it gives information on the range of correlations. Furthermore, it can be measured above T_c where relaxation time problems are not severe, so one really obtains equilibrium results. Data on several systems [12] give strong evidence for a power law divergence

$$\chi_{nl} \sim (T - T_c)^{-\gamma} \tag{7}$$

with an exponent γ in range $\gamma \sim 2.0 - 3.2$ depending on which range of reduced temperature was used in the analysis. It is interesting to speculate whether real glasses have a length analogous to ξ_{SG} which grows as the temperature is reduced through the glass forming region and which would diverge if there is ultimately a true glass transition.

5. Models

Theorists use the observed universality of spin glass behavior to study simple models with the necessary ingredients of randomness and frustration. The most popular is due to Edwards and Anderson [13]. The Hamiltonian is given by

$$H = - \sum_{<i,j>} J_{ij} S_i S_j \tag{8}$$

where the spin S_i, chosen to be Ising here, lie on each site of a regular lattice. The interactions J_{ij}, which are nearest neighbor in the simplest versions of the model, are independent random variables. To get a "pure" spin glass, with no bias towards ferromagnetism or antiferromagnetism, one takes a symmetric distribution. A Gaussian distribution and a binary distribution (where $J_{ij} = \pm 1$) are two popular choices. One can easily generalize this model to Heisenberg spins, to longer range interactions, for example with a standard deviation which falls off as R_{ij}^{-3} to imitate the RKKY coupling, and to asymmetric distributions which are needed to study the change from spin glass to ferromagnetism, including reentrant effects discussed in Section 7. Other choices of J_{ij} give the neural network model studied by Hopfield [4].

6. Results

Apart from results for an infinite range model proposed by Sherrington and Kirkpatrick [14], very little in this field is known analytically, so a variety of numerical approaches have been tried. First of all, I will discuss what is known above T_c.

Monte Carlo simulations [15, 16], domain wall renormalization group studies [17] and high temperature series expansions [18] all agree that there is a phase transition in a short range Ising spin glass. In particular, the Monte Carlo results for χ_{SG} look very similar [19] to experimental data on χ_{nl}. Furthermore, the exponent γ seems [15, 18] to be close to 3, in fair agreement with experiment. Unfortunately, this fair agreement may be a coincidence since similar calculations [19, 20] on Heisenberg models which should be more appropriate for most real systems, find $T_c = 0$. This paradox has led to renewed interest in the role of anisotropy and also the range of interactions. Presumably an anisotropic Heisenberg model will behave, close to T_c, like an Ising system, but the expected crossover behavior has not been seen to my knowledge. For models where $T_c = 0$ with short range interactions, it has been argued [21] that RKKY interactions lead to different behavior and are at, rather than below, their lower critical dimension. This would mean, for instance, that χ_{nl} diverges exponentially as $T \to 0$, rather than with a power law which is found for short range Heisenberg models. An exponential divergence as $T \to 0$ may be hard to differentiate from a power law divergence at finite T. Recent simulations [22] indeed find a more rapid increase in χ_{SG} with RKKY interactions than occurred in the short range case [19] but, from the data, one cannot say for sure that one is at the lower critical dimension. By contrast, Chakravarti and Dasgupta [23] find a less rapid increase in ξ_{SG} as $T \to 0$ for an RKKY model than for the short range case [19]. Clearly more work is needed to clarify the role of anisotropy and the range of interaction so one will be able to compare in detail with experiment.

Next we will discuss current theories of the state below T_c. Fisher and Huse [23] and subsequently Bray and Moore [24] have argued that the remanent magnetization should decay with a fractional power of $\ln t$, which seems somewhat different from experiment. However, the theory, as usual, is for an Ising model whereas real systems are much more Heisenberg-like. Since one has very slow relaxation the system is never in true equilibrium below T_c. One might therefore ask what theoretical predictions, which generally assume an initial equilibrium state, can be meaningfully compared with experiment. It has been argued, [23] for the remanent magnetization, m_R, for example, that waiting a time t_w at a temperature below T_c before removing the field and observing m_R at subsequent times t

will give equilibrium and hence reproducible results for $t \lesssim t_w$. Only for $t > t_w$ will one be affected by the initial state at t_w being not in equilibrium.

One curious result is that the surface tension (free energy per unit area of an interface) vanishes. Physically this is because of cancellations of contributions from interactions of different sign. Similarly, the relative orientation of spins a large distance apart is a delicate balance of contributions from many different paths through the lattice. This balance is completely changed if one makes a very small change in temperature so "the relative sign of spins separated by sufficiently long distances changes randomly with any change in temperature" [23]. We see that a spin glass is therefore not just an "anti-ferromagnet in disguise" with an usually complicated staggered ordering. Competing interactions really lead to different physics.

7. Reentrant Spin Glasses

As one changes the relative concentration of atomic species in spin glass materials, one often finds a range of concentration where the successive phases are paramagnet, ferromagnet, and finally spin glass as the temperature is lowered. These are called reentrant spin glasses because the system reenters a type of disordered phase at low T. This is somewhat counterintuitive because it implies that the ferromagnet has more entropy than the spin glass. Although widely observed, this behavior is not understood theoretically. It does not occur in the infinite range model, for example. Recent studies [25] of two-dimensional Ising systems with various types of competing interactions also fail to show the effect. Presumably one needs (a) to look at three dimensions, or (b) have Heisenberg spins, or (c) include dipole-dipole interactions, or a combination of these. Unfortunately, they are rather hard to include in a satisfactory theory. The possible importance of transverse fluctuations, which of course can only occur for vector spin models, has been discussed by a number of authors [26].

8. Conclusions.

We have seen that experiment and theory agree on the existence of a phase transition, though the role of RKKY interactions and anisotropy remains to be clarified. Below T_c the dynamics awaits a detailed theory. Novel aspects of the theory below T_c are hard to test, though it would be very interesting to do so. Finally, the observed reentrant phase diagrams remain a mystery.

Acknowledgements.

My recent work in the field has benefited considerably from a stimulating collaboration with Joseph Reger. Support from the NSF under grant DMR 84-19536 is gratefully acknowledged.

References

1. K. Binder and A. P. Young, Rev. Mod. Phys. **58,** 801 (1986).
2. J. L. van Hemmen and I. Morgenstern, Eds., *Glassy Dynamics and Optimization* (Springer, Berlin, in press), 1987.
3. S. Kirkpatrick, C. D. Gelatt Jr., and M. P. Vecchi, *Science* **220,** 671 (1983).
4. J. J. Hopfield, Proc. Natl.. Acad. Sci., U. S. A., **79.** 2554 (1982).
5. D. Sherrington in "Proceedings of the Heidelberg Colloquium on Spin Glasses", edited by J. L. van Hemmen and I. Morgenstern, *Lecture Notes in Physics,* Vol. 192 (Springer Berlin), p. 348.
6. H. Maletta and W. Felsch, Z. Phys. B. **37,** 55 (1979).
7. L. E. Wenger, C. A. M. Mulder, A. J. van Duyneveldt and M. Hardiman, Phys. Lett. A. *87,* 439.
8. U. T. Hochli, Phys. Rev. Lett. **48,** 1494 (1982).

9. A. Loidl, R. Feile and K. Knorr, Phys. Rev. Lett, **48,** 1263 (1982).
10. V. Cannella and J. A. Mydosh, Phys. Rev. B **6,** 4220 (1972).
11. R. V. Chamberlin, G. Mazurkevich and R. Orbach, Phys. Rev. Lett. **52,** 867 (1984).
12. P. Monod and H. Bouchiat, J.Phys (Paris) Lett. **43,** 145 (1982); B. Barbara, A. P. Malozemoff and Y. Imry, J. Appl. Phys. **53,** 7672 (1982); Y. Yeshurun and H. Sompolinsky, Phys. Rev. B. **26,** 1487 (1982); R. Omari, J. J. Prejean and J. Souletie, J. Phys. (Paris) **45,** 1809 (1983); H. Bouchiat, Phys. Rev. B **23,** 1375 (1986).
13. S. F. Edwards and P. W. Anderson, J. Phys. F **5,** 965 (1975).
14. D. Sherrington and S. Kirkpatrick, Phys. Rev. Lett., **35,** 1972 (1975).
15. R. N. Bhatt and A. P. Young, Phys. Rev. Lett., **54,** 928 (1985).
16. A. T. Ogielski and I. Morgenstern, Phys. Rev. Lett., **54,** 928 (1985).
17. A. J. Bray and M. A. Moore, Phys. Rev. B **31,** 631 (1985); W. L. McMillan, Phys. Rev. B **31,** 340 (1985).
18. R. R. P. Singh and S. Chakravarty, Phys. Rev. Lett., **57,** 245 (1985).
19. J. A. Olive, A. P. Young and D. Sherrington, Phys. Rev. B. **34,** 6341 (1986).
20. B. W. Morris, S. G. Colborne, M. A. Moore, A. J. Bray and J. Canisius, J. Phys. C., **19,** 1157 (1986); W. L. McMillan, Phys. Rev. B **31,** 342 (1985); J. R. Banavar and M. Cieplak, Phys. Rev. B **26,** 2662 (1982).
21. A. J. Bray, M. A. Moore and A. P. Young, Phys. Rev. Lett., **56,** 2641 (1986).
22. J. D. Reger and A. P. Young (unpublished).
23. D. S. Fisher and D. Huse, Phys. Rev. Lett., **56,** 1601 (1986).
24. A. J. Bray and M. A. Moore in "Glassy Dynamics and Optimization", edited by J . L. van Hemmen and I. Morgenstern (Springer, Berlin, in press).
25. J. D. Reger and A. P. Young (unpublished).
26. W. M. Saslow and G. Parker, Phys. Rev. Lett., **56,** 1074 (1986); G. S. Aeppli, S. M. Shapiro, H. Maletta, R. J. Birgeneau and H. S. Chen, J. Appl. Phys. **55,** 1628 (1984) and references therein.

Neural Networks: A Tutorial

E. Domany

Department of Electronics, Weizman Institute of Science,
76100 Rehovot, Israel

The subject of Neural Networks does not fit naturally into the general context of a meeting on "Competing Interactions & Microstructures: Statics and Dynamics". Therefore, I will try to explain what Neural Networks are, why are (some) physicists interested in them, and why is it not completely absurd to have a talk on them in such a meeting. A much extended version of my strongly biased views on the subject is given in a review article [1], which also contains a more detailed list of references. Here I present only an extended abstract that summarizes the main points of my talk.

The ultimate goal of research in the field of Neural Network is to gain understanding of how the brain works. I believe that so far not much progress towards that goal has been achieved by research on Neural Networks. Most of this research is oriented towards constructing model networks, that consist of many intricately coupled elementary units (cells, formal neurons). These networks are expected to perform various tasks that mimic some higher neural functions. In my talk I limit my attention to networks that model a memory. Model memories are set by the external world in some initial state. Subsequently the networks develops in time; its state changes according to some dynamic rule, untill a final state, to serve as output, is obtained. Since the network consists of a large number of strongly coupled elements, the resulting dynamics is usually complex. The main potential contribution of physicists to the field is, in my opinion, to help to understand such complex dynamics.

An important step in this direction was taken by HOPFIELD [2], who introduced a simple, fully connected network with symmetric intercell couplings. The dynamic rule of the network was chosen so that it could be viewed as relaxation to equilibrium of an Ising spin-glass. A memory-state is a (key) pattern, i.e. one of the 2^N states of the network; one wishes to choose the bonds or couplings so that all key patterns are stable states. Each such stable state may have a sizeable domain of attraction, i.e. many nearby initial states will flow to the key pattern under the dynamic rule. This property is viewed as error-correction, content addressability, etc. Since many bonds determine the stable states of the system, destruction of even a sizeable fraction of them may leave the stable states relatively intact; i.e. the memory is robust. These are attractive properties of the Hopfield model, and it is interesting to study the extent to which these properties depend on various assumptions taken, such as symmetric bonds, functional uniformity (i.e. the entire network serves as input, processing and output) and interpretation of recall as a steady activity pattern associated with a stable state. To this end a layered neural network was introduced [3]. All bonds of this network are unidirectional; it has no feed-back, with information flowing only forward, from input units via processing layers to output. Nevertheless, many of the attractive properties of the Hopfield model do survive; in particular, the most important property of analytic solvability. Whereas for the Hopfield model the equilibrium statitical mechanics is exactly soluble [4], for the feed-forward network discussed one can analytically solve the dynamics [5]. That is, if the overlap of the input with a key pattern is known, we can calculate the overlap with the associated key patterns on the processing and output layers.

Feed-forward networks have a long and interesting history [6]. The single layer Perceptron invented by ROSENBLATT [7] was highly acclaimed for its performance, and in particular its learning algorithm, that was proven to converge to a solution of the problem at hand (provided such a solution did exist). Subsequently MINSKY and PAPERT have shown [8], that the set of problems solvable by the single layer perceptron is severely restricted. More recently it was demonstrated [6] that many of the insoluble problems can, in fact, be solved by multi-layer perceptrons, such as the one described above. However, for these networks there is no known learning algorithm that has an associated convergence theorem. I view the goal of finding such a learning rule the central (short range) challenge of the field.

<u>References</u>

1. E. Domany, J. Stat. Phys.,in print.
2. J.J. Hopfield, Proc. Natl. Acad. Sci. USA, <u>79,</u> 2554 (1982).
3. E. Domany, R. Meir and W. Kinzel, Europhys. Lett. <u>2,</u> 175 (1987)
4. D.J. Amit, H. Gutfreund and H. Sompolinsky, Ann. Phys. <u>173,</u> 30 (1987).
5. R. Meir and E. Domany, Phys. Rev. Lett. <u>59,</u> 359 (1987).
6. D. E. Rumelhart and J. L. McClelland: <u>Parallel Distributed Processing:Explorations in the Microstructure of Cognition,</u> 2 vols. The MIT Press Cambridge Mass. 1986.
7. F. Rosenblatt, Principles of <u>Neurodynamics,</u> Spartan, Washington D.C. 1961.
8. M. Minsky and S. Papert, <u>Perceptrons,</u> MIT Press, Cambridge Mass. 1969.

Ordering Kinetics in Quasi-One-Dimensional Systems and Polymer Melts

K. Kawasaki

Department of Physics, Faculty of Science,
Kyushu University 33, Fukuoka 812, Japan

Abstract

Existence of the two universality classes in quasi-one dimensional ordering kinetics of layered Ising-like magnets is demonstrated. In polymer melts, loops of correlated reptating chains are found to be important for morphology dynamics. We also argue for apparent violation of the monomer density conservation law in polymer blends.

1. Introduction

When a system in a stable high temperature disordered phase is quenched into thermodynamically unstable states, initially microscopic thermal fluctuations grow in their amplitudes and at the same time develop (often random) characteristic mesoscale spatial structures. Simple analytic treatments of this problem are possible only for very short times after quench where amplitudes of fluctuations are so small that linearization is legitimate. In reality, however, the growth process is so rapid that this linear regime characterized by the exponential growth law is rarely realized /1/. The important exceptional case is the phase separation kinetics of polymer blends where entanglement effects enormously slow down the process /2,3/. Even here deviations from the linear behavior become appreciable after certain times, say, 30 minuites /3/.

The most striking feature that emerges in the nonlinear regime of ordering kinetics is the fact that the order parameter fluctuation appears to have a single characteristic length which increases in time (scaling behavior) /1/. This is expressed in terms of fluctuation spectrum defined by

$$I_{\underline{k}}(t) = \langle |S_{\underline{k}}|^2 \rangle_t \tag{1.1}$$

where S_k is the Fourier component of the local order parameter fluctuation and the angular bracket with the suffix t is the average at time t after the quench. The scaling behavior for a d-dimensional system is then expressed as

$$I_{\underline{k}}(t) / \int I_{\underline{k}'}(t) d\underline{k}' = \ell(t)^d \, F(k\ell(t)) \tag{1.2}$$

where $\ell(t)$ is the characteristic length and F(x) is the scaling function. Our theoretical understanding of ordering kinetics is still far from complete, despite rapid progress in recent years, owing to many causes such as the problem concerning the choice of a starting model (e.g. continuum versus discrete lattice models), difficulties of analyzing given model systems where nonlinearity is essential, and absence of general theoretical approach which is universally applicable, such as the renormalization group for critical phenomena, etc. It appears that more accumulation of individual case studies is still needed before deeper and thorough understanding of ordering kinetics can be achieved.

A particular approach which holds promise for understanding of scaling is to focus on time evolution of random domain walls, which are formed sometime after quench when local phase separation process is completed /4/. This in a way is an extension of the nucleation picture beyond metastable states. There are several known cases where scaling naturally emerges from analyses of suitable domain wall equations of motion /4/. Indeed it now appears to be widely accepted that validity of the scaling is closely tied to emergence of some kind of domain wall states after local phase separation is completed, although this may not be the whole story.

In the next section we present a particular case study where ordering proceeds through movements and annihilations of almost flat but undulating domain walls. This is followed by discussions on ordering kinetics of polymer blends and block copolymer melts, which are receiving increasing attention recently but still defy theoretists' challenge.

2. Quasi-One-Dimensional System

The system we have in mind is an Ising magnet with strong spatial anisotropy of uniaxial type where domain walls formed in a state quenched below the critical temperature are almost perpendicular to the symmetry axis (x-directions). Domain walls are labelled consecutively by their intersections with the x-axis from negative to positive x-directions. The configuration of the i-th domain wall is then specified by the function $x_i(\underline{r}_\perp)$ giving the x coordinate of the domain wall at $\underline{r}_\perp$, the position vector in the plane perpendicular to the x-axis.

The purely dissipative domain wall equation of motion takes the following general form (the arguments t are often suppressed here and after for simplicity):

$$\kappa \frac{\partial}{\partial t} x_i(\underline{r}_\perp) = - \frac{\delta H\{x\}}{\delta x_i(\underline{r}_\perp)} \tag{2.1}$$

where κ is the friction constant of the wall motion and $H\{x\}$ is the free energy functional of the domain wall system. Derivation of equations like (2.1) from more basic equations like dissipative field equation for the local order parameter was described elsewhere /4/.

For $H\{x\}$ we consider two kinds of contributions; one is the sum of the free energies of individual domains proportional to $[\nabla_\perp x_i(\underline{r}_\perp)]^2$ where $\nabla_\perp$ is the gradient in the plane perpendicular to the x-axis and unimportant constant terms are dropped. Here we have assumed $|\nabla_\perp x_i| \ll 1$ reflecting strong anisotropy. This energy contains increases of both the domain wall area and the anisotropy energy associated with tilt of the wall. Another contribution to $H\{x\}$ contains interaction energies $V(x_{i+1}-x_i)$ of adjacent domain walls. Thus we take

$$H\{x\} = \sum_i \int dr \, \{\frac{\sigma}{2} [\nabla_\perp x_i(\underline{r}_\perp)]^2 + V(x_{i+1}(\underline{r}_\perp)-x_i(\underline{r}_\perp))\} \tag{2.2}$$

Eq. (2.1) then assumes the following form:

$$\kappa \frac{\partial}{\partial t} x_i(\underline{r}_\perp) = \sigma \nabla_\perp^2 x_i(\underline{r}_\perp) + \sum_i \{V'(x_{i+1}(\underline{r}_\perp)-x_i(\underline{r}_\perp)) - V'(x_i(\underline{r}_\perp)-x_{i-1}(\underline{r}_\perp))\} \tag{2.3}$$

where primes denote differentiations. This equation must be supplemented by the condition that whenever portions of adjacent domain walls touch, portions annilate each other.

In the absence of long-range interactions such as those due to elasticity, V(z) is normally very weak and is short-ranged and behaves for the present case as

$$V(z) = -V_0\, e^{-z/\xi}, \qquad (z \geq 0) \tag{2.4}$$

where ξ has the size of the domain wall thickness.

It is interesting to note that there exists in nature a system which can be regarded as an almost ideal realization of the model described here. This is a layered Ising antiferromagnet of composition $Rb_2Co_{1-x}Mg_xF_4$ with x=0.3 extensively studied by H. Ikeda /5/. If the system is quenched from a temperature slightly above the Neel temperature to that slightly below, almost one-dimensional domain growth of the kind described here takes place. Note that this is a rather slow process requiring movements of large sections of almost parallel domain walls and can take many hours. This is in contrast to real one-dimensional magnetic chains with fast single spin flip processes /6/.

When the anisotropy is absent, the weak interactions in (2.3) can be safely ignored, which are masked by the first term of (2.3). In this case the domain wall equation simply reduces to the familiar curvature-driven interface equation discussed, for example, by Allén and Cahn among others /7/.

If the anisotropy were chosen to be infinite from the outset, x_i no longer depends on $\underline{r}\perp$ and we find

$$\kappa \frac{d}{dt} x_i = V'(x_{i+1}-x_i) - V'(x_i-x_{i-1}) \tag{2.5}$$

and the problem becomes one-dimensional.

Let us now assume that V(z) is negative with positive derivative for $x \geq 0$ (attractive force) and tends to zero rapidly in a short distance, say, ξ, like (2.4) and we focus on the late time behavior where the distance between adjacent domain walls (now kinks and antikinks) is much greater than ξ so that the domains are well-defined. Here we notice that our dynamical system (2.5) is very peculiar. In well-defined domain wall states the distance of the size of domain wall thickness (here ξ) should be immaterial. Yet the force V'(z) acting between adjacent domain walls radically changes within the distance of $O(\xi)$. To see a consequence of this circumstance we consider a domain wall state with the domains of sizes $z_1 < z_2 < z_3 < \cdots$ such that $z_{i+1}-z_i \gg \xi$ /8/. (This last condition is not valid in the thermodynamic limit of infinite system size with the continuous domain size distribution. Here in fact we are assuming that the system can be divided into subsystems of finite sizes, in each of which this condition is valid, and yet the different subsystems are statistically independent.) These domains need not be ordered on the x-axis, of course. Since the force acting between the kink-antikink pair bordering the smallest domain overwhelms others, we first need to consider only shrinking and annihilation of the smallest domain of the size z_i which takes place during the time interval $T(z_i)$. Here T(z) is the time required for an isolated domain of size to be annihilated and is given, solving the equation of motion, by

$$T(z) = \frac{\kappa}{2} \int_0^z \frac{dz}{V'(z)} \underset{=}{\sim} \frac{\kappa\xi^2}{2V_0}\, e^{z/\xi} \tag{2.6}$$

where the second member corresponds to the case of (2.4) with $z \gg \xi$. The shrinking distance Δz_i of other domains with $i \geq 2$ can be found not to exceed the value determined by the equation

$$T(z_1) = T(z_i) - T(z_i-\Delta z_i) \tag{2.7}$$

For the special case (2.4) we have

$$\Delta z_i < \xi \exp[(z_1 - z_i)/\xi) \ll \xi \tag{2.8}$$

which indeed is a very small number and can be safely neglected as was anticipated before. After annihilation of the smallest domain, shrinking and annihilation of the smallest of the remaining domains start to take place and the process is repeated until there remains at most a single domain. From this argument we expect that during the interim period the domain size distribution will develop a sharp cut-off structure with no domain with sizes smaller than a time-dependent cut-off $z_c(t)$. Since (2.6) is rapidly decreasing in z, (2.6) itself gives the time dependence of the cut-off:

$$z_c(t) = T^{-1}(t) \cong \xi \ln t \tag{2.9}$$

where $T^{-1}(t)$ is the inverse function of T(z) and the second member is for the exponential force law.

The general feature of the domain size distribution and its temporal growth has indeed been confirmed some time ago in the computer simulation of the exponentially attracting kink-antikink systems with random initial distribution of kinks and antikinks by Nagai and Kawasaki /8/ although the statistics was not good enough to verify scaling. Nevertheless we believe that the scaling holds and the obtained scaled domain size distribution is universal, independent of the specific form of $-V'(z)$, as long as it is attractive and short-ranged although the growth law (2.9) is not universal. This is because the actual procedure of constructing domain wall states as outlined above is independent of the specific force law acting between kinks and antikinks.

Using the domain size distribution function thus obtained, we can also derive $I_k(t)$, (1.1), if we assume statistical independence of different domains. We can now test the prediction made for this one-dimensional model by Ikeda's experimental results for the neutron scattering structure factor $I_k(t)$. Indeed the scaling behavior (1.2) was verified with $\ell(t)$ given by $z_c(t)$, (2.9). However, the properly normalized theoretical scaling function F(x)/F(0) (see (1.2)) was found to fall far below the experimental one for $x > 1$ (see Fig.2 of /9/). The primary cause of the discrepancy is absence of the domains with sizes smaller than the cut-off $z_c(t)$, indicating inadequacy of the one-dimensional model (2.5). Indeed, transition from the original model (2.3) to the one dimensional model (2.5) is delicate. When the anisotropy increases indefinitely, the domain walls tend to be more flat reducing $\nabla_\perp^2 x_i(r_\perp)$ but at the same time the coefficient σ in front increases indefinitely, and it is unclear that the product of the two can be neglected in the limit. In fact, for the attractive interaction between domain walls the two models (2.3) and (2.5) turn out to belong to different universality classes of ordering kinetics. This stems from the instability of flat domain walls against long wave length undulatory perturbations. Intuitively, whenever two portions of adjacent domain walls are curved and come close together, the attractive force between the two portions is enhanced, which results in further curving of the walls. This is in sharp contrast to the cases with repulsive domain wall interactions where no instability against undulation exists and the first term on the r.h.s. of (2.3) should be irrelevant.

One might in general expect that the model (2.3) is more difficult to analyze than (2.5), However, it transpires that as far as the universal aspects such as scaling are concerned, quite the opposite is true. Indeed we can transform the model (2.3) into a much simpler model system which is readily analyzed. (The rest of this section should be read together with /9/.) We focus on events occurring in a fictitious tube lying along the x-axis whose diameter is roughly the threshold wave length of undulatory instability. The portions of domain walls inside the cylinder can then be regarded as almost flat. If this cylinder were isolated from the rest of the system, we would have the one-dimensional situation described by (2.5) with a cut-off domain size distribution. However, if many such identical cylinders (with hexagonal cross sections in this case) were put together to form the entire system with connected domain walls, we encounter a quite different situation. Whenever domain wall annihilation takes place in any of the cylinders, the hole generated in the pair of domain walls spreads out in all the directions in the $\underline{r}_\perp$ plane very rapidly to reduce the domain wall energy. Such an event can occur in any of the cylinders and is very frequent. If we are focussing on any one of these cylinders whose domain size distributions are assumed to be uncorrelated, we find that adjacent domain walls are annihilated at the rate independent of the wall distance. Then it is a simple matter to write down the kinetic equation for the domain size distribution function. This equation has the following exact scaling solution for the normalized domain size distribution function g(z,t);

$$g(z,t) = [2\pi\, z\bar{z}(t)]^{-1/2} \exp[-z/2\bar{z}(t)] \tag{2.10}$$

Here $\bar{z}(t)$ is the average domain size whose asymptotic long time behavior is given by

$$\bar{z}(t) = \nu\,\xi \ln t \tag{2.11}$$

ν being some numerical coefficient. The scattering structure function is computed from (2.10) neglecting again the statistical correlations of different domains, and is compared with Ikeda's experimental result. This time the agreement turned out to be excellent, as shown in Fig.2 of /9/.

Again we believe that the scaled domain size distribution function obtained from (2.10) as well as the scaled scattering structure function are universal although the growth law (2.11) is not. However the universality class of this case is different from the one for the one-dimensional case that was discussed earlier. The cross-over between the two universality classes takes place when the diameter of the cylinder, that is, the wave length of instability threshold against undulation, becomes comparable to the linear dimension of the system perpendicular to the x-axis. The critical wavelength grows in time fast by some power law in contrast to (2.11). The situation here still remains to be elucidated both theoretically and experimentally. Possible importance of domain wall pinning due to impurities etc. in the one-dimensional situation is an additional complicating factor in this problem.

3. Ordering Kinetics in Polymer Melts

Study of ordering kinetics in polymer melts has some distinct aspects:

(i) Highly entangled mess of polymer chains varies very slowly in time permitting study of various stages of kinetics ranging from the early linear regime to the late stage with well-defined domain structures /10/.

(ii) Variety of polymer systems (polymer blends, block copolymers, star polymers, etc) show many different kinds of ordered structures (mesophases) /11/.

(iii) Complexity of dynamics of entangled polymer chains presents a big challenge to theorists.

Here we briefly describe our recent attempts at morphology dynamics of polymer blends and block copolymer melts done in collaboration with K. Sekimoto /12/.

First of all we must stress that the large size of polymer chains precludes uncritical use of kinetic models invented primarily for systems of small molecules such as the time-dependent Ginzburg-Landau equation;

$$\frac{\partial}{\partial t} S(\underline{r}) = L\,\nabla^2 \frac{\delta H\{S\}}{\delta S(\underline{r})} + \zeta(\underline{r}) \tag{3.1}$$

where H{S} is the appropriate free energy functional for the polymer system of the type considered by Leibler and others /13/, L is a kinetic coefficient, $S(\underline{r})$ the order parameter, here the local monomer concentration of one of the species which is conserved, and ξ is the thermal noise satisfying the usual fluctuation-dissipation relation. In fact (3.1) closely resembles the usual diffusion equation which relies on the picture that diffusing particles move by jumps over microscopic (thus macroscopically infinitesimal) distances. Use of (3.1) is especially not warranted to describe kinetics of micro phase separation (i.e. formation of mesophases) in block copolymer systems where one is concerned with microstructures of the scales of molecular size /13/. Thus clearly a more fundamental approach is called for.

The dynamics of highly entangled polymer chains still remains to be one of the most difficult many-body problems. Considerable insight has been gained since introduction of the reptation concept /14,15/. Athough inadequacies of the original reptation model are being disclosed, the many successes of the model so far achieved in explaining main features of viscoelastic properties and other dynamical properties of polymer melts warrant attempts at phase separation kinetics on the basis of this model for the cases where polymer chains have similar chain lengths /16/.

Our approach starts from constructing a general dynamical equation for polymer melt morphology on the basis of the reptation model that replaces (3.1). The reptation model is a kind of mean field model in which a particular chain under consideration can move in a tube which represents the effects of entanglements with other chains surrounding the one under consideration. Polymer ends can move out of the tube to create a new portion of the tube or can be retracted. The part of the tube evacuated by the chain is destroyed. In this way a polymer chain can change its configuration within the time called the reptation time proportional to N^3 where N is the polymerization index. In applying this picture to the cases with more than one species of monomers with repulsive interactions between monomers of different species we must take into account the fact that the reptation rate is "biased", that is, the rate is affected by the inhomogeneous environment of the tube /17/. This is a kind of local form of the mean field theory, Focussing on the late stage kinetics where polymer concentration changes very slowly in time compared with the reptation process, we find the following equations of motion for the local monomer number densities $\rho_K(\underline{r})$ where K=A or B specifies the monomer species:

$$\kappa_c \frac{\partial}{\partial t} \rho_K(\underline{r}) = - \int d\underline{r}' [\Lambda^{KK}(\underline{rr}') \mu_K(\underline{r}') + \Lambda^{K\overline{K}}(\underline{rr}') \mu_{\overline{K}}(\underline{r}')] \tag{3.2}$$

where κ_c is the friction constant of the reptative motion proportional to the chain length, $\overline{K}$ specifies the species different from K, and $\mu_K(r) \equiv \delta H\{\rho_A,\rho_B\}/\delta\rho_K(\underline{r})$ is, apart from the sign, the driving force for concentration change. The thermal noise is omitted for simplicity. The Λ's are the nonlocal kinetic coefficients defined as follows:

For the blend of n_A and n_B chains, each consisting of N_K monomers of the species K=A and B, we have $\Lambda^{K\overline{K}}(\underline{rr}')=0$ and

$$\Lambda^{KK}(\underline{rr}') = 2p_K(\underline{r})\, \delta(\underline{r}-\underline{r}') - 2P_{OK}(\underline{rr}') \tag{3.3}$$

where $P_{OK}(\underline{rr}')$ denotes the probability that the two ends of a chain can be found ar $\underline{r}$ and $\underline{r}'$ and $p_K(\underline{r})$ is the probability that one of the ends is at $\underline{r}$ so that

$$p_K(\underline{r}) = \int P_{OK}(\underline{rr}')d\underline{r}' \tag{3.4}$$

$p_K(\underline{r})$ is further normalized by

$$\int p_K(\underline{r})d\underline{r} = n_K \tag{3.5}$$

These probabilities are to be calculated in the local equilibrium state with given inhomogeneous concentration.

For the block copolymer system consisting of n identical linear chains, each consisting of N_A monomers of the species A connected to N_B monomers of the species B, we have

$$\Lambda^{KK}(\underline{rr}') = [p_K(\underline{r}) + p_J(\underline{r})]\delta(\underline{r}-\underline{r}') - P_{\bar{K}J}(\underline{rr}') - P_{JK}(\underline{rr}') \tag{3.6a}$$

$$\Lambda^{K\bar{K}}(\underline{rr}') = P_{KJ}(\underline{rr}') + P_{J\bar{K}}(\underline{rr}') - P_{K\bar{K}}(\underline{rr}') - p_J(\underline{r})\delta(\underline{r}-\underline{r}') \tag{3.6b}$$

The meaning of the P's and the p's here is similar to that of blends where J now denotes the junction point of the A- and B- portions of a chain and n_K should be replaced by n. The equations of motion (3.2) are rather easy to interpret. The terms with $\delta(\underline{r}-\underline{r}')$ in the Λ's describe concentration changes at the same locations where the forces (the μ's) act. Other terms arise from concentration changes at polymer ends or junction points that result from reptative motions by forces acting at other locations. Note the Onsager reciprocity for the Λ's implies that the force μ_K can effectively act only at polymer ends and junction points involving the species K. This is illustrated in Fig.1 for the case of block copolymers.

Fig. 1. Diagrams of Λ^{AA} (left) and Λ^{AB} (right). The solid and wavy lines are the A and B blocks of a copolymer chain respectively. The white and black circles are the end and junction points, respectively, and the arrows directed to and out of the circles represent the forces and the concentration changes, respectively.

The conservation laws are satisfied by each of the Λ's:

$$\int \Lambda^{KK}(\underline{rr}')d\underline{r} = \int \Lambda^{K\bar{K}}(\underline{rr}')d\underline{r} = 0 \tag{3.7}$$

which is reflected by the sign relations of various terms in (3.6).

Now, the treatment described above in which each polymer chain moves independently of the other is too simplified to deal with inhomogeneous states. In fact, since the system under consideration is highly incompressible, motions of individual chains must be coordinated so as to maintain the total monomer density uniform. This is formally taken care of by adding to all the μ's a function of $\underline{r}$, which is then determined so as to maintain the constant uniform total monomer density /16/. Now, only one of the concentrations, say ρ_A, is independent and we obtain

$$\kappa_c \frac{\partial}{\partial t} \rho_A(\underline{r}) = - \int d\underline{r}' \, \Lambda(\underline{rr}')[\mu_A(\underline{r}')-\mu_B(\underline{r}')] \tag{3.8}$$

where $\Lambda(\underline{rr}')$ is related to the Λ's by the following equivalent matrix equations in the $\underline{r}$-space:

$$\Lambda = \Lambda^{AA} - (\Lambda^{AA}+\Lambda^{AB})\cdot\Lambda_T^{-1}\cdot(\Lambda^{AA}+\Lambda^{BA})$$

$$= -\Lambda^{AB} + (\Lambda^{AB}+\Lambda^{AA})\cdot\Lambda_T^{-1}\cdot(\Lambda^{BB}+\Lambda^{AB}) \qquad (3.9)$$

together with two more equations obtained by interchanging A and B in the above, where Λ_T is the sum of all the Λ's. These equivalent forms for Λ show that $\Lambda(\underline{rr}')$ is nonvanishing only when both types of monomers are present at both the locations $\underline{r}$ and $\underline{r}'$. Since we have seen that the $\Lambda(\underline{rr}')$'s with superfixes exist only for both $\underline{r}$ and $\underline{r}'$ at a junction point or a polymer end, we must hence conclude that $\bar{\Lambda}(\underline{rr}')$ is nonvanishing only when both $\underline{r}$ and $\underline{r}'$ are at a junction point or a point where an end of the A-block (or of the A-type polymer chain) and an end of the B-block (or of the B-type polymer chain) meet where the expressions in parentheses refer to the polymer blend. Furthermore, the expression (3.9) can be further analyzed algebraically both for blends and copolymers and can be brought to the forms which permit pictorial interpretation in terms of the reptation processes during morphology change. The analyses and the final results are too lengthy to be reproduced here and we only present diagrams showing the correlated reptative motions of many chains. Note that the incompressibility condition does not permit isolated movement of a chain end inside the system but must be coupled with a compensating move of another chain end at the same location. Such a motion is called the end-end collision and is represented by a double white circle in Fig.2.

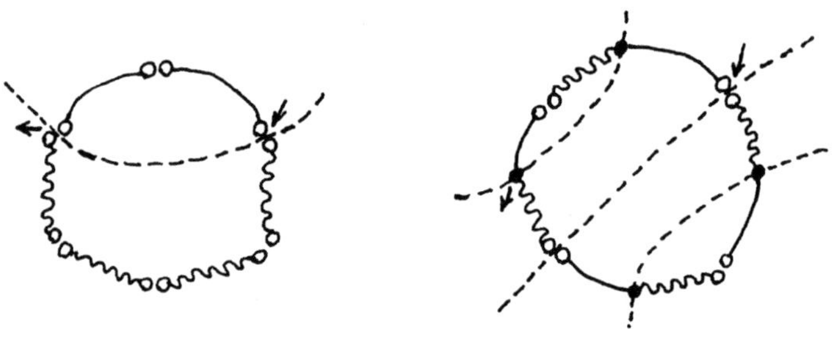

Fig. 2 Examples of diagrams contributing to the concentration change in the polymer blend (left) and in the block copolymer (right). Dashed lines represent domain walls in SSL (strong segregation limit).

Correlated reptative motions of many chains always form a closed loop inside the system and the line can terminate only at the system boundary. In this sense there is a close analogy with lines of singularities in ordered media such as dislocation and disclination lines. The loop of similar sort has been considered by PAKULA /18/ constructing a microscopic lattice model for computer similation, where "defects" (or localized slacks) of the line which Pakula calls kinks play major roles in dynamics. Our chains are more like primitive chains in the sense of Doi and Edwards where these "defects" are already averaged out, and thus chain ends play major roles in dynamics.

Consequences of our theory of morphology dynamics are yet to be worked out. We must mention that the growth rate of concentration fluctuation in the linear regime of phase-separating polymer blend agrees with that of PINCUS /16/. The corresponding result for block copolymers can be readily worked out and will be given in Appendix. One prediction we have made is concerned with coarsening in the late stage spinodal decomposition of polymer blend. As is shown in Fig.2, the concentration change takes place through effective jumps of monomers over distances which can become much greater than the radius of gyration of chains. This will lead to apparent violation of the conservation law if we focus only on the finite region whose size is smaller than the effective jump distance.

Therefore, we may expect the existence of a regime where the average domain size of a phase separating polymer blend grows as square root of the time, which is characteristic of the domain growth in systems with non-conserved scalar order parameter /19/. An available experimental evidence /20/ is suggestive of but not yet conclusive for the predicted behavior.

The author acknowledges T. Nagai, K. Sekimoto, T. Ohta and J. Ogawa for valuable discussions on various aspects of the works reported here.

Appendix

Here we merely present the Fourier transform $\Lambda(\underline{k})$ of $\Lambda(\underline{r}-\underline{r}')=\Lambda(\underline{r},\underline{r}')$ appearing in (3.8) for states with very small spatially inhomogeneous fluctuations. This is useful since $\Lambda(\underline{k})$ for polymer blends has been often used as a substitute for $\Lambda(\underline{k})$ of the block copolymers. Thus we need to evaluate quantities appearing in (3.3)-(3.6) and (3.9) in spatially homogeneous state of polymer blends and block copolymers, which are represented as ideal Gaussian chains. Then (3.5) implies $p_k=c_k\equiv n_k/\underline{V}$ for polymer blends, V being the system volume. For block copolymers, $p_K=p_J=c\equiv n/\underline{V}$. The joint probability densities are also known /22/ and are given in terms of their Fourier transforms $\overline{P}_{OK}(\underline{k})$, etc. by

$$\overline{P}_{KO}(\underline{k}) = \overline{P}_{OK}(\underline{k}) = c_K[1-\eta_k(N_K)] \qquad \text{(blend)} \tag{A.1}$$

$$\left.\begin{aligned} \overline{P}_{KO}(\underline{k}) &= \overline{P}_{KJ}(\underline{k}) = c[1-\eta_k(N_A)] \\ \overline{P}_{\overline{K}J}(\underline{k}) &= \overline{P}_{J\overline{K}}(\underline{k}) = c[1-\eta_k(N_B)] \\ \overline{P}_{K\overline{K}}(\underline{k}) &= \overline{P}_{\overline{K}K}(\underline{k}) = c[1-\eta_k(N_A+N_B)] \end{aligned}\right\} \qquad \text{(copolymer)} \tag{A.2}$$

where

$$\eta_k(N) \equiv 1 - \exp(-\frac{1}{6}Nb^2k^2) \tag{A.3}$$

with b the effective bond length of the chain assumed to be the same for the two monomer species. Substituting these into (3.9) for blends and copolymers we find

$$\Lambda(k) = \Lambda_{bl}(k) \equiv 2c_A\eta_k(N_A)c_B\eta_k(N_B)/[c_A\eta_k(N_A)+c_B\eta_k(N_B)] \qquad \text{(blend)} \tag{A.4}$$

$$\Lambda(k) = \Lambda_{co}(k) \equiv 2c\eta_k(N_A)\eta_k(N_B)[1-\frac{1}{4}\eta_k(N_A)\eta_k(N_B)]/\eta_k(N_A+N_B) \qquad \text{(copolymer)} \tag{A.5}$$

(A.4) coincides with Pincus' result /16/ but (A.5) is new. We now discuss the limiting cases of these results when $N_A \sim N_B \sim N$.

(i) $k^2 \ll 6/Nb^2$

Here (A.4) and (A.5) coincide if expressed in terms of the monomer densities $\rho_K=N_Kc_K$ ($c_K=c$ for copolymers):

$$\Lambda(k) \cong \frac{1}{3}(bk)^2 \rho_A\rho_B/(\rho_A+\rho_B) \tag{A.6}$$

(ii) $k^2 \gg 6/Nb^2$

$$\Lambda_{bl}(k) \cong 2c_Ac_B/(c_A+c_B), \qquad \Lambda_{co}(k) \cong \frac{3}{2}c \tag{A.7}$$

Thus we see no important difference in the behaviors of $\Lambda_{bl}(k)$ and $\Lambda_{co}(k)$ although $\mu_A - \mu_B$ can behave quite differently.

(iii) $k^2 \sim 6/Nb^2$

$$\Lambda_{bl} \sim \Lambda_{co} \sim \rho_0/N \tag{A.7}$$

If this is combined with the fact that the wave number dependent osmotic compressibility behaves as N^2 in this medium wave number region, the wave number dependent relaxation time behaves as N^3 just like the reptation relaxation time. Hence we may expect that the incompressibility restriction is not that strong as to destroy the usual unrestricted reptation picture which has successfully explained many dynmaical behaviors of polymer melts. /14,15/.

Note Added

It was pointed out that for block copolymers in the strong segregation limit the shear strain in the lamellar structure cannot be relaxed by simple reptation since junction points are fixed to domain boundaries /23/. Thus our theory for block copolymers may have to be modified in the strong segregation limit by taking this fact into account.

References

1. The field was reviewed, for instance, in J.D. Gunton, M. San Miguel and P.S. Sahni, In Phase Transitions and Critical Phenomena, vol.8, ed. by C. Domb and J.L. Lebowitz, (Academic Press, New York, 1983).
   K. Binder and D.W. Heermann, In Scaling Phenomena in Disordered Systems, ed. by R. Pynn and A. Skjeltorp (Plenum, New York, 1985).
   H. Furukawa, Advances in Phys. 34, 703 (1985)
2. H.L. Snyder, P. Meakin and S. Reich, Macromolecules 16 757 (1983)
3. T. Hashimoto, J. Kumaki and H. Kawai, Macromolecules 16 641 (1983)
4. K. Kawasaki and T. Ohta, Physica 118A, 175 (1983)
   K. Kawasaki, Ann. Phys. (N.Y.) 154, 319 (1984).
5. H. Ikeda, J. Phys. C: Solid State Phys. 16, 3563 (1983), ibid 19 L535 (1986)
6. J. Villain, Physica B79, 1 (1975)
7. S.M. Allen and J.W. Cahn, Acta Metall. 27, 1085 (1979)
8. T. Nagai and K. Kawasaki, Physica 120A, 587 (1983)
   K. Kawasaki and T. Nagai, Physica 121A, 175 (1983)
   K. Kawasaki, In Computer Analysis for Life Science, ed. by C. Kawabata and A.R. Bishop (Ohmsha, Tokyo, 1986)
9. K. Kawasaki and T. Nagai, J. Phys. C: Solid State Phys. 19 L551 (1986)
10. Reference /2,3/ and also:
    H.L. Snyder, P. Meakin, J. Chem. Phys. 79, 5588 (1983)
    T. Hashimoto, M. Itakura and N. Shimidzu , J. Chem. Phys. 85, 677 (1986)
11. I. Goodman, ed. Developments in Block Copolymers 1 (Applied Science Publishers, London, 1982)
12. K. Kawasaki and K. Sekimoto, Physica A (in press), Europhys Lett. (in press) and a paper in preparation
13. L. Leibler, Macromolecules 13, 1602 (1980)
    T. Ohta and K. Kawasaki, Macromolecules 19, 2621 (1986)
14. P.G. de Gennes, Scaling Concepts in Polymer Physics (Cornell University Press, Ithaca, 1979)
15. M. Doi and S.F. Edwards, The Theory of Polymer Dynamics (Clarendon Press, Oxford, 1986)
16. P.G. de Gennes, J. Chem, Phys. 72, 4756 (1980)
    P. Pincus, J. Chem. Phys. 75, 1996 (1981)
    K. Binder, J. Chem. Phys. 79, 6387 (1983)
17. G.W. Slater and J. Noolandi, Phys. Rev. Lett. 55, 1579 (1985)
18. T. Pakula, Macromolecules 20, 679 (1987).

19. Reference /7/ and also
J.W. Cahn and S.M. Allen, J. de Physique, Colloque 7(C), 54 (1977)
20. T. Izumitani, M. Takenaka and T. Hashimoto, Polym. Prepr. Jpn. 35, 2974, 2978 (1986)
21. T. Hashimoto, Macromolecules 20, 465 (1987)
G.H. Frederickson, J. Chem. Phys. 85, 5306 (1986)
A. Onuki, RIFP preprint (1987)
22. E. Helfand and Z.R. Wasserman, in /11/
23. F.S. Bates, Macromolecules 17, 2607 (1984)
A. Onuki, a talk at RIFP research meeting (1987)

Effects of Impurities on Domain Growth

D.J. Srolovitz[1;2], *G.S. Grest*[3], *G.N. Hassold*[2], *and R. Eykholt*[1;4]

[1]Los Alamos National Laboratory, Los Alamos, NM 87545, USA
[2]Department of Materials Science and Engineering, University of Michigan, Ann Arbor, MI 48109, USA
[3]Exxon Research and Engineering Co., Annandale, NJ 08801, USA
[4]Department of Physics, Colorado State University, Fort Collins, CO 80523, USA

1. Introduction

The kinetics of domain growth in simple spin models quenched from a high temperature ($T>>T_c$) to a low temperature ($T<T_c$) have received a great deal of attention in the last 10 years. These spin systems provide a simple model for ordering processes in real materials (e.g., grain growth [1,2], spinodal decomposition [3], etc.). Such studies generally show that the correlation length in the system grows with time as t^n, where n is the temporal growth exponent. In cases with nonconservative dynamics, n is generally found to be 1/2. Occasional difficulties in obtaining this value (and the corresponding value of 1/3 for conservative systems) in Monte Carlo simulations are often attributable to finite system sizes (i.e., correlation length not sufficiently larger than lattice size). In many experimental systems, however, a growth exponent less than expected is observed [4]. Frequently, these low exponents are attributable to impurities in the experimental system. These impurities can take the form of second phase particles, atomic impurities, or lattice defects. At sufficiently low temperatures, these impurities are usually immobile and can be viewed as static or quenched with respect to the motion of domain walls. On the other hand, at sufficiently high temperatures where the impurity mobility is significant, the impurities can diffuse along with the moving domain walls. In either case, the impurities impede domain wall motion, resulting in slower growth kinetics.

In the present report, we consider the effects of both static and diffusing impurities on domain growth kinetics. In particular, we employ Monte Carlo simulations for nonconservative (Glauber) dynamics to examine the effects of quenched impurities on domain growth in the Potts model with varying degeneracy Q ($2\leq Q\leq 48$) [5,6]. The effects of diffusing impurities are examined within the framework of the Ising model (i.e., Potts model with Q=2) as a function of impurity diffusivity [7]. Finally, a theoretical analysis of the diffusing-impurity results is presented [8]. Due to the limited space available, however, this paper presents only a review of our recent work, to which the interested reader is referred for more details [5-8].

2. Simulation Procedure

The Hamiltonian describing the Q-component ferromagnetic Potts model is written as

$$H = -J \sum_{NN} \delta_{S_i S_j} \qquad (1)$$

where S_i is the state of the spin on site i ($1 \le S_i \le Q$), δ_{AB} is the Kronecker δ function, and J is a positive constant. The summation in Eq. (1) is taken over all nearest neighbor (NN) spins on a two-dimensional lattice. Since, in the present simulation, the order parameter is not conserved, spin-flip (Glauber) dynamics are employed. The Monte Carlo procedure employed was made more efficient by adoption of the technique known as the "n-fold way" [9]. In order to account for impurities, the normal Potts model is modified to allow $0 \le S_i \le Q$, where $S_i=0$ corresponds to an impurity site [5,6].

The model was initialized by randomly placing Nc impurities with $S_i=0$ on the lattice, where N is the number of lattice sites and c is the concentration of quenched-impurities. The remaining sites were assigned a random value between 1 and Q. During the course of the quenched impurity simulations the impurity sites were not updated and hence both the impurity positions and concentration were time independent. The surface (line) energy of an impurity interacting with a spin on a NN site is the same as the domain-boundary energy between two sites with $S_i \neq S_j \neq 0$. The static impurity simulations were performed on a 200x200 triangular lattice for Q>4, on a 400x400 triangular lattice for Q = 3 or 4, and on a 500x500 triangular lattice for Q=2 (i.e., the Ising model). For Q<48, results of ten configurations were averaged for each value of Q and c. For Q=48, five simulations were performed for c≥0.02 and two for smaller values of c.

The diffusing-impurity simulations [7] were all performed on the Ising model on the square lattice. It is convenient to write the Ising Hamiltonian as

$$H = -J \sum S_i S_j \qquad (2)$$

where the spins are now $S_i=\pm 1$. As before, impurities correspond to $S_i=0$. The nonzero spins were updated using spin-flip dynamics. The impurity spins, on the other hand, were updated using spin-exchange (Kawasaki) dynamics, since the concentration of impurity sites is conserved. The same spin-updating probability (W) was employed for both impurity and non-impurity spins, i.e.

$$W = (D/2)[1+\tanh(-\Delta E/k_B T)] \qquad (3)$$

where ΔE is the difference in the energy of the system following and prior to a spin update and $k_B T$ is the thermal energy. The parameter D was chosen as unity for the non-impurity spins, and was assigned a value less than or equal to 1 for the impurity spins. Since D determines how often impurity and non-impurity site exchanges are attempted, D is proportional to the impurity diffusivity. The diffusing-impurity simulations were performed on 250x250 square lattices at $T=0.08J/k_B$. Simulations were performed for $0 \le D \le 1$ and $0.003 \le c \le 0.1$. Due to the large number of simulations required to examine this two-dimensional parameter space, only two simulations were performed for each set of conditions.

Several different methods are available for characterizing the degree of order in the system as a function of time. We have found [7] that the mean chord length, L, the inverse perimeter density, R, and the inverse square root of the second moment of the structure factor, $2\pi k_2^{-1/2}$, all yield approximately the same domain-growth exponent. However, the inverse perimeter density ⟨R⟩ was the most efficient to calculate and, hence, will be employed here. Noting that, at low temperature, the perimeter density is inversely proportional to the excess energy (over the ground-state energy E_o), we write

$$\langle R \rangle = -E_o/[E(t)-E_o] \qquad (4)$$

where E(t) is the energy of the system at time t. Since the ground-state energy of the non-impurity spins is a function of the instantaneous impurity configuration, the time dependence of E_o must be carefully accounted for.

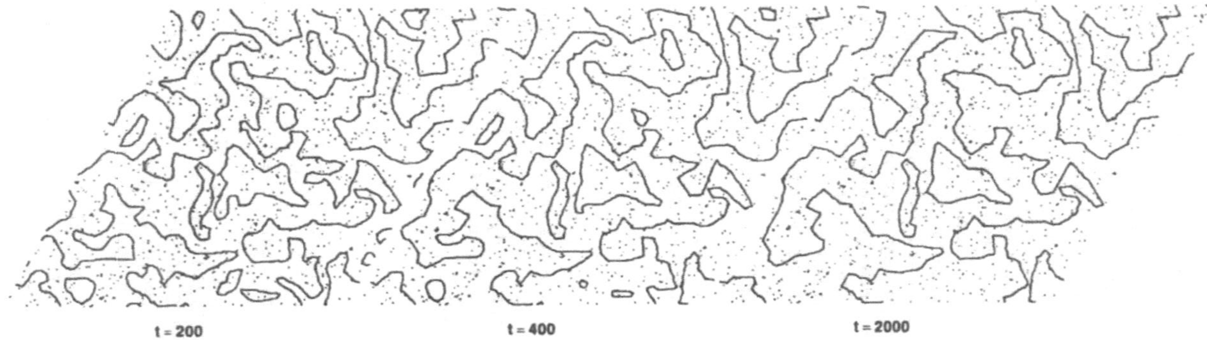

1. Evolution of the domain structure for the nonconserved Ising model quenched to T=0 on a 500x500 triangular lattice with c=0.005.

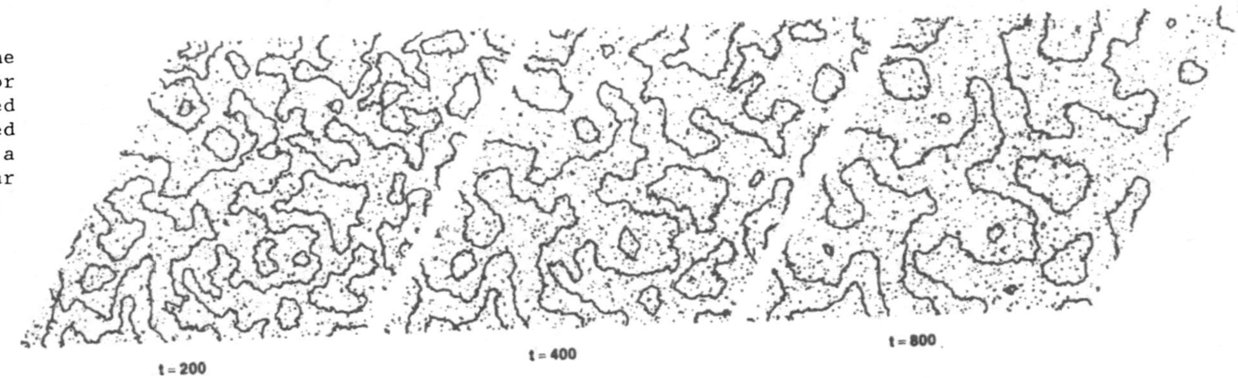

4. Evolution of the domain structure for the nonconserved Ising model quenched to $T=1.3J/k_B$ on a 500x500 triangular lattice with c=0.01.

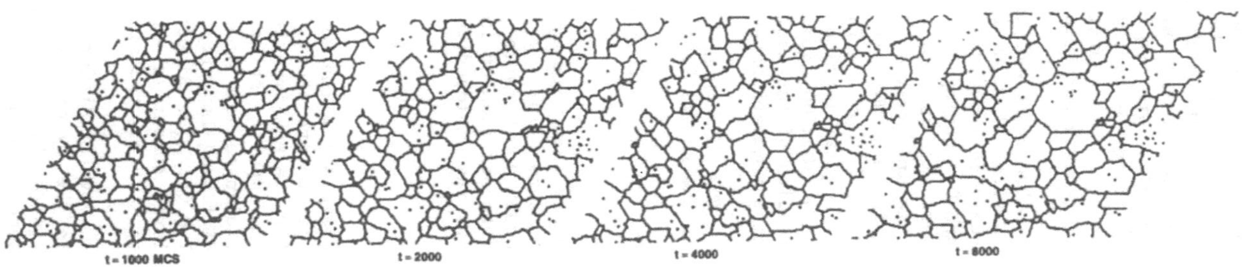

6. Evolution of the domain structure for the nonconserved Q=12 Potts model quenched to T=0 on a 200x200 triangular lattice with c=0.01.

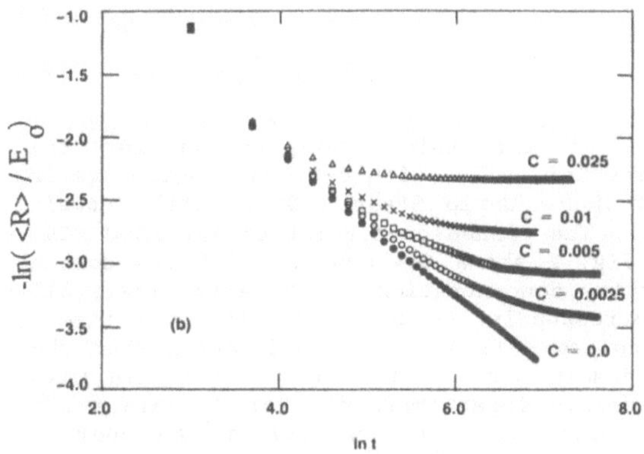

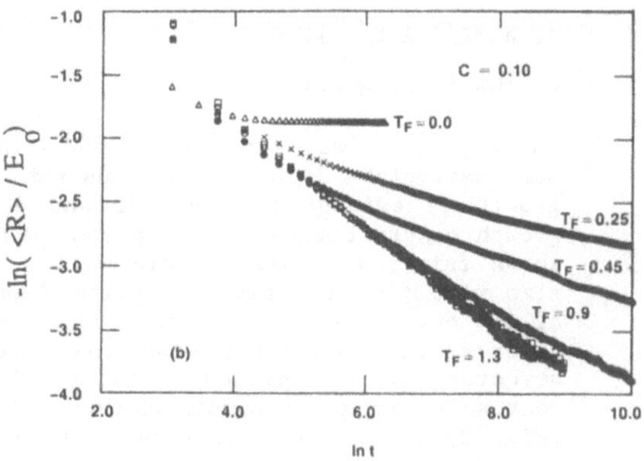

2. Normalized domain size vs time for the Ising model quenched to T=0. The data are averaged over 10 runs on a 500x500 lattice.

5. Normalized domain size vs time for the Ising model quenched to different temperatures for c=0.10.

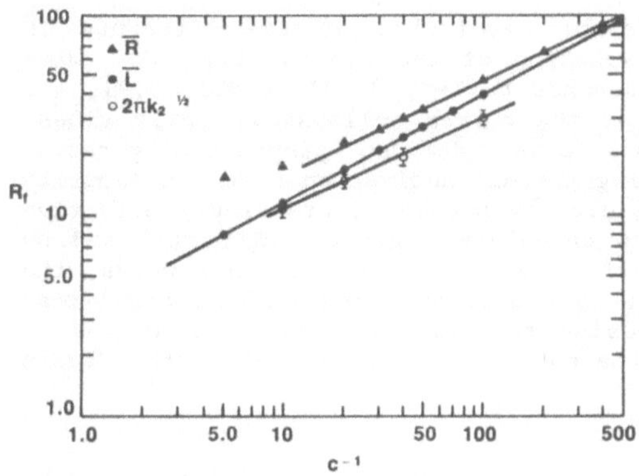

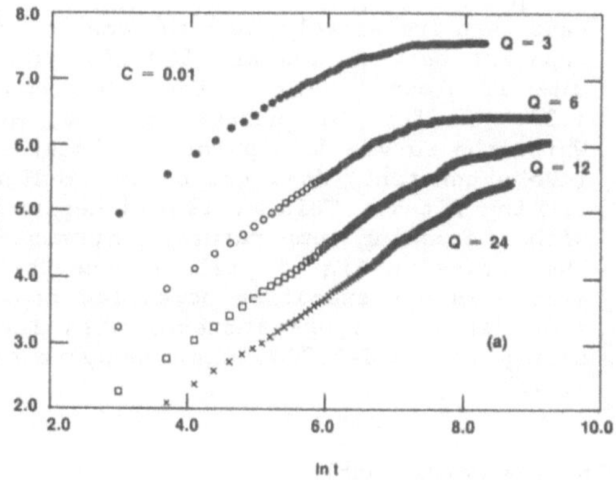

3. Log-log plot of the pinned domain size, R_f, vs the inverse concentration, c^{-1}. The three different measures are defined in the text.

7. The average domain area vs time for T=0 quenches of the Q = 3, 6, 12, and 24 Potts model with c=0.01.

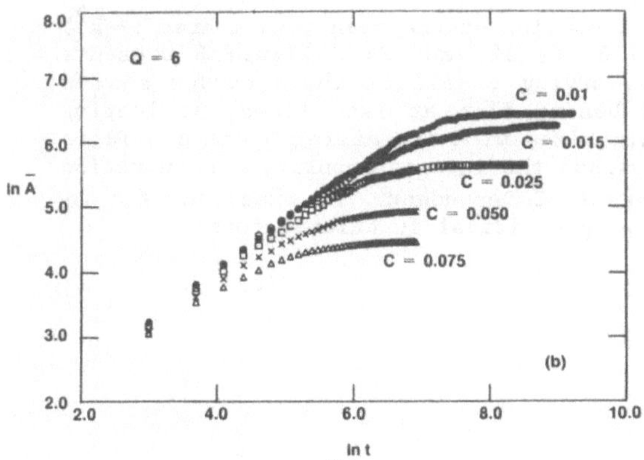

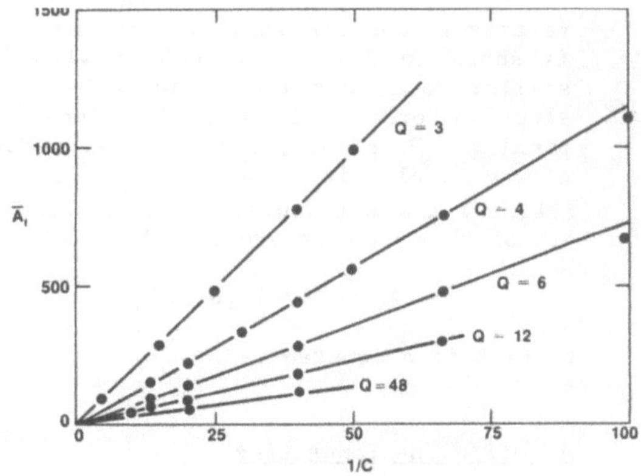

8. The average domain area vs time for T=0 quenches of the Q = 6 Potts model with c = 0.01, 0.015, 0.025, 0.050, and 0.075.

9. The pinned domain area, A_f, vs the inverse impurity concentration, 1/c, for Q = 3, 4, 6, 12, and 48.

3. Static Impurities

A. The Ising Model

In Fig. 1, we show the evolution of the spin configuration for the nonconservative Ising model quenched from $T\approx\infty$ to $T\approx0$ with $c=0.005$. While domain growth is evident between 200 and 400 Monte Carlo Steps (MCS), little domain growth occurs between 400 and 2000 MCS. The pinned configuration (at 2000 MCS) shows fairly irregular domain walls. Fig. 2 shows the evolution of the domain size with time for five different impurity concentrations. At early times, all of the curves fall on the same line, corresponding to $\langle R\rangle\sim t^{1/2}$. At later times, however, the curves corresponding to nonzero c flatten out, indicating that the structure becomes pinned. The final domain size R_f is seen to increase with decreasing impurity concentration. A double logarithmic plot of R_f versus 1/c (Fig. 3) shows that the final, pinned domain size is inversely proportional to the square root of the impurity concentration. Such scaling relations collapse all of the data [5] onto a single curve obtained by plotting $\ln(R(t)/t^{1/2})$ versus $\ln(ct)$.

The microstructural evolution corresponding to a quench to $T\sim0.7T_c$ with $c=0.01$ is shown in Fig. 4. Unlike for the quenches to T=0, the domain walls in this case are relatively smooth and considerable evolution of the structure is apparent between 400 and 800 MCS. The evolution of the domain size, $\langle R\rangle$, with time is shown in Fig. 5 for $c=0.1$ and quenches to T=0, 0.25J, 0.45J, 0.9J, and 1.3J. As for the quenches to T=0, all of the curves collapse at early times. While the curves do separate at later times, a well-defined, pinned configuration ($\langle R\rangle$ = constant) does not occur, indicating thermal activation over the impurity pinning sites. This activated depinning clearly becomes increasingly effective with increasing temperature. Attempts to calculate a growth exponent based on the curves in Fig. 5 yield exponents which vary with time: in other words, the system is not executing power-law growth. The exponents measured at the latest times in the figure increase with increasing temperature (from zero at T=0 to nearly 1/2 at T=1.3J), again supporting the notion of thermally activated domain growth.

B. The Potts Model

In addition to the Q=2 (Ising) limit discussed above, the effects of impurities on domain growth in the Potts model were examined for Q = 3, 6, 12, 24, and 48 [6]. Figure 6 shows the evolution of the microstructure for a quench to T=0 with c=0.01. Unlike in the Ising model, where domain walls never cross, for $Q\geq3$ (in two dimensions), the domain walls meet at three-fold vertices. This results in relatively compact domains. The temporal evolution of the mean domain area ($A\sim R^2$) is shown in Fig. 7 for c=0.01 with Q = 3, 6, 12, and 24. Figure 8 presents similar data, but for fixed Q (=4) and varying c. All of these curves show a slope of order 0.5 at early times and become flat at late times, indicating pinning. Pinning occurs at smaller domain sizes with increasing Q and c. Plots of the final, pinned domain size (area) versus the inverse impurity concentration (Fig. 9) are all linear, with a slope that is Q dependent. An excellent fit to all of this data is provided by the following empirical functional form:

$$A_f = (1/c)\ [3+B/(Q-1)^{3/2}] \qquad (5)$$

where B is a constant.

4. Diffusing Impurities

When the impurities are free to diffuse [7], three new effects can occur: 1) the impurities can diffuse to domain walls, 2) the impurities can diffuse along with moving domain walls, and 3) the impurities can cluster. All of these features can be seen by comparing Figs. 10 and 11, which show the temporal evolution of an Ising model on a square lattice quenched from $T=\infty$ to $T=0$ with $c=0.01$ and with $D = 0$ or 1, respectively. The most notable effect is that, when the impurity diffusivity is finite, the final or pinned domain size is smaller than when the impurities are static ($D=0$). Additionally, in the pinned configuration of the $D=0$ simulation relatively few impurities are observed in the center of the domains (i.e., away from domain walls). The effect of variations of D (diffusivity), at fixed c, on the rate of domain growth is indicated in Fig. 12. This figure clearly shows that increasing diffusivity leads to earlier pinning at smaller domain sizes. Similar results were obtained for all nonzero impurity concentrations simulated.

A great deal can be learned by examining the time dependence of the density of impurities on domain walls (see Fig. 13). While the fraction of impurities on domain walls generally shows a decrease with time, the density of impurities along the domain walls (i.e., the number of impurities per unit domain wall length) increases with time, since $\langle R\rangle \sim t^{1/2}$, at least at early times. These curves appear to show a break that occurs at later times with decreasing D. This is most clearly seen in the $D=0.3$ curve, where the edge fraction initially decreases with time and then increases. At early times, the domains are small and the domain-wall curvature is large, and, hence, the domain-wall velocity is large. When the domain wall velocity is large, the impurities are not able to diffuse at a sufficiently high velocity to keep up with the domain wall. However, at later times, the domain size has grown, the domain-wall curvature and velocity have decreased, and, hence, the impurities can diffuse at rates sufficient to allow the domain walls to sweep impurities. The slower the domain wall, the more impurities it collects, and, as a result, it moves even slower. This feedback mechanism results in the catastrophic pinning of domain walls.

5. Analytic Results

To provide some theoretical insight into the above data, we [8] have performed an analytical analysis of the interaction between moving domain walls and diffusing impurities for a realistic form of the impurity--domain-wall interaction energy. This interaction energy is derived within the framework of the modified sine-Gordon model with either misfit impurities (coupling to the gradient of the order parameter) or elastic-modulus impurities (coupling directly to the order parameter). This interaction potential is then employed in calculating the steady-state impurity-concentration profile about a moving domain wall for arbitrary domain wall velocity and impurity diffusivity. At low velocities, the concentration profile is nearly symmetric about the position of the domain wall. As the velocity increases, the amplitude of the concentration profile decreases and the profile becomes increasingly asymmetric. The drag on the wall exerted by the impurities is given by

$$F_{drag} = \int_{-\infty}^{\infty} c(u)\ [-dE(u)/du]\ du \qquad (6)$$

where c(u) is the steady-state impurity concentration, E is the impurity--domain-wall interaction energy, and u is the separation of the impurity from the domain wall.

Eq. (6) can be evaluated analytically and then inverted to yield the steady state relation between the applied force on a domain wall and its resultant velocity. In the high-diffusivity/weak-interaction limit, this reduces to a linear force-velocity relation at small driving forces, an inflection at larger forces followed by another linear region, and, finally, saturation to an asymptotic velocity corresponding to the speed of sound in the material (Fig. 14a). The inflection in the curve is due to a transition from impurity-limited domain-wall motion to a regime in which the diffusing impurities are essentially

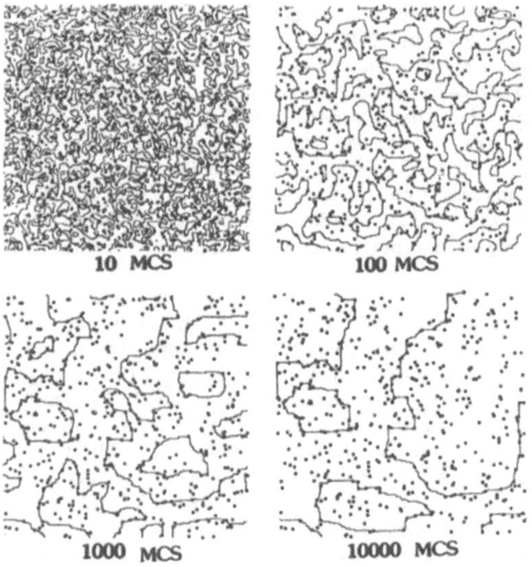

10. Evolution of the domain structure for the nonconserved Ising model quenched to $T=0.04J/k_B$ on a 200x200 triangular lattice with c=0.01 and D=0.

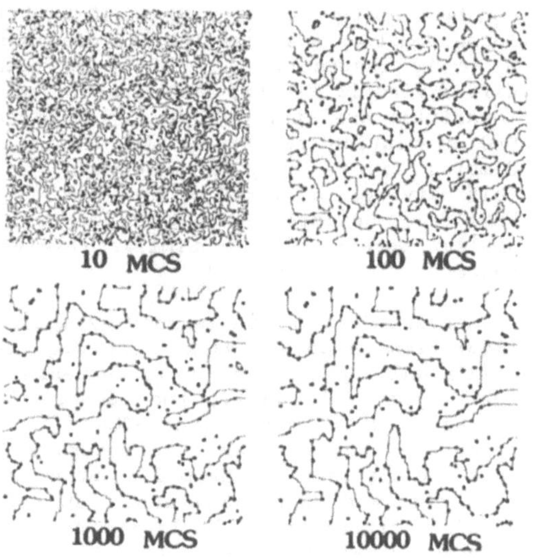

11. Evolution of the domain structure for the nonconserved Ising model quenched to $T=0.04J/k_B$ on a 200x200 triangular lattice with c=0.01 and D=1.

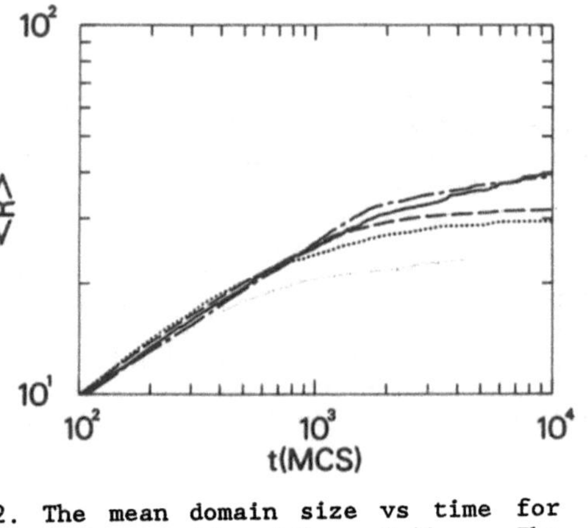

12. The mean domain size vs time for the Ising model with c=0.01. The dotted, dashed, chain-dotted, and solid curves correspond to D=1,0.3,0.1,0.03,0, respectively.

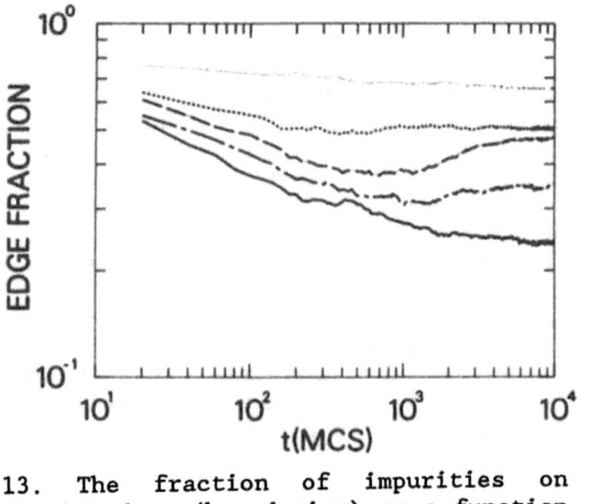

13. The fraction of impurities on domain edges (boundaries) as a function of time for c=0.01. The curves correspond to the impurity diffusivities considered in Fig. 12.

incapable of keeping up with the domain wall. For strong-impurity--domain-wall interactions and lower impurity diffusivity, this inflection turns into a bifurcation with both high- and low-velocity branches (Fig. 14b). As the domain wall is quasi-statically accelerated from rest, it moves along the low-velocity branch until a critical force is applied, after which the domain wall travels at a velocity determined by the upper branch. Similarly, when the domain wall is slowed from high velocity it traverses the upper branch and then discontinuously moves to the lower one at a lower critical applied force. Relatively simple

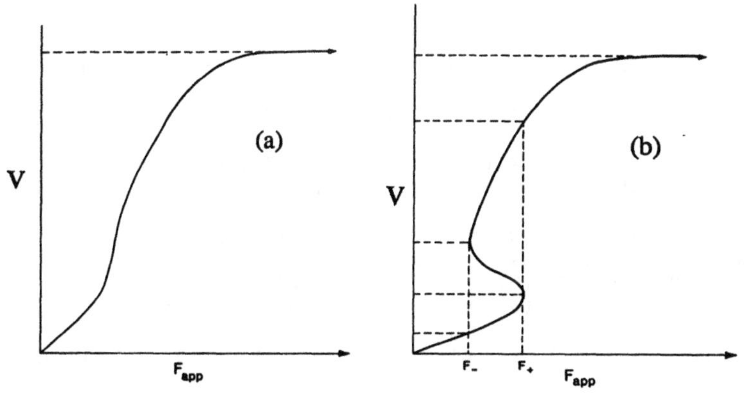

14. Qualitative forms of the domain-wall velocity vs applied force. The transition from impurity-dominated behavior to impurity-free behavior can be either (a) gradual or (b) hysteretic.

expressions for the critical points on the hysteresis loop have been obtained in different limits.

Application of these results to domain growth in the presence of impurities confirms the qualitative picture described above (at the end of Section 4). Following the quench to low temperature, the domain-wall velocity is initially high due to the large-curvature driving force at small domain size and then slows as the domain size increases (decreasing driving force). If conditions are such that the force-velocity relation depicted in Fig. 14b is applicable, the domain-wall velocity will decrease slowly during domain growth until its velocity falls below a critical value where it is captured by the impurities and slowed very drastically, if not effectively pinned. Since different domain walls will be moving with different velocities during domain growth, the pinning of the system is expected to occur one domain wall at a time until no walls are free. If the temperature is such that significant domain-wall velocities are possible in the regime where the impurities are diffusing along with the walls (i.e., the lower branch), then further domain growth can occur, although with an effectively enhanced damping. It is interesting to note that, once the system finds itself in the regime where the impurities can keep up with the domain walls, the effective bulk concentration of impurities is reduced (i.e., a non-negligible fraction of the impurities is on the walls), and a moving wall feels a smaller impurity drag than expected based on the total impurity concentration. It is expected that this will lead to a rescaling of the steady state.

6. Conclusions

Recent Monte Carlo simulations on the effects of impurities on domain growth [5-7] show that, at low temperatures, there is a rather abrupt transition from normal domain growth (i.e., $R{\sim}t^{1/2}$) to a pinned state. The mean, pinned domain size decreases with increasing degeneracy and with increasing impurity concentration as $R_f{\sim}c^{-1/2}$. This transition is less abrupt and, with increasing temperature, it occurs at larger domain size and later times. When the impurities are free to diffuse, they slow domain growth by diffusing to and diffusing along with moving domain walls. Increasing diffusivity results in decreased final domain sizes, at least at low temperatures. A one-dimensional analytical analysis [8] shows that the presence of diffusing impurities can lead to nonlinearities and hysteresis in the relationship between the driving force on a domain wall and the resultant domain-wall velocity and, hence, non-classical domain growth exponents.

References

1. M. P. Anderson, D. J. Srolovitz, G. S. Grest, and P. S. Sahni: Acta Metall. 32, 783 (1984).
2. D. J. Srolovitz, M. P. Anderson, P. S. Sahni, and G. S. Grest: Acta Metall. 32, 793 (1984).
3. J. L. Lebowitz, J. Marro, and M. H. Kalos: Acta Metall. 30, 297 (1982).
4. F. Haessner: Recrystallization of Metallic Materials, (Riederer Verlag, Berlin 1978).
5. G. S. Grest and D. J. Srolovitz: Phys. Rev. B 32, 3014 (1985).
6. D. J. Srolovitz and G. S. Grest: Phys. Rev. B 32, 3021 (1985).
7. D. J. Srolovitz and G. N. Hassold: Phys. Rev. B 35, 6902 (1987).
8. D. J. Srolovitz, R. Eykholt, D. M. Barnett, and J. P. Hirth: Phys. Rev. B 35, 6107 (1987).
9. A. B. Bortz, M. H. Kalos, and J. L. Lebowitz: J. Comput. Phys. 17, 10 (1975).

Introduction to Growth Oscillations

R. Savit

Physics Department, University of Michigan,
Ann Arbor, MI 48109, USA

1. Introduction

Recent developments in the study of growth and aggregation have underscored the incredibly rich and varied properties that can be obtained from very simple dynamical algorithms. One such remarkable aspect of non-equilibrium growth is the existence of oscillatory behavior in a variety of physical properties of a growing cluster [1]. These oscillations, which should occur in a wide range of growth models, arise as the result of an induced incommensuration between two length scales. One length scale is a statically defined length corresponding to the size of the accreting particles, or if the model is defined on a lattice, to the lattice spacing. The second length scale is dynamical in origin. In the simplest models, in which growth events occur at discrete times, this second length scale can be thought of as the average distance which the growing interface moves per time step. (Generalizations to growth systems for which time is not discrete will be discussed below.) It is the beating between these two length scales that causes the oscillations.

Growth oscillations can be observed in the velocity of propagation of a growing interface, as well as in the density of the resulting aggregate. The conditions necessary for the existence of these oscillations are very general. Although there are exceptions (see below), growth oscillations may generically be expected in systems in which the following two conditions are met: 1. Growth takes place by the accretion of discrete particles or lumps of material and 2. Growth occurs at a fairly well-defined interface.

As we shall show, growth oscillations typically have a highly multiperiodic character, and, in fact, can usually be expected to manifest quasiperiodic behavior [2]. Moreover, depending on the precise conditions of the growth, the wavelengths of the oscillations can vary over an enormous range, from distances of a few particles sizes to hundreds, or even thousands of particle diameters. Thus the oscillations can range from the micro- to the meso-, or even macroscopic regime.

In the following section I will first explain the origin of these growth oscillations in simple physical terms. In section 3 I will present evidence for their existence in a simple physical model of growth, a version of ballistic aggregation with a finite density of raining particles. Next, in section 4, I will introduce another model of growth [3] in which we also find growth oscillations. This second model is related to the Broadbent, Hammersley, Leath, Alexandrowicz (BHLA) algorithm for the generation of percolation clusters [4], but is somewhat simpler. In addition to their intrinsic interest these models are useful to study in that their algorithms can be expressed in a straightforward way as functional stochastic iterative maps. The maps can then be approximated in a number of ways. Studying the analytic approximations reveals some very interesting properties of growth oscillations. In particular, we will see that the oscillations play a role analogous to limit cycles in simple 0-dimensional iterative maps, such as the logistic map [5].

Furthermore, this way of formulating the problem suggests an interesting scenario concerning the origin of random fractal growth. Following this, in section 5, I will show why growth oscillations are generically expected to be quasi-periodic, and I will end with a brief summary and conclusions.

2. Origin of Growth Oscillations

Imagine a cluster which grows according to some rules. We suppose that the cluster grows by the addition of particles of finite size. Alternatively, we may imagine that the growth takes place on a lattice. For simplicity, let us suppose that the cluster grows in one direction (say, the z-direction) from an initially flat substrate normal to z. If the growth is stochastic, then there will generally be irregularities along the top surface of the cluster, and indeed, the growing interface may be rough, although there will still be a well-defined average interfacial position. The cluster itself may also contain holes so that the density, while having a non-zero average value, is not necessarily completely uniform.

To make the discussion as simple as possible, it may be useful to have a specific model in mind. Let us consider, therefore, finite density ballistic aggregation [6]. The model in two dimensions is defined on a square lattice as follows (generalizations to higher dimensions will be obvious): To begin, place a line of L seed particles at adjacent sites along the horizontal axis. Particles are added to the cluster at discrete times according to the following algorithm: Let h(i,n) be the height above the substrate (of seed particles) of the uppermost particle in column i at time n. If at time n+1 a particle falls in column i, it will stick at a height determined by

$$h(i,n+1)=\max[h(i-1,n),\ h(i,n)+1,\ h(i+1,n)]. \tag{1a}$$

If at time n+1 no particle falls in column i, then

$$h(i,n+1)=h(i,n). \tag{1b}$$

A particle will fall in a given column at a given time, independently and randomly with a probability p, where p is a fixed control parameter. Thus, at each time step pL new particles, on average, are added to the cluster. To complete the description of the model, we can specify periodic boundary conditions for the substrate so that we can think of the L seed particles as being wrapped around the surface of a cylinder.

Now, let us define $P_n(y)$, which is the probability that a particle will stick to the cluster at a site whose vertical coordinate is z=y at time n. Since the interface has some non-zero width, we expect $P_n(y)$ to have some bell-like shape, the details of which are not important for us now. We now ask the following question: Given the curve $P_n(y)$, what is $P_{n+1}(y)$? Ignoring for the moment possible broadening of the interface due to roughening, we expect that $P_{n+1}(y)$ will look just like $P_n(y)$, only translated along the y axis a distance given by the distance of propagation of the interface in one time step, as shown in Fig. 1. But this is not what is observed. The actual situation is represented in Fig. 2. Here we see that the shape of the functions $P_n(y)$, (and in particular their heights) oscillate as a function of time. The apparent period of these oscillations is observed to depend on p, growing as $p \rightarrow 1$.

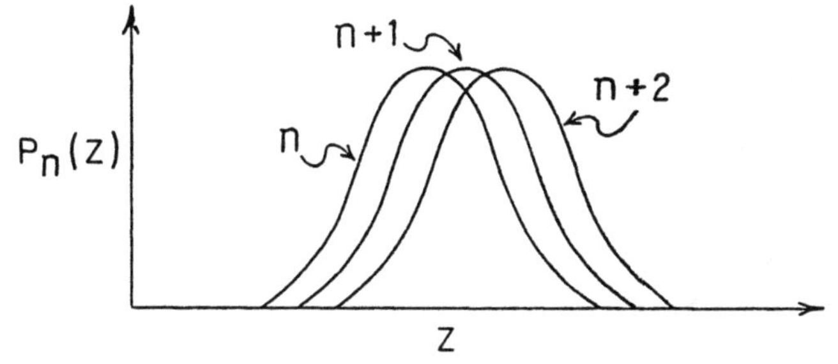

Fig. 1. Schematic representation of the translation of a uniform interface with time

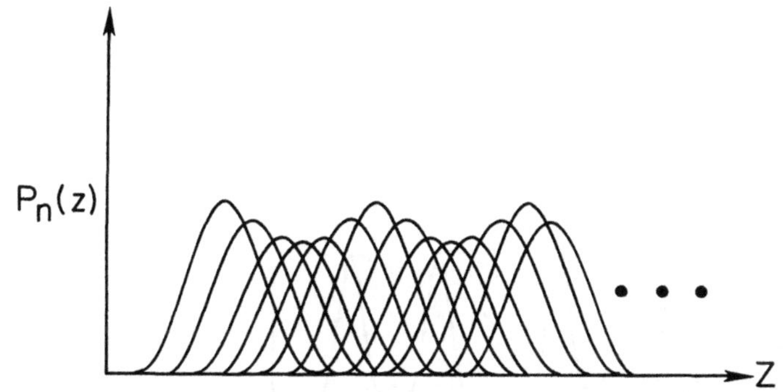

Fig. 2. Schematic representation of an interface which oscillates with time

How do we account for this unexpected behavior? The important point is to remember that there is an inherent discreteness in the growth process. For example, in the ballistic aggregation model defined earlier growth occurs on a lattice. Thus, z is not a continuous variable and can take on only integer values. $P_n(y)$ is therefore defined only for integer values of y. The implications of this observation can be understood in the following way [7]. Let us suppose for the moment that the growth is continuous and that z is a continuous variable. Then, the function $P_n(z)$ is a continuous function defining a kind of meta-interface. Suppose that as time goes by this meta-interface does not change its shape, but translates along the z-direction with each time step. In general, the meta-interface will move a fraction of a lattice spacing in each time step. If we now recall that z only takes on integer values, we have the situation depicted in Fig. 3. Although the meta-interface does not change its shape with time, the integer values of z intersect the curve at relatively different places at different times since the interface does not move an integer number of lattice spacing per time step. For example, if the average position of the interface moves .95 lattice spacings per time step, it will require 20 time steps before the meta-interface is in the same relative position with respect to the underlying lattice. Thus, the shape and maximum height of $P_n(y)$ will appear to oscillate with a period of about 20 time steps.

In Fig. 4 I have plotted $P_n(y)$ as a function of y for all values of n. (This graph happens to be associated with a growth model which is discussed in section 4, below, but other growth models yield qualitatively similar graphs.) The solid lines pass through $P_n(y)$ for a given value of n. It is interesting to note that the values of $P_n(y)$ lie naturally

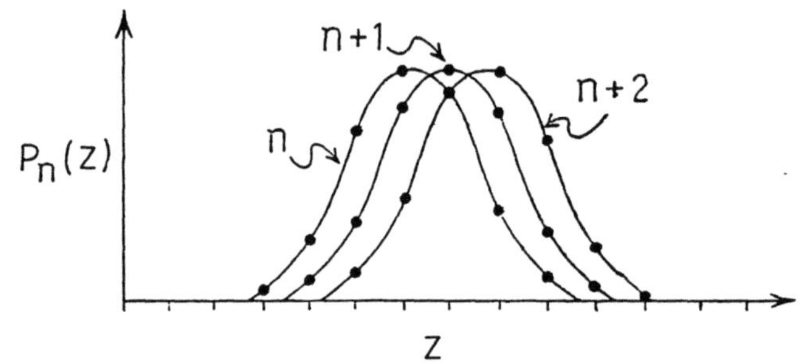

Fig. 3. Intersection of lattice points with a smooth meta-interface which moves a fraction of a lattice spacing per time step

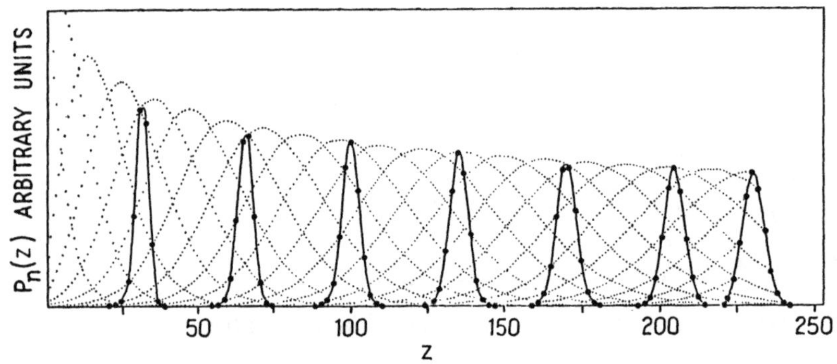

Fig. 4. $P_n(z)$ as a function of integer z for all values of n. Each solid line passes through the values of $P_n(z)$ for some fixed value of n

on broad curves. It can be shown [8] that the kth curve from the left of the graph is the curve of values of $P_{y+k}(y)$. The distance between maxima of the broad curves is approximately the fundamental period of the oscillations described above.

Oscillations in $P_n(y)$ will result in oscillations in physical quantities. For example, the density of the cluster averaged over directions normal to the direction of growth is just given by

$$\rho(z) = \sum_n P_n(z) \tag{2}$$

while the velocity of the interface as a function of time is

$$v(n) = \sum_z z[P_n(z)-P_{n-1}(z)]. \tag{3}$$

Since these quantities are just sums and differences of the $P_n(y)$, it should be clear that they will incorporate the oscillations contained in $P_n(y)$. On the other hand, because of the evident subsidiary structure in $P_n(y)$, we may expect the spectrum of these physical quantities to contain more than just a single frequency. The spectra of several physical quantities have been studied in a number of different models. I now want to turn to a description of one example.

3. Oscillations in Finite Density Ballistic Aggregation

From the description of the origin of growth oscillations in the previous section, it should be clear that such oscillations are to be expected at least in growth models defined on a lattice, for which the growth takes place at discrete times. (In fact, as we will discuss below, we believe these strict conditions can be relaxed so that growth oscillations should occur in a wider class of systems.) To verify this, we have studied large single clusters grown according to the finite density ballistic aggregation algorithm described above [6,9]. Since this model is stochastic, it also serves as an arena in which we can learn something about the effects of noise on the presence of growth oscillations. Although it is more difficult computationally, we chose to study large single clusters rather than averaging over the results of many small clusters, since the former is a situation of more common experimental interest. (Results of averages over many small clusters have also been performed [1,7,8].)

As we shall see, the typical amplitude of the oscillations is rather small, requiring great care to eliminate spurious computational effects. We generated three different finite density ballistic aggregation clusters with substrates of 10^7, 5×10^6 and 6×10^5 particles respectively, and with a value of the control parameter, p=0.9. Each cluster was grown for 1200 generations using a different random number generator to avoid spurious correlations. We then analyzed the density and growth velocity of these clusters for the presence of oscillations. We found evidence for growth oscillations, consistent with our expectations, in all the spectra we analyzed. One typical spectrum is shown in Fig. 5. This is the Fourier transform of the quantity

$$D(n) = \langle d(n)\rangle - u_0 n \tag{4}$$

where $\langle d(n)\rangle$ is the average over the cluster interface of

$$d(i,n)=\max[\, h(i-1,n),\, h(i,n)+1,\, h(i+1,n)] \tag{5}$$

which just represent the zone of potential growth at the next time step. The Fourier transform is performed for a set of data taken over a time interval, M, which excludes some transients due to the initial conditions. To eliminate the very strong peak in the Fourier transform near n=0, we have subtracted in (4) the effect of the constant, average background velocity, u_0, which is defined as

$$u_0 = \frac{1}{M}\sum_{n=t}^{t+M}[\,\langle d(n)\rangle-\langle d(n-1)\rangle] \equiv \frac{1}{M}\sum_{n=t}^{t+M}u(n). \tag{6}$$

There are several important things to notice about Fig. 5. First, it is clear that there is a definite signal for an oscillation with $n\approx0.05$, which corresponds to a period of about 20 time steps. This fundamental period of oscillation is what we would expect from the argument given in section 2: On the basis of that argument, we expect the wavelength of the most prominent oscillation to be given by $\lambda_t=(1-u_0)^{-1}$ [6]. For finite density ballistic

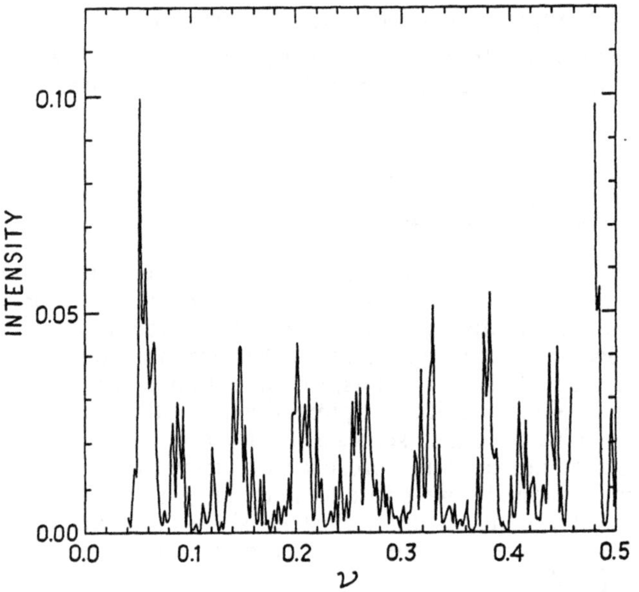

Fig. 5. Power spectrum of D(n) in the range $588 \leq \nu \leq 1100$, for finite density ballistic aggregation for a sample with a substrate of 107 lattice sites and p=0.9.

aggregation with p=0.9, $u_0 \approx 0.95$, so that $\lambda_t \approx 20$. Second, although there is a very clear oscillatory signal in this spectrum the effect .is small. In fact, $[u(n)-u_{0]}/u_0 \approx 10^{-4}$ or 10^{-5}. Third, there are clear groups of peaks of varying heights which are associated in part with harmonics of the peak at $\nu \approx 0.05$. Finally, despite the fact that some of the oscillatory signal can be associated with the fundamental frequency of oscillation and its harmonics, it is clear that the spectrum is a good deal more complicated. In the next section we will discuss some deterministic approximations to another growth model which manifests oscillations, and we will show that the oscillations are highly multiperiodic. Furthermore, in section 5 we will argue that, in general, the spectra of the oscillations should manifest quasiperiodic behavior. Returning to Fig. 5, it is not difficult to see that this spectrum is reminiscent of that of a quasiperiodic system whose spectra have a dense delta-function structure.

4. Percolation Related Models and Their Deterministic Approximations

I have demonstrated the existence of growth oscillations in a stochastic growth model of considerable physical interest. I have also argued that these oscillations depend on only a few very general conditions, and should occur in a wide range of systems. I now want to briefly describe growth oscillations in another model which is related to the BHLA algorithm for generating percolation clusters [4]. As stated earlier, studying this model is useful for several reasons. First, it gives added support to the ubiquity of growth oscillations. Second, it is a model for which it is straightforward to construct a deterministic approximation. This approximation, in turn is useful for at least two reasons. Its numerical solution shows very clearly the rich multiperiodic structure of growth oscillations. In addition, the form of the equation suggests an intriguing speculation about the origin of random fractal growth and its relation to growth oscillations.

Let us begin by defining the model [3]. Consider a square lattice (the generalization to higher dimensions is obvious), and place a seed particle at the origin. During the first time step, each nearest neighbor of the seed site is filled independently and randomly with a probability, p. At the next time step, each empty site which is a nearest neighbor of at least one site which was filled during the first time step is filled independently and randomly with the same probability, p. During the third time step, each nearest neighbor of a site that was filled during the second time step is occupied independently and randomly with probability p, etc. Each site is sampled at most once during any time step even if it has more than one neighbor filled during the previous time step. This model differs from the BHLA algorithm in that sites that are not filled at some time are not blocked, and can be filled at a later time.

The clusters grown according to this algorithm are finite density, compact clusters for p greater than $p_c \approx 0.54$. The shape of the clusters depends on p, but they are roughly diamond shaped with rounded corners which intersect the lattice axes. For large enough p the clusters are faceted along their edges. For $p=p_c$, the clusters are fractals, probably with the same Hausdorff dimension as ordinary percolation clusters. A more detailed discussion of the morphology of this model can be found in [3].

The structure of the oscillations in this model has been studied in [7,8], and is quite complicated due to the anisotropic morphology. If we consider a ray extending from the center of the cluster, the amplitude and period of the oscillations depends on the orientation of the ray, being most pronounced along a lattice axis.

To make further progress, it is useful to express the growth algorithm in an analytic form. This can be done as follows:

$$s(i,n+1)=q(i,n)\left[1-\prod_j(1-s(j,n))\right]\prod_{m=1}^{n}(1-s(i,m)) \qquad (7)$$

The variable s(i,n) is one if and only if site i is occupied exactly at time n. Otherwise it is zero. The product over j in the second factor on the right hand side is a product over all nearest neighbors of the site i. The variables q(i,n) take on values of one or zero randomly and independently, but such that the average value is p. Equation (7) is a stochastic functional iterative map which bears a strong formal resemblence to ordinary zero-dimensional maps, such as the logistic map. In particular, we see in (7) competing factors (the second and third factor on the right hand side, which play a role analogous to the f and (1-f) factors in the logistic map. In the logistic map, limit cycles are due to the competition between these two factors. Similarly, in our case growth oscillations are caused by the competition between the last two factors in (7) [7,8].

Unfortunately, (7) is very difficult to analyze, both because of its intrinsic structure, and because of the complication of noise. However, some additional insight can be gained by constructing a kind of mean-field approximation to (7). We do this by simply replacing each stochastic variable by its ensemble average. Thus we have the following deterministic equation:

$$P(i,n+1)= p\left[1-\prod_j(1-P(j,n))\right]\prod_{m=1}^{n}(1-P(i,m)) \qquad (8)$$

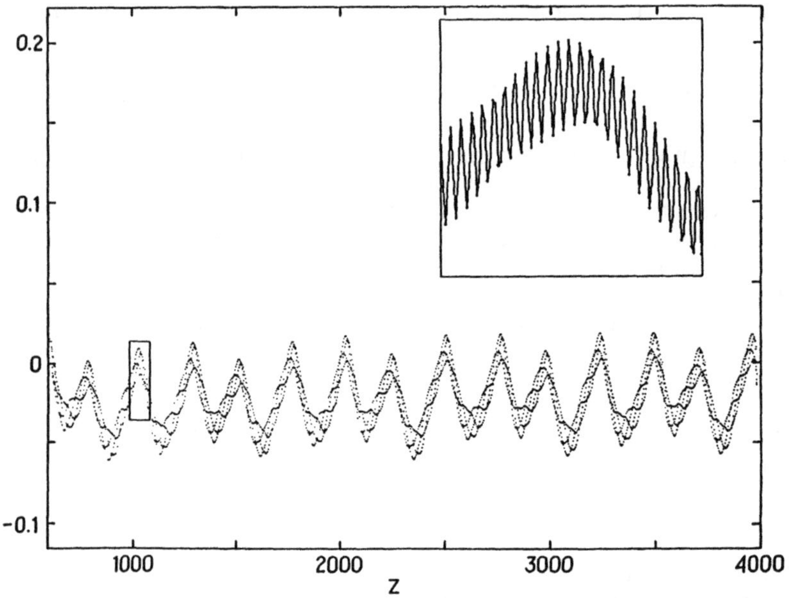

Fig. 6. Multiperiodic density oscillations for a deterministic one-dimensional growth model. See [8] for details

where P(i,n) is the probability to occupy site i exactly at time n, and p=<q(i,n)>. Various versions of (8) have been studied numerically [8] and show an amazingly rich multiperiodic structure which is exquisitely sensitive to the value of p. An example of the oscillations obtained for the density in a one-dimensional version of (8) is shown in Fig. 6.

5. Quasiperiodic Behavior in Growth Oscillations

It is clear that growth oscillations both in stochastic and deterministic models are highly multiperiodic. In fact, it not difficult to see that, in general, they are quasiperiodic [2]. To do this, let's go back to the picture of the propagation of a smooth meta-interface which we presented in section 2. Suppose that this interface propagates in some direction, x. For simplicity, let us ignore the effects of broadening of the interface and suppose that the interface is defined by a curve, f(x-vt), where v is a constant average background velocity. The functions $P_n(y)$ are just the values of f for integer time,n, and integer values of position, y: $P_n(y)=f(y-vn)$. This can be pictorially summarized using a version of the projection method developed for the study of quasicrystals [10]. Figure 7 shows a lattice of space-time points. The slope of the straight line is v. The curve f(x-vt) is drawn for t=0. This curve is smoothly translated along the straight line; and the values of $P_n(y)$ are the values of f where it intersects the lattice points. Physical quantities are easily derived from this picture. For example, the density as a function of y is obtained by summing the $P_n(y)$ horizontally across the plane.

Since there are so many more irrational than rational numbers, we expect that, unless some specific dynamics forbids it, v in general will be irrational. Thus, on the basis of

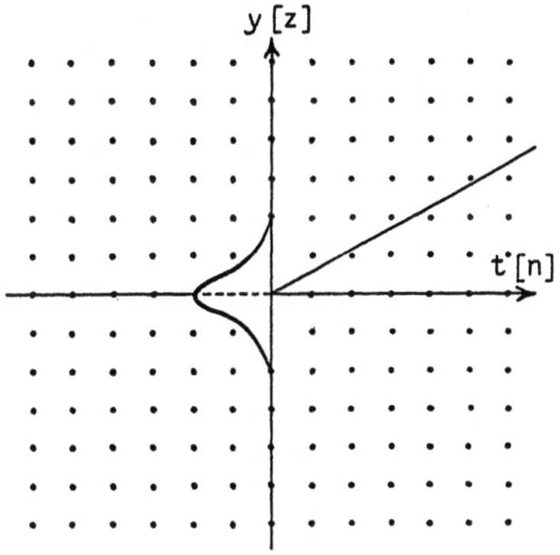

Fig. 7. Geometric representation of the quasicrystal projection method as applied to growth oscillations

the analysis developed for quasicrystals, it is clear that the oscillatory structure will have a quasiperiodic character. Moreover, using this approach, it is straightforward to identify the contribution that is associated with the most prominent peak in the Fourier spectra. For example, for an interface, f, that is a smooth decreasing function of its argument, it can be shown [2], that the largest peak in the power spectrum of D(n) occurs at ν=1-v, as implied by the simple physical arguments of sections 2 and 3. The large (in principle infinite) number of subsidiary peaks of the power spectra can also be deduced from the simple construction represented in Fig. 7. Of course, since the interface of most physical systems of interest is rough, the interface broadens with time, and cannot be described by a meta-interface which retains its shape. We have been able to show [2] that the primary effect of this broadening is just to double the peaks in the power spectra of various physical quantities, with the separation between peaks proportional to the rate of broadening.

6. Summary and Conclusions

In this talk I have tried to demonstrate the nature and generality of growth oscillations. These oscillations should be present in a wide range of growth models, and will appear in a variety of physical quantities. The oscillations are an intrinsically dynamical phenomenon, and result from a dynamically induced incommensuration between two length scales; a static length characterizing the size of the aggregating particles, and a dynamical length associated with the kinetic properties of the growth. These two lengths beat against each other in the process of growth and cause the oscillations. We expect that growth oscillations will occur generically in systems in which 1. growth proceeds by the addition of discrete lumps or particles of material of finite size and 2. growth occurs at a moderately well-defined interface. (The necessity for these two conditions should be clear from the discussion of section 2.)

The specific models we have discussed in this paper were defined on a lattice, and the growth took place in discrete time. One may ask whether these ingredients are necessary for the existence of the oscillations. We believe that models defined in continuous space and time will also manifest oscillations of the type described here. If the growth process consists of the addition of particles of finite size, a static length scale will still exist. Aggregation of finite size particles in continuous space will introduce other sources of noise (for example, if we perform ballistic aggregation in the continuum, new particles may not be aligned directly above the particles to which they are sticking) making the oscillations more difficult to observe, but, in principle, they should still be present. In addition, if the growth process takes place in continuous time, oscillations should still exist. It must be remembered that a growth event (eg. the addition of a particle to the cluster) occurs (for the purposes of our discussion) at a discrete moment in time, even though the river of time is continuously flowing. Thus, depending on the details of the dynamics, it is possible for the growing system to behave as if time is discrete. For example, if the dispersion in times between growth events is not too large, then the growth will proceed as if time was discrete.

There is another point of view to take which also argues against the necessity for externally imposed discrete time. It is clear that growth oscillations ultimately depend on two competing length scales. Growth processes defined, as above, in discrete time (or effectively discrete time) clearly have the potential for generating a dynamical length scale. However, one can easily see the emergence of a dynamical length, which is in general incommensurate with particle size in another way: If the geometry of the growing interface is such that the average penetration depth for a new particle is some fraction of the size of the particle, then we can expect a situation in which the local geometry of the interface oscillates with a period which is determined by the ratio of the particle size and the penetration depth. This oscillation in the local geometry will clearly affect physical quantities such as the density and velocity of growth. Notice that in this argument there is no reference at all to time, obviating a fortiori the need for discrete time.

Growth oscillations have a very rich multiperiodic structure, which in general we expect to be quasiperiodic. We have shown how to understand that quasiperiodicity, and how to deduce the positions of the most prominent peaks in the power spectra. Although the magnitude of the oscillations is typically rather small, they should still be observable in carefully controlled growth experiments using MBE or physical vapor deposition techniques. The most promising experiments in which to observe the oscillations are those in which there is relatively little relaxation or surface diffusion during growth. Such relaxation will destroy the effects of the dynamical length scale, and hence will destroy the oscillations.

One issue that, in my mind, does not have a completely satisfactory explanation concerns the amplitude of the oscillations. It would be very helpful to understand in a simple physical way, how the growing system generates a scale 10^{-4} or 10^{-5} times smaller than other quantities in the system. In this regard it should be noted that in a system with a rough interface, the amplitude of the oscillations may decrease as the size of the system and the width of the interface increases. Generally, we expect this decrease in amplitude to be power behaved [1], with an exponent related to the roughening exponent of the interface. Nevertheless, in some cases, the signal to noise ratio for the power spectrum peaks associated with the growth oscillations can still improve as the cluster grows. See [1,7] for details.

One final point concerns a speculation about the relation of growth oscillations to random fractal growth [1,8]. Consider a growth process such as the one described in section 3, in which the growth algorithm can be expressed as a functional stochastic iterative map. We argued there that growth oscillations are analogous to limit cycles in simpler single variable maps, such as the logistic map. We observe that as the control parameter, p, of our system increases, the fundamental period of the oscillations increases. On the other hand, as $p \to p_c$, the oscillations decrease in period. When $p = p_c$, the oscillations disappear. Simultaneously, the cluster becomes fractal-like. In the logistic map, as the control parameter is decreased, the limit cycles decrease in period and finally, the solution of the map becomes a fixed point. Thus, random fractal growth is apparently controlled by the fixed point of a functional stochastic iterative map. Growth oscillations in these maps are the analogues of limit cycles in the logistic map, and the scaling dimensions and other properties of the critical fractal are determined by the way in which the fixed point is approached as the control parameter approaches its critical value. We have demonstrated these ideas in a very simple one dimensional model of growth [8]. Work on the general problem is in progress.

This work was supported by the U. S. Department of Energy under grant number DE-FG02-85ER45189.

References

1. Z. Cheng and R. Savit: J. Phys. A **19**, L973 (1986).
2. R. Savit and R. K. P. Zia: J. Phys. A (Lett.) **20**, L987 (1987).
3. R. Savit and R. Ziff: Phys. Rev. Lett. **55**, 2515 (1985).
4. R. Broadbent and J. M. Hammersley: Proc. Cambridge Philos. Soc. **53**, 6239 (1957); P. L. Leath, Phys. Rev. B **14**, 5046 (1976); Z. Alexandrowicz: Phys. Lett. **80**A, 284 (1980).
5. M. J. Feigenbaum: J. Stat. Phys. **19**, 25 (1978); **21**, 699 (1979).
6. Z. Cheng, L. Jacobs, D. Kessler, and R. Savit: J. Phys. A (Lett.) **20**, L1095 (1987).
7. R. Baiod, Z. Cheng, and R. Savit: Phys. Rev. B **34**, 7764 (1986).
8. Z. Cheng, R. Baiod, and R. Savit: Phys. Rev. A **35**, 313 (1987).
9. Z. Cheng, L. Jacobs, D. Kessler, and R. Savit: University of Michigan preprint, submitted for publication (1987).
10. H. G. de Bruijn: Ned. Akad. Weten. Prod. Ser. A **43**, 39 and 53 (1981); R. K. P. Zia and W. Dallas: J. Phys. A **18**, L341 (1985); V. Elser: Acta Cryst. A **42**, 36 (1986); M. Duneau and A. Katz: J. Phys. (Paris) **47**, 181 (1986).

Index of Contributors